# ESSAYS IN BIOINFORMATICS

**NATO Science Series**

A series presenting the results of scientific meetings supported under the NATO Science Programme.

The series is published by IOS Press and Springer Science and Business Media in conjunction with the NATO Public Diplomacy Division.

*Sub-Series*

| | | |
|---|---|---|
| I. | Life and Behavioural Sciences | IOS Press |
| II. | Mathematics, Physics and Chemistry | Springer Science and Business Media |
| III. | Computer and Systems Sciences | IOS Press |
| IV. | Earth and Environmental Sciences | Springer Science and Business Media |
| V. | Science and Technology Policy | IOS Press |

The NATO Science Series continues the series of books published formerly as the NATO ASI Series.

The NATO Science Programme offers support for collaboration in civil science between scientists of countries of the Euro-Atlantic Partnership Council. The types of scientific meeting generally supported are "Advanced Study Institutes" and "Advanced Research Workshops", although other types of meeting are supported from time to time. The NATO Science Series collects together the results of these meetings. The meetings are co-organized by scientists from NATO countries and scientists from NATO's Partner countries – countries of the CIS and Central and Eastern Europe.

**Advanced Study Institutes** are high-level tutorial courses offering in-depth study of latest advances in a field.
**Advanced Research Workshops** are expert meetings aimed at critical assessment of a field, and identification of directions for future action.

As a consequence of the restructuring of the NATO Science Programme in 1999, the NATO Science Series has been re-organized and there are currently five sub-series as noted above. Please consult the following web sites for information on previous volumes published in the series, as well as details of earlier sub-series:

http://www.nato.int/science
http://www.springeronline.nl
http://www.iospress.nl
http://www.wtv-books.de/nato_pco.htm

Series I. Life and Behavioural Sciences – Vol. 368 ISSN: 1566-7693

# Essays in Bioinformatics

Edited by

David S. Moss

*School of Crystallography, Birkbeck College, London, UK*

Sibila Jelaska

*Department of Molecular Biology, Faculty of Science, Zagreb, Croatia*

and

Sándor Pongor

*International Centre for Genetic Engineering and Biotechnology, Padriciano, Trieste, Italy*

IOS Press

Amsterdam • Berlin • Oxford • Tokyo • Washington, DC

Published in cooperation with NATO Public Diplomacy Division

Proceedings of the NATO Advanced Study Institute on Introduction to Bioinformatics
Dubrovnik, Croatia
19–23 May 2003



ISBN 1-58603-539-8
Library of Congress Control Number: 2005930249

*Publisher*
IOS Press
Nieuwe Hemweg 6B
1013 BG Amsterdam
Netherlands
fax: +31 20 687 0019
e-mail: order@iospress.nl

*Distributor in the UK and Ireland*
IOS Press/Lavis Marketing
73 Lime Walk
Headington
Oxford OX3 7AD
England
fax: +44 1865 750079

*Distributor in the USA and Canada*
IOS Press, Inc.
4502 Rachael Manor Drive
Fairfax, VA 22032
USA
fax: +1 703 323 3668
e-mail: iosbooks@iospress.com



PRINTED IN THE NETHERLANDS

*Essays in Bioinformatics*
*D.S. Moss et al. (Eds.)*
*IOS Press, 2005*



# Foreword

When, as President of the Committee on International Co-operation of the Croatian Academy of Sciences and Arts, I visited the Royal Society and the British Academy in July 2000, I had in mind that the co-operation with the Royal Society be expanded as much as possible and, for the first time, that preparations be made for signing an Agreement on Co-operation with the British Academy. As usually, I could not help visiting the Birkbeck College of the University of London so well-known to me since the time Professor J.D. Bernal was there. I owe that visit mostly to Professor Alan Mackay, FRS, to whom I am tied by many years of friendship. It was on that occasion that in a conversation with David Moss, Professor of Biomolecular Structures, and his co-worker Dr. Clare Sansom, the idea was conceived to organize the postgraduate course in bioinformatics, this newly emerging interdisciplinary research area as the interface between biological and computational sciences, primarily aimed at research students from Central and Eastern Europe.

During the visit to the Royal Society, Alan and I met Professor Brian Heap, Vice-President and Foreign Secretary of the Royal Society at that time, and his collaborators. Professor Brian Heap supported our efforts on the condition that the Royal Society and the Croatian Academy of Sciences and Arts acted as initiators, while the Birkbeck College in London and the Faculty of Science in Zagreb took over organization. However, this was not the only part of activities which were agreed upon.

In view of the Agreement on Co-operation concluded between the Royal Society and the Croatian Academy of Sciences and Arts, Dr. Clare Sansom several times visited Zagreb and the International University Centre (IUC) in Dubrovnik where the course was intended to be organized. The realization of the course would be hardly thinkable without her persistence and wish for success. However, Professor Sibila Jelaska, Department of Molecular Biology, Faculty of Science, Zagreb, and Professor David S. Moss, School of Crystallography, Birkbeck College, supervised the course as its co-directors. It is a special pleasure to me that Dr. Kristian Vlahovicek, a former research student of mine, also greatly contributed to the organization of the course.

The course aroused far more interest among young researchers than it had been expected so that the number of participants had to be limited due to objective reasons (lack of room and mostly lack of computers in the IUC). Eight lecturers from five countries and 23 students from some ten countries took part in the course. The success was surprising, students enjoyed the course and learnt a lot finally marking the course with the average score on the Good/Excellent boundary.

Last but not least, the organization of the course was facilitated by the financial support of the NATO within the NATO Science Programme. The course was also sponsored by the Faculty of Science of the University of Zagreb and PLIVA, Zagreb, the largest Croatian pharmaceutical industry. For Croatian participants generous financial support was obtained from the Ministry of Science and Technology of the Republic of Croatia. Gratitude is due to the International University Centre, the organizers of the course, to all lecturers and participants.

All students would like such advanced courses to be continued in future. Let us act according to their wishes.

Professor Emeritus Boris Kamenar
Zagreb, July 2004

# Editors' Note

The course *Introduction to Bioinformatics* was held in Dubrovnik, Croatia, between 19th and 23th of May, 2003. The chapters of this proceedings volume were written by the lecturers of this course as well as by other authors recommended by them. The chapters were compiled so as to cover a wide range of subjects, from historical and theoretical background to practical applications. At the end of the course, the students were asked to write a paper on how they use what they learned during the course. These are added to this volume as an appendix. We hope that you will find much to interest you in this work. The assistance of Ms. Elena Stubel and Dr. Clare Sansom in preparing the manuscripts for publication is gratefully acknowledged.

Professor David S. Moss
School of Crystallography
Birkbeck College
London, UK

Professor Sibila Jelaska
Department of Molecular Biology
Faculty of Science
Zagreb, Croatia

Professor Sándor Pongor
International Centre for Genetic Engineering and Biotechnology
Padriciano
Trieste, Italy

# Contents

**Appendix. Student Papers**

*Essays in Bioinformatics*
*D.S. Moss et al. (Eds.)*
*IOS Press, 2005*




# Biology and Informatics

Alan L. MACKAY
*School of Crystallography, Birkbeck College, University of London[1], Malet Street, London WC1E 7HX*
*(The Inter-University Centre, Dubrovnik, 19-24 May 2003)*

**Abstract.** The advent of modern bioinformatics is the result of a long succession of scientific discoveries and paradigm changes in chemistry and biology. This chapter provides an introduction to the pertinent events in these diverse fields.

## Introduction

"Is it not a wonder that anyone can bring himself to believe that a number of solid and separate particles by their chance collisions and moved only by the force of their own weight could bring into being so marvellous and beautiful a world?" Marcus Tullius Cicero[2] (106-43 BC), *"On the Nature of the Gods"*

"Molecular Biology is the confluence of information and conformation" John Kendrew, (1965)

"How does so little information control so much behaviour?" Richard L. Gregory, in *Towards a Theoretical Biology*, (ed. C. H. Waddington), (1969)

"In less than a generation we have witnessed a radical, irreversible, world-wide transformation in the way that science is organised, managed and performed." John Ziman, *"Real science: what it is, and what it means"*, CUP 2000, (p.67).

## 1. Atomism

The city of Dubrovnik, earlier The Republic of Ragusa, in which this workshop was held, is a historic place and we have to mention its most famous scientist, Roger Joseph Boscovich[3] (1711-1787, FRS (1761)), who was born here and who looked after its interests, although he usually resided elsewhere. He worked mostly on astronomy, but he was an atomist and had proposed an important theory of point atoms, between which were mutual forces with a number of minima at different distances, running from a strong repulsion at very short distances, to an inverse square attraction like gravitation at very long distances[4]. This removed difficulties about what happens at the discontinuity of the surface of a billiard-ball kind of atom. Boscovich, who was a Jesuit, lived a generation after Newton and influenced Maxwell and Kelvin among others. He aimed to understand the properties of things in terms of their structure and his main work was called *"Philosophiae naturalis theoria reducta ad unicum legem virium in natura existentium"*, (Vienna, 1758, etc.) [Physics reduced to a single law of the forces existing in nature]. Also in Dubrovnik, Marin Getaldic[5] (1568-1626), a century earlier, appears as a pioneer of the algebraic geometry which is the basis of computer graphics. Already two and three hundred years ago European scientists were remarkably closely in touch with each other.

A generation before Boscovich, Newton, having determined "the motions of the planets, the comets, the Moon and the sea", was unfortunately unable to determine the remaining structure of the world from the same propositions because, as Newton said:

"I suspect that they may all depend upon certain forces by which the particles of the bodies, by some causes hitherto unknown, are either mutually impelled towards one another, and cohere in regular figures, or are repelled and recede from one another. These forces being unknown, philosophers have hitherto attempted the search of Nature in vain; but I hope the principles laid down will afford some light either to this or some truer method of philosophy". *(Preface to the Principia).*

But neither Newton nor Boscovich had the present-day experimental access to the atomic and molecular level necessary for the understanding of chemistry and biochemistry.

Boscovich was just one in the long tradition of atomism, which had started with Leucippus and Democritos, and which was promoted by Lucretius in his Latin poem, "On the Nature of the Universe" which sought to explain everything in terms of atoms[6]. Lucretius specifically claimed that mind and spirit are both also made of atoms. Atomism has long been a difficulty for the Vatican, most recently in connection with allergy to gluten and its implications for the doctrine of transubstantiation, but Lucretius' programme is steadily becoming reality.

Atoms became visible after the proof of their arrangement in crystals of sodium chloride by Lawrence Bragg and William Bragg, following the discovery of X-ray diffraction in 1912 by Laue, Friedrich and Knipping. Explaining everything in terms of atoms then became a major feature of modern science, especially of molecular biology[7].

## 2. Towards a theoretical biology

In the 1920s and 1930s around the laboratory of F. Gowland Hopkins ("the father of biochemistry") in Cambridge, there flourished the Club for Theoretical Biology, which was a most important source of ideas about molecular biology. The key idea was that the three-dimensional structure of molecules determined their behaviour. The group included Joseph Needham, Conrad Waddington, Desmond Bernal, Lancelot Whyte[8], and others. They aimed to make biology a real science like physics, where there were interactions to and fro between theory and experiment, and to understand the origin and processes of life. They also had radical political ideas. Theoretical biology was a new concept. Darwin had formulated the principles of evolution by natural selection, but now there was a prospect of elucidating the mechanisms of heredity, which appeared to operate at the atomic level.

Needham made a proposal (1935) for an Institute of Physico-chemical Morphology (to the Rockefeller Foundation, through Warren Weaver) but this was not funded, although Weaver and Astbury (independently) had coined the expession 'Molecular biology'. Needham as an embryologist had to ask how shape and the unfolding of shape in the embryo following a programme, was determined by the hereditary material.

Already at that time (1931) Bernal[9] had recognised that in order to be replicated, the hereditary material, then thought to be protein, had to be a linear structure. It was not demonstrated until 1944 that genes were nucleic acids and not protein (although associated with protein)[10]. In 1934 Bernal showed that if a crystal of pepsin were kept in its mother liquor, the diffraction pattern had information out to inter-atomic dimensions, that is, a protein molecule has every atom in its proper place. This was an epoch-making discovery. The concept of a mystical protoplasm thus collapsed. Proteins had a structure which could be investigated by physical methods, chiefly X-ray crystal structure analysis. Bernal and Astbury agreed to divide the new world between themselves, Bernal taking the globular proteins and Astbury the fibrous. There are several excellent studies of the development of molecular biology[11]. Schroedinger's well-known book "What is life" (1945)[12] appeared rather late in the day.

C. H. Waddington (1905-1975), a biologist, one of the Cambridge Club, after the war organised a series of influential seminars under the title of "Towards a Theoretical Biology" (1968, 1969, 1971) which brought together varied people who digested the revolution in computing, information, the structure of molecules, genetics and the origin of life. He himself promoted the concept of the "epigenetic landscape" as a way of visualising the development of an organism as genes were switched on and off to make choices between various paths. Almost all the people concerned with protein structure had exceptionally well-developed abilities for spatial visualisation (now shown by PET scanning to correspond to physical development of structures in the brain. Information and thought really have a material basis as Lucretius had suspected).

## 3. Hierarchy

The great success of X-ray crystal structure analysis in providing the shapes of molecules, has obscured the fact that most materials are not crystalline, although almost everything gives useful X-ray diffraction patterns. Crystallisation is a test for purity, but crystals are exceptional in that one rule takes one from the atomic level of 1 Angstrom (0.1nm) right up to 10 cm. The span of operation of this rule is unusually great. The recent discovery of quasi-crystals has led to a profound re-assessment, leading in the direction of hierarchy, of the laws of crystallography.

Biological structures are distinctively hierarchic with perhaps six levels of organisation with much smaller spans, each with its characteristic rules of ordering. These levels overlap to a greater or lesser extent. Properties at one scale are determined by structure at that scale, but may be critically influenced by certain detailed configurations in the level below for which the level above forms an average climate.

Levels of organisation or integration were clearly recognised by, for example, Joseph Needham[13], representing the thought of the Club for Theoretical Biology.

## 4. Information theory and the computer. Information and material structure

The concept of information began to appear in the 1920s. Not surprisingly, information theory began with the question: "How much should you pay for your telegraph message and how fast would it go?" At first it was so much a word but then newspaper correspondents began to make up pseudo-words like "Pariswise urgentmost". Theory began to be developed for questions of military cryptography, as the story of the Enigma machine has revealed. The Colossus computer[14] was built for cryptography at Bletchley Park. Questions of bandwidth arose. How much information could be transmitted over a land-line? The first Atlantic cable could only carry a few bits per second. Nyquist (1924), Kolmogorov and Hartley (1928), Claude Shannon, Louis Brillouin, Warren Weaver, Leo Szilard, Norbert Wiener were all concerned with the foundation of information theory[15].

John Tukey invented the word "bit" for a binary digit and Shannon used the word "entropy" for information content. as $-\Sigma\, p_i \log p_i$. (where $p_i$ is the fractional probability of the i-th kind of character. There is still great confusion as to the entropy content of meaningless and meaningful information. Shannon's example was of printed English text and he showed that about half the information is arbitrary, that is, is "meaning", and half is redundancy due to the intrinsic structure of the language which every native speaker knows. This redundancy can be used to correct mistakes in transmission. The Huffmann algorithm for compression[16] is based on a knowledge of the relative probabilities of different symbols, measured over a particular text. Since Shannon, the analogy between DNA and protein sequences and natural languages has been pervasive.

Information theory was developed in dialogue with the construction and use of computers which have made both the examination of the arrangement of atoms and the operation of data-bases possible. "Cyberspace" was invented and colonised the literary world[17].

Donald Booth at Birkbeck, recruited by Bernal to make a computer for crystallography, invented the floppy disc[18], using a primitive speech recorder with a magnetic disc, but he discarded it, and toyed with the machine translation of natural languages, an idea, which emerged in discussions with Warren Weaver. The Cambridge Crystal Structure Database was begun in an attic at Birkbeck College, originally on cards, before being established in Cambridge. Its creation was due to Olga Kennard and J. D. Bernal (who had far earlier been concerned with the development of Structure Reports (originally Strukturbericht) collecting all data on the arrangement of atoms in crystals.

Gregory Chaitin proposed that the amount of information in a structure could be defined in terms of the shortest computer programme necessary to generate it. The number of operations necessary to sort a sequence of N numbers into an arbitrary order is N log N ("operation" needs more careful definition).

## 5. Cellular automata

Robert May alerted us to the fact that there were many "simple mathematical models with very complicated dynamics", although the immensely creative J. B. S. Haldane had noted this more graphically in 1932[19]. In particular, finite difference equations, for example $x_{t+1} = f[x_t]$, have results which cannot be predicted far ahead better than by simply iterating the process. It also emerges that eventually the finite accuracy of all computing processes, including those in nature, will render the outcome indefinite and unpredictable. This kind of equation can be extended to two or three (or more) dimensions, the equations may be coupled or non-linear, so that the complexity increases. Stephen Wolfram[20] has developed certain classes of "cellular automata" in such detail that classification is possible. Intriguing and unpredictable patterns may emerge[21]. It is immediately clear that patterns in nature, particularly those in biological systems produced by the switching on and off of genes which synthesise proteins, must be physically analogous to such mathematical phenomena. Now even the classical mechanical problems of Newton, the pendulum and the solar system are seen to be weakly chaotic.

## 6. Structural molecular biology. Proteins and nucleic acids

Desmond Bernal had the good fortune to be the right man in the right place at the right time. In February 1945, before returning to Birkbeck after the war, Bernal produced a plan *"to set up a research centre for the study of the structure and properties of large molecules by all available physical and chemical methods"*. This was based directly on the thinking of the Cambridge club and was effectively the charter for the Birkbeck Laboratory, set up in 21-22 Torrington Square, which Bernal headed from 1938 to about 1964. In the 1950s

Aaron Klug, Rosalind Franklin, Kenneth Holmes and others contributed greatly to the establishment of molecular biology. I do not need to list their enormous achievements.

If we take a large molecule, for example the protein lysozyme, it contains C, H, O, N, S atoms in definite numbers and so should appear as a region in the phase diagram of this 5-component system. It would probably be in a meta-stable energy minimum. However, this is clearly unrealistic and lysozyme is much better considered as being specified by a number which represents its amino-acid sequence, which is effectively its address in phase space. Given the sequence, lysozyme can be now be made by adding the right amino residues in the right order. That is, it has a description. Information is stored in such meta-stable systems.

The proteins of life are a very special and minutely small subset of all possible amino-acid sequences characterised by being able to fold up into a unique configuration.

## 7. The double helix

Darwin and Mendel recognised the discrete nature of the hereditary substance but could get no further without access to the levels below those provided by optical microscopy.

The 50th anniversary[22] of the spatial structure of the DNA double helix has ensured that the circumstances of the discovery should now be well-known. I remember opening the copy of Nature for 25 April 1953 and reading the three papers disclosing the double helix, Crick and Watson's paper ending with the sentence: *"It has not escaped our notice that the specific pairing we have postulated immediately suggests a possible copying mechanism for the genetic material"* and thinking, yes, of course, it must be something like that. An immediate entry to the mechanism of heredity had opened, just as van't Hoff's vision in 1874 of the tetrahedral carbon atom had opened up organic chemistry, making it clear that is was the arrangement of atoms in three-dimensional space that was the determining factor for molecules (although Pasteur had demonstrated optical enantiomorphism in 1848, picking left- and right-handed crystals, and this implied spatial structure).

## 8. Dialectics

The key theme of this workshop is the relationship between information and structure. The more we look into it, the more complicated it gets. A very informative modern survey of the interaction between *"nature and nurture"* has been provided by Matt Ridley[26]. The conflict has been fought at all levels from the molecular to the politics of agriculture and education. The Lysenko affair in the Soviet Union was one acute manifestation, but there are still deepening conflicts with religious views.

The basic idea, that one structure should be a description of another, but both composed of atoms subject to the same laws of chemistry, has been revolutionary[23].

```
Real Space                         Representational Space

 Phenotype          ↔                 Genotype

             →    Projection    →
↑                                        ↓
Comparison                        Manipulation
↑                                        ↓
     ←           Restoration        ←
```

Earlier philosophical systems analysed economics and society as equilibrium systems, in many cases fixed by the unchanging dogmas of sacred texts. Change can now also be handled explicitly. Newton and Leibnitz, with the differential calculus, provided the tools for physics and Hegel introduced the idea of dialectics into philosophy[24]. In science there are many new ways of handling change, for example, the epigenetic landscape of C. H. Waddington for biology, all kinds of computer simulations of systems ranging from the Solar system (found to be weakly chaotic) to the British economy. Arthur Winfree, a pioneer in dealing with non-linear systems, gave his book the intriguing title "The geometry of biological time" indicating that changes in time and space were intimately mixed (as also in relativity) although Joseph Needham had much earlier written: "form is simply a short time-slice of a single spatio-temporal entity". The sudden changes in such systems have been illustrated in the 'catastrophe theory' of René Thom which has been expounded by Christopher Zeeman for social as well as for physical systems. They find that there are only seven types of geometrical singularities in the configuration space. These are by way of being mathematisations of the "double bind" kind of situation which philosophers describe where one can get out of a knot only by jumping to some other position.

## 9. Experimental techniques

Of course the whole progress of bioinformatics has depended on the development of experimental methods and their implementation, both facilitated by the advent of computer hardware and of appropriate algorithms. Structural studies stand on X-ray crystal structure analysis, electron microscopy, atomic force microscopy and nuclear magnetic resonance and all their variants.

Fred Sanger in Cambridge quietly provided the methods for sequencing both proteins (1949-55) and DNA, which are the absolute fundamentals for bioinformatics, but the flood of sequence data is a result of the industrial-scale implementation of sequencing methods on a huge scale[33].

Numerous automated biochemical techniques for identification and, for example, for combinatorial chemistry, have become essential.

Computer handling of gigantic data-banks and computer modelling of the conformations of proteins and the expected chemical properties of molecules are now central to bioinformatics.

## 10. Genomics

The key problems[25] include:

- The structure of proteins, protein folding, the operation of proteins.
- DNA, its sequence, replication, transcription, its interaction with proteins, the switching of genes.
- The ribosome.
- The structure and operation of chromosomes, meosis, mitosis, replication, mutation, variation. Extra-chromosomal nuclei acid.
- Phylogenetics, evolution, speciation.

The nematode worm, *Caenorhabditis elegans,* with 302 nerve cells, was the essential link, chosen by Sydney Brenner, between behaviour, molecules and genetics.

## 11. Genetic and financial engineering

The nature/nurture interaction means that the results of the expression of genes as proteins depend on the environment in which they are expressed[26]. Analogously, the consequences of the developments of biotechnology depend on the social system within which they are expressed. There are huge possibilities of good or evil. Thus, scientists cannot be unconcerned with politics and must act responsibly. We may be sure that some people, somewhere[27], are thinking about the uses of bioinformatics for military, selfish and destructive ends[28]. Social control of applications of genomics cannot be left to oligarchies[29]. This means that scientists in genomics must work to create an informed public and this implies an opposition to secrecy[30].

Today, information, especially that relating to molecular structure and genetic sequences, is being enclosed, as land was enclosed in the 18th and 19th centuries in Britain, and is becoming private property[31] (as are also computer components, algorithms and methods[32]). The Human Genome Project has generated acute conflicts in the "Republic of Science" and more generally[33]. Huge data-banks of the DNA information on individuals are being built up for social, political and military purposes. Even the ownership of ordinary standard English words and phrases are being claimed by arrogant companies and litigation[34] absorbs a large proportion of the social product, especially in the USA.

With the development of socio-biology, through the efforts of E. O. Wilson, D. S. Wilson, R. Dawkins, J. Goodall and many others, the extension of biological ideas, from the collective behaviour in insect societies to the noösphere, is making progress towards understanding the behaviour and evolution of individuals, groups and species. “Memes” have been proposed as units of social structure[35] circulating in the world of information. At the insect level some quantitative confirmation of numerical predictions has been achieved. Such topics should eventually be included in bioinformatics as part of the dialogue between information and matter.

## 12. Lucretius

In due course I and you and everyone else will cease to operate as living systems. The atoms will disperse and all that will be left will be traces of information distributed round the world. There will be bits of genetic sequence continuing in descendants and in relatives, some genetic information may be recoverable from organic specimens, there will be items in the Internet and in documents of all kinds. There will also be transient memories residing in others. It will all be a matter of chance as to what survives of us, but it will be information recorded in various kinds of matter.

I must draw your attention again to Lucretius' book: *De Rerum Natura*, [on the nature of things]. It reached us from antiquity in only a single manuscript copy and, with the development of printing at the end of the fifteenth century, it was reprinted and translated and generally circulated so that, by chance, this remarkable philosophic outlook has survived to our own times[36] and remains a source of inspiration and consolation for us even two thousand years later.

It is information transmitted in code from our ancestors and it is this coding into language which distinguishes the human species from all others. Lucretius had said "I set out to loose the mind from the knots of religion"[37]. His book’s great merit is that it sought to give a complete, unitary picture of the universe, free from prevailing superstitions. Also, we might note, prophetically perhaps, in view of the concern over AIDS and SARS (Severe

Acute Respiratory Syndrome) that Lucretius ended his book with a description of the social chaos which occurred with the plague in Athens. I commend it to you as the foundation of bioinformatics.

## 13. The Present Crisis

This workshop takes place at a critical time in human history[38]. Science and technology have changed the world[39]. We cannot avoid the political significance of bioinformatics and indeed the militarisation of science must be one of our major concerns[40].

The human race faces the possibility of various catastrophes, from oligarchies to chaos, as well as natural disasters, most of its own making. In particular, the growth of the world population cannot continue indefinitely at its present rate. The only way in which these can be avoided is by knowledge and the intelligent application of knowledge[41]. Thus it is vital to build a world-wide network of people who understand each other, who have each other's confidence, who can operate in their own societies, and who will be able to inject their special knowledge into the decision-making centres[42] and thus to influence the course of history. The social parts of our meetings, begun in Dubrovnik, are at least as important as the technical parts.

## Notes

[1][London] … "the quick forge and working-house of thought" *W. Shakespeare, King Henry V, 5:23.*

[2]Cicero was supposed to have been the editor of Lucretius' m/s "De rerum natura".

[3]L. L. Whyte (ed.) *"Roger Joseph Boscovich",* Allen and Unwin, London, 1961.

[4]G. Malescio, "Intermolecular potentials – past, present and future", Nature materials, **2**, 501-503, (2003).

[5]"De resolutione et compositione mathematica", Rome (1630). Getaldic also made a concave parabolic mirror, 70 cm in diameter, which is now in the London Maritime Museum (inventory NAV 0928), and probably also a reflecting telescope.

[6]Lucretius, [Titus Lucretius Carus], "On the Nature of the Universe", (trans. R. E. Latham), Penguin, revised edition 1994. In this book, the concept of "swerve", [clinamen] which has so worried the classical commentators, can perhaps be understood retrospectively in terms of chaos theory where the progress of an idealised game of billiards cannot be forecast more than a few impacts ahead.

[7]A. L. Mackay, "Generalised crystallography", Structural Chemistry, **13,** (3/4), 217-222, (August 2002). http://sinapse.arc2.ucla.edu/Mackay02.pdf

[8]The editor of the work on Boscovich and a writer on atomism.

[9]"The facts of genetics demand, as J.B.S. Haldane has pointed out, that, at some stage in mitosis, the individual molecules in a chromosome must be exactly duplicated. A complete molecule can be duplicated in three ways. If it is solid and three dimensional only a supernatural agency, a divine copyist, can, entering its inner complexity, reproduce it in detail. If we prefer a natural solution, we must imagine the molecule stretched out either in a plane or along a line. In either case the simpler constituent molecules have only to arrange themselves one by one on their identical partners in the original molecule, and then become linked to each other by the absorption of suitable quanta from radiation or from second order collisions. That such autocatalysis is possible is indicated by recent work in Russia and America, where the regular atomic arrays of metallic catalysts are shown to operate like laceworker's frames on which simple organic molecules settle to be joined into larger aggregates. A two-dimensional reproduction of this kind is impossible, owing to the fact that the constituent amino acids in nature are not symmetrical, but exist in right or left hand forms. Two-dimensional reproduction would lead to mirror image molecules, which are not found in nature. There remains then only one dimensional reproduction. At the moment of reproduction, but not necessarily at any other time, the molecule of the protein must be imagined as a pseudo-linear, associating itself, element by element, with identical groups, related by an axis instead of a plane of symmetry, and thus preserving only right – or only left handed symmetry. This hypothesis is clearly indicated by Astbury's explanation of Svedburg's numbers. Svedburg has established that most natural proteins consist of M Wt 34,000 or multiples 2, 3, or 6 times that number. This gives us the confidence to treat all protein molecules, regardless of their

complex constitution, as belonging to one natural species. It is impossible to claim that these ideas are anything but preliminary guesses, but they have the advantage of being susceptible to experimental test."
J. D. Bernal (1931) [Int. Congress of the History of Science. Bernal Archive, Cambridge. A4.7 Box 22, by courtesy of Andrew Brown].
[10]Philip Ball, "Portrait of a molecule", Nature, 421, 421-422, (23 January 2003).
[11]H. F. Judson, *"The Eighth Day of Creation"*, Simon and Schuster, New York, 1979.
Robert Olby, "The Path to the Double Helix ", Macmillan, London,
Nature, **421** , (6921), (23 Jan. 2003) [special supplement for the 50$^{th}$ anniversary of the double helix]
[12]Schroedinger used the term "aperiodic crystal" which later entered the discussion of quasi-crystals after 1985. He said: "We believe a gene – or perhaps the whole of the chromosome fibre – to be an aperiodic solid".
[13]J. Needham, *"Order and Life"*, (1936) Reprinted MIT Press, 1968. [Dedicated to the Theoretical Biology Club.]
[14]The Colossus computer, all copies of which were destroyed after the war on Churchill's orders, is now being rebuilt at Bletchley Park as an historic monument.
[15]L. Brillouin, "Science and Information Theory", New York, 1956.
C. E. Shannon and Warren Weaver, *"The Mathematical Theory of Communication"*, University of Illinois Press, (1949).
D. M. Mackay, "Quantal Aspects of Scientific Information", Phil. Mag., 41, (1950) and Proc. First London Symposium on Information Theory, (1950)
[16]A. L. Mackay, "Optimisation of the genetic code", Nature, **216**, 159-160, (1967).
[17]"Cyberspace. A consensual hallucination experienced daily by billions of legitimate operators, in every nation, by children being taught mathematical concepts... A graphical representation of data abstracted from the banks of every computer in the human cystem. Unthinkable complexity. Lines of light ranged in the non-space of the mind, clusters and constellations of data. Like city lights, receeding... " William Gibson (ca. 1982).
[18]A. D. Booth, "A magnetic digital storage system", Electronic Engineering (July, 1949)
[19]J. B. S. Haldane (1892-1964) wrote, very presciently,
"Even in a non-mathematician like myself, some differential equations evoke fairly violent physical sensations similar to those described by Sappho and Catullus when viewing their mistresses. Personally, however, I obtain an even greater 'kick' from finite difference equations, which are perhaps more like those which an up-to-date materialist would use to describe human behaviour". Haldane was, indeed "an up-to-date materialist"! *"The Inequality of Man"*, (1932), Penguin, (1937), p. 39.
Robert M. May, Nature, **261**, 459-, (10 June 1976).
(http://nedwww,ipac.caltech.edu/level5/Sept01/May/May_contents.html)
See also: A. L. Mackay, Physics Bulletin, 495-497, (Nov. 1976) and Izv. Jugoslav. Centra za Krist., **10**, 25-36, (1975). (http://www.cryst.bbk.ac.uk/surfaces/zagreb.html). J. W. Galloway, Physics Bulletin, **34**, 161-164, (1983).
[20]Stephen Wolfram, *"A New Kind of Science"*, Wolfram Media, 2002.
[21]For example, P. Ball, *"The self-made tapestry: Pattern formation in Nature"*, Oxford, (1999).
[22]Nature, **421**, (23 January 2003).
[23]A. L. Mackay, "From 'The Dialectics of Nature' to the inorganic gene", Foundations of Chemistry, **1**, (1), 43-56, (1999).
[24]I have discussed this at greater length in "From the 'Dialectics of Nature' to the inorganic gene", Foundations of Chemistry, **1**, 43-56, (1999).
[25]A. M. Lesk, *"Introduction to Bioinformatics"*, Oxford University Press, 2002.
(www.oup.com/uk/lesk/bioinf/)
[26]Matt Ridley, *"Genome: the autobiography of a species in 23 chapters"*, Fourth Estate, London, 1999.
*"Nature via Nurture"*, Fourth Estate, London, 2003.
[27]Military uses of bioinformatics are discussed in: Tom Mangold and Jeff Goldberg, "Plague Wars", Macmillan, London, 1999.
[28]Concerns about anthrax illustrate this. The USSR had a serious accident releasing anthrax; the USA also had a dramatic terrorist attack associated with its own weapons programme; much earlier Churchill wished to use anthrax, tested on Gruinard Island in the North of Scotland, against the German civil population; Iraq too, had sought to develop anthrax.
[29]In Britain already two million DNA profiles are held in police records.
[30]If you are a scientist at an American research university like mine, you know what to do if you think you've hit on some technique or bit of knowledge that might have commercial potential. You go online to the university's technology transfer office, download an invention and technology disclosure form, and fill in the details. You have to do that because all such intellectual property (IP) discovered by this university's employees belongs to the university. If the local bureaucrats think there's something in it, they will file a

provisional patent and. after formally offering it to any government agency that funded the research – which usually declines – they will start hawking the IP about to see if any entrepreneurs or companies want to license it. Priority in your IP is protected at this stage, and you can now go ahead and publish if you wish, but eventually you may proceed to full (or utility) patent, where property rights are wrapped up more securely, and, while IP lawyers make fortunes from litigation about who in fact owns the property, basically the matter is now in the domain of formal law. If the university does manage to license the IP, you will get perhaps 35 per cent of the royalty stream. Or, if that's not enough for you, you can cut yourself free from academia and take your chances with the venture capitalists as an independent entrepreneur. - Steven Shaplin, (University of California at San Diego), London Review of Books, 6 March 2003, p.14.

[31]"Monsanto aim to control the world food supply", [London, Channel 4 TV. "DNA the story of life", 19:00, 15 March 2003] see also for example the website www.cryptome.com for the current applications of surveillance technology.

[32]L. Cranswick, "The potential power of 'software patents' to destroy crystallographic software", Crystallography News, (84), (March 2003). http://www.ccp14.ac.uk/maths/software-patents/

[33]J. Sulston and Georgina Ferry, "The Common Thread: A story of science, politics, ethics and the human genome" Bantam, (2002).
see the review of this by Robin McKie, The Observer 3 Feb. 2002. at www.guardianunlimited.co.uk/ (search for McKie).
Apparently J. D. Watson told Sulston "Venter wanted to own the whole genome the way Hitler wanted to own the world".

[34]There are some 900,000 lawyers in the USA. In Japan, with a different social structure there are only 18,000. Science is now done with lawyers looking over your shoulder.

[35]Richard Semon (Munich) had proposed "mnemes to be the preserving principle in the interaction of organic events" and this idea was promoted by Ernst Haeckel.

[36]Karl Marx, as a young man, wrote his doctoral thesis (presented *in absentia* at the University of Jena) on a comparison of the philosophies of Democritos and Epicurus.

[37]"religionum animum nodis exsolvere pergo"; I. 932.

[38]R. Brenner, "Towards the precipice: the crisis in the US economy", London Review of Books, 25, (3), (6 Feb. 2003); Chalmers Johnson, "Who's in Charge" (Review of Daniel Ellsberg, "Secrets: A Memoir of Vietnam and the Pentagon Papers), (LRB same number: see the London Review of Books website www.lrb.co.uk ). E. Hobsbawm, "Age of Extremes: The short twentieth century 1914-1991", London, (1994).

[39]See, for example, Chapter III of "The Theory and Practice of Oligarchical Collectivism" by Emmanuel Goldstein, (1949).

[40]War also is being privatised as Eisenhower's 'military-industrial complex'. In 2001 expenditure on military research and development was: (in millions of dollars) USA 39,340; (total EU 9,100;) Britain 3,986; France 3,145; Germany 1,286; Italy 291; Spain 174; Canada 121; Netherlands 65; Turkey 50. (Economist, 3/5/03). The total US expenditure on defence is about 340,000 per annum.

[41]M. L. Sifry and C. Cerf, "The Iraq War Reader: History Documents, Opinions, Simon and Schuster, New York, 2003.

[42]As I write (in London in July 2003) the crisis over the death of the principal British scientific expert on biological warfare, who found that the scientific situation was misrepresented by political leaders, exhibits the problems of the relationship between science and politicians. "What is truth said jesting Pilate, and would not wait for an answer" Francis Bacon (1561-1626).

*Essays in Bioinformatics*
*D.S. Moss et al. (Eds.)*
*IOS Press, 2005*




# Concepts of Similarity in Bioinformatics

Vilmos ÁGOSTON[1], László KAJÁN[2], Oliviero CARUGO[2,3], Zoltán HEGEDÜS[1], Kristian VLAHOVICEK[2,4] and Sándor PONGOR[2]

[1]*Bioinformatics Group, Biological Research Center, Hungarian Academy of Sciences, Temesvári krt. 62, 6726 Szeged, Hungary*

[2]*Protein Structure and Bioinformatics Group, International Centre for Genetic Engineering and Biotechnology, Area Science Park, 34012 Trieste, Italy*

[3]*Department of General Chemistry, Pavia University, viale Taramelli 12, 27100 Pavia, Italy*

[4]*Molecular Biology Department, Biology Division, Faculty of Science, University of Zagreb, 10000 Zagreb, Croatia*

**Abstract.** The key problem of bioinformatics is the prediction of properties, such as structure or function, based on similarity This chapter reviews the concepts and tools of similarity analysis used in various fields of bioinformatics.

## Introduction

The concept of similarity is fundamental in the study of macromolecular structures, genomes, proteomes and metabolic pathways. Similar objects are often assumed to take part in similar mechanism, or to carry out a similar function. Similarity, on the other hand is a highly intuitive concept, and its use in various fields – such as the comparison of sequences or of 3-D structures – is quite different. For students of molecular biology it is sometimes difficult to find straightforward definitions of the basic concepts that originate from as diverse fields as cognitive psychology, systems science as well as various branches of mathematics. The motivation of this review is to provide a – not necessarily complete - compendium of useful concepts and definitions and to show the commonalities underlying the various applications. We will use three main forms of representations: sequences, 3-D structures and graphs. The discussion will be based on an entity-relationship description of macromolecular structures [1], as applied to the description of small molecules [2] as well as biological objects used in genome analysis [3].

Most concepts of molecular similarity have been proposed in applied contexts that are so numerous that an exhaustive coverage would detract from our focus on the underlying mathematical spaces. In particular, machine learning methodologies used in bioinformatics [4, 5], such as neural networks [6] and support vector machines [7] are based on specific concepts that in our view cannot be adequately described in the framework of a general discussion. Similarly, we could not include a practice-oriented overview of applications such as the comparison of sequences, 3D structures and genomes (a review on these topics will be published elsewhere [8]). Several fields that are gaining importance in bioinformatics, such as the analysis text similarities [9], could not be incorporated because of space limitations. Although a significant amount of research is thus excluded from this overview, a broad, and we hope to show, integrated body of research remains.

The primary focus of this work is to present a set of useful definitions pertinent to the similarity analysis of macromolecular structures, meant as reference material for advanced bioinformatics courses. *Section 2* describes the basic concepts used in macromolecular similarity analysis, pointing out, whenever possible, the parallel concepts in other fields. *Section 3* focuses on four distinct mathematical relationships, each of which constitutes a possible definition of similarity: equivalence, matching, partial ordering, and proximity.

## 1. Basic concepts

### *1.1 Model, description, analysis*

When we speak about molecules, what we mean are not physical entities, rather abstract models of reality. It is useful to distinguish three concepts underlying molecular data:

The *models* are the conceptual structures or mental representations used to store information on molecules. These models never incorporate all of information available on a given macromolecule – the mere listing of the atoms and bonds in a macromolecule would be beyond the reach of human memory – rather we deal with a set of models of varying complexity, each describing a certain aspect of the molecular structure, such as linear sequence, domain topology, active site contacts, etc.

Various formal and/or narrative *descriptions* of the data constitute the backbone of molecular databases. We can imagine the descriptions as the mathematical representation of a particular model. Similarity measures are calculated between descriptions (and not between models).

The *analysis* covers everything we do with molecular data in such fields as molecular modelling, prediction, classification, similarity search, visualization etc.

For example we may start noticing a new regularity when classifying the existing molecular descriptions (*analysis*). If this new feature "makes sense" (e.g. it points to a meaningful subclass of the objects) we may include this into our abstract *model*, and we may proceed to construct a new kind of description that includes the new feature. In a further round of analysis we may find new examples that contain the feature in question, in addition we may experiment with new feature candidates analogous or similar to the previously found features. As this cycle is repeated, the models and the descriptions undergo an evolutionary change, and in fact this is how databases develop [10].

### *1.2 Entities, relationships, structure and function*

In the first approximation, bioinformatics is concerned with the structure of protein and DNA molecules that fulfil functions in a series of interdependent systems such as pathways, cells, tissues, organs and organisms. This complex scenario can be best described with the concepts of systems theory (**Figure 1**).

According to systems theory [11, 12], a system is a group of interacting elements functioning as a whole and distinguishable from its environment by recognizable boundaries Molecules can be regarded as such systems. Generally speaking, *structure* is fixed state of a system, and the study of a system usually starts with its characteristic structures that are recurrent in space or time. As structures are detected by recurrence, the symmetries (internal repetitions) are integral parts of structural descriptions. Using the terms of the previous paragraphs, systems are conceptual models of reality, while structures are descriptions.

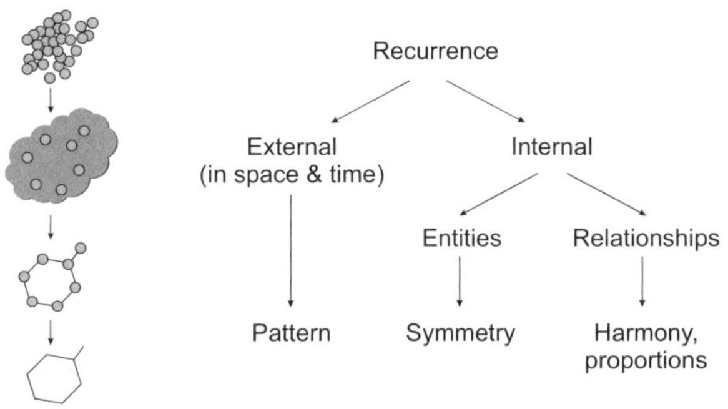


**Figure 1.** Simplified overview of concepts underlying structural descriptions.

Descriptions rely on elements (entities) and binary relationships between them [1, 13] (Table 1).

In the case of molecules, both the elements (substructures) and the relationships can be described in terms of systems of categories. The categories and the relations between them can be formalized into ontologies, which include the definitions of the elements as well as the operations that are possible within the system (**Figure 2**). Ontologies give itemized descriptions each functions and roles a molecule can fulfil, so it is a logically coherent world description. Entity-relationship-descriptions are generally applicable and can be extended to such concepts as similarity groups, vicinities and networks (**Figure 2**).

**Table 1.** Examples of models and descriptions

| System | Entities | Relationships |
|---|---|---|
| a) Conceptual models of natural systems | | |
| Molecules | Atoms | Atomic interactions (chemical bonds) |
| Assemblies | Proteins, DNA | Molecular contacts |
| Pathways | Enzymes | Chemical reactions (substrates/products) |
| Genetic networks | Genes | Co-regulation |
| b) Structural descriptions | | |
| Protein structure | Atoms | Chemical bonds |
| Protein structure | Secondary structures | Sequential and topological vicinity |
| Folds | $C_\alpha$ atoms | Peptide bond |
| Protein sequence | Amino acid | Sequential vicinity |

Elements and relationships can be described not only in terms of categories, but we can assign to them property descriptors, such as physicochemical, chemical descriptors. In terms of contents, there are two kinds of properties in proteins and DNA that deserve special attention. i) The *position* of an element (nucleotide, atom) can be defined either within the molecular chain (sequential position, with respect to the N-terminus, etc.) or in as 3-D coordinates. ii) The *function* is a property or role that can be defined in the context of a higher level. E.g. "protease" is a function defined either in an *in vitro* (e.g. action on a certain substrate) or *in vivo* environment (e.g. role in complement activation). In addition to these two main classes, there are a whole list of properties that can be assigned to entities and relationships within a model. In terms of mathematical form, the descriptors of the properties can be continuous, discrete or binary variables, even statements in human language.

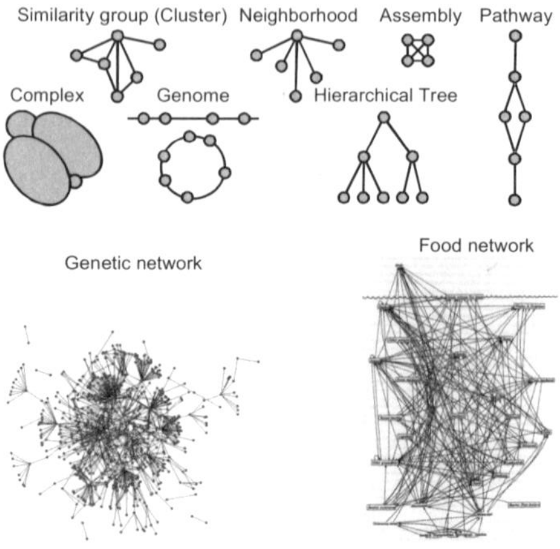


**Figure 2.** Molecular structures can be represented as entities and relationships [1, 13]. Implicit to a structure is the description of the underlying concepts (entities and relationships as well as their properties), which can be summarized in an ontology [14]. The same principle can be easily extended to genomic and "systems biology" applications.

Entity/relationship models have been used in psychology as well. Erich Goldmeier's "Similarity of visually perceived forms" defines similarity in terms of partial identities that may include a varying proportion of entities and relationships [15, 16]. If we apply this definition to molecular graphs such as shown in Figure 2, we arrive to a plausible definition: Two molecular graphs are similar if they have a common sub-graph (Figure 3).

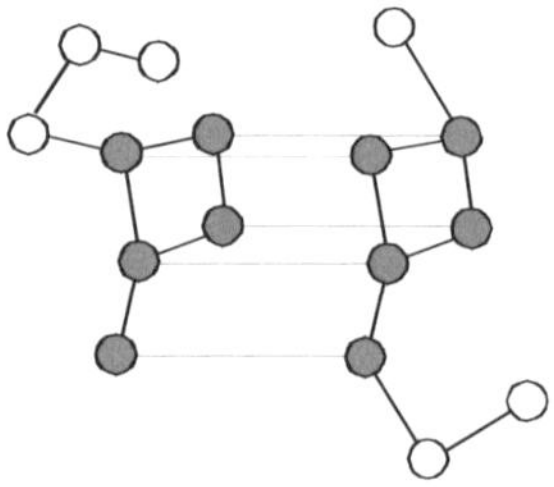

**Figure 3.** Molecular similarity as sub-graph isomorphism. Similarity of structures can be defined as a common sub-graph shared by two entity-relationship descriptions.

Dedré Gentner [17] drew a map classifying the similarities of narrative descriptions (Figure 4a), which can be extended without difficulties to the description of protein structures (Figure 4b). For example, molecular descriptions are considered identical if they consist of the same substructures and relationships. If two descriptions only share the substructures but not the relationship, they are identical in terms of composition only. If the relationships are identical, but not the substructures, we speak about equivalent topology. Alpha-helices (and other protein secondary structure elements) are examples for this kind of partial identity, since in this case the identity of amino acid residues (i.e. the entities) is immaterial. All identities and similarities are true only at the given level of description (e.g. backbone conformation, amino acid composition, etc.).

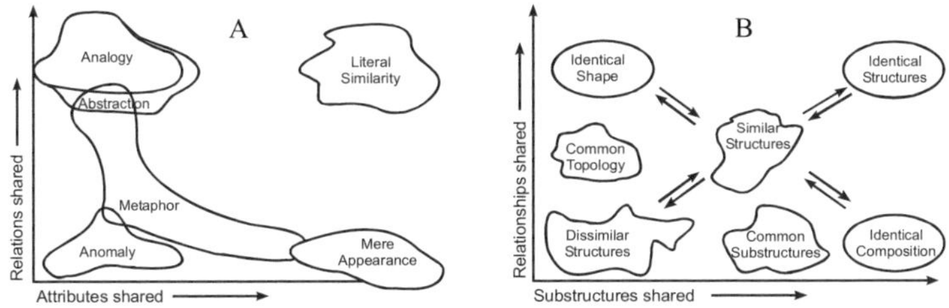


**Figure 4**. Identity, different kinds of similarity and non/identity can be pictured as regions in a plot of shared entities vs. shared relationships. This representation was developed by Dedré Gentner for narrative descriptions [17] (A), but can be extended to molecular descriptions as well (B).

Figures 4 implies that similarity of two molecules can be captured if we can define equivalencies between their constituents, i.e. if we match the similar parts of the two descriptions to each other. Finding common substructures relies on matching, and some numerical parameter of matching is used in most cases as a measure of similarity. For example, two 3D structures are obviously similar if more than 90% of their alpha carbons can be superposed. We mention that matching is used not only for establishing similarity, but also for finding complementarity, such as surface-complementarity used in molecular docking, or strand-complementarity used in the analysis of anti-sense RNA.

Based on the above concept we can define two further concepts, similarity groups and functional units. The similarity group is such a group of molecules that are connected by structural similarity. This similarity can be local or global (see 2.3) or it can be general or specific (section 3.3). Biologically important similarity groups, such as those of protein domains belong to the latter class, as all group members are characterised by a common sequence-description or a common fold-description.

Functional units denote a group of molecule that jointly fulfil a biological function. Enzymes, regulators and substrates of a metabolic pathway are examples of functional unit. Members of a functional unit are similar in their common function, but they do not need to be structurally similar. This is thus a contextual similarity, as opposed to the structural similarity.

### *1.3 Elements of molecular descriptions*

#### *1.3.1 Focusing of descriptions*

The entity-relationship framework and the underlying category definitions can be used to construct a very large number of description that can focus on various aspects of a molecular model [13]. One of the practical ways of generating simplified descriptions is to concentrate on parts of a molecule that are important for the actual goal of the analysis. Starting from a generalized theoretical model containing detailed descriptions of all entities and relationships in various forms, one can derive simplified descriptions by omitting some of the descriptors. For example, a hydrophobicity plot is a description of protein structure wherein the entities are amino acid residues described in terms of only two parameters, the sequence position and the residue hydrophobicity index. On the other hand fold descriptions include only the $C_\alpha$ atoms of a protein, while surface descriptions include only those atoms in contact with the environment (solvent). But we may choose to use higher categories, such as domain-units instead or amino acid residues. TOPS cartoons are simplified description in which the entities are secondary structural elements; the relationships are topological links describing sequential or spatial vicinities.

**Table 2.** An example of simplified descriptions

| Model | Descriptor | |
|---|---|---|
| | Position | Hydrophobicity |
| Hydrophobicity plot | + | Real number |
| Hydrophobic segments | + | Discrete (0 or 1) |
| Average hydrophobicity | - | Real number |
| Hydrophobic character | - | "Hydrophobic"/"Hydrophilic" |

Another avenue of fine-tuning consists in decreasing the detail - the resolution - of the descriptors (Table 2). For example, residue hydrophobicity can be described in quantitative terms, using a hydrophobicity scale (with continuous variable represented as a real number) or qualitatively (discrete variable, represented as 0 or 1 or with categories "hydrophobic" and "hydrophilic").

The intuitive concept of *resolution* also refers to the number of categories used in a given description. An amino acid composition is a vector in a 20-dimensional space, and since most proteins contain all of the amino acids, all the components of the vector are non-zero. On the other hand, we have 400 dipeptides and 8000 tripeptides. In a tripeptide-based composition, however, many (or most) of the components would be zero or 1. Very high-resolution descriptions are highly characteristic "fingerprints" that can be used to identify individual structures. For example, mass spectra are efficiently identified by the presence/absence of their constituent peaks, and similarly, small molecular structures can be retrieved from databases using queries constructed from their constituent fragments. On the other hand, high-resolution fingerprints cannot be easily generalized to similar molecules, so the resolution of the descriptions has to be optimized so as to include the right scope of similar descriptions.

### *1.3.2 Kinds of descriptors*

Descriptors can be categorized according to their contents. On the one hand we have various levels, such as atoms, residues, secondary structure element, domain etc. Whether we talk about DNA or about proteins, there is an apparent lowest level that is not divided into further categories. For example, structural biology is rarely concerned with particles below the atomic level, while molecular biologists use nucleotides and amino acids as the lowest level. Higher-order units can be built up from the lower levels. In most cases the higher units are non/overlapping, i.e. one atom can be part only with one residue. On the other hand we use overlapping fragment descriptions as well, for example nucleotide sequences can be described in terms of overlapping di- or trinucleotide words, protein 3D structures can be described as peptide fragments.

We use the term "*structured descriptions*" for those descriptions that contain both entities and relationships. Protein 3-D structures and sequences are such descriptions even though the relationships are not explicitly included in the actual descriptions found in databases. For example, the atoms are named in PDB files, but the connectivity of atoms in amino acids is not part of the database, it rather has to be included in the program reading the database entries. If a description contains only entities or only relationships, we term it an "*unstructured description*". Examples include amino acid composition (only entities) and $C_\alpha$ distance-distributions (only relationships).

Finally, descriptors can be classified also depending on what they refer to. Descriptors referring to an entire molecule are global descriptors, such as a protein function. Local descriptors, such as the role of a domain within the protein are local descriptors.

### *1.4 Overview of macromolecular descriptions*

Based on the concepts introduced in the preceding sections we can now attempt to classify the molecular descriptions. One simple classification distinguishes 1D, 2D and 3D descriptions. 1D descriptions, such as sequences and hydrophobicity plots, are residue-based, and include only the chain-topology. 2D descriptions are graph-like and include relations in addition to the chain topology (e.g. helical circle and helical net diagrams provide a symbolic view of the 3D arrangements). 3D descriptions are those in which Cartesian coordinates are included among the descriptors.

A more detailed classification is possible according to the mathematical machinery. This classification essentially follows that of Johnson set up for small molecules [2, 18].

The most complete description is a generalized labelled graph in which both the vertices, and the edges can be provided with arbitrary labels such as numbers, vectors, names even statements in human language. Labels can be attached to individual entities or to groups of them (such as segments of a polypeptide chain). This is a hypothetical, multi-level description that is best approximated by a well-annotated 3D database record that is cross-referenced to (possibly all) the available biological databases. Such variable-level descriptions are rarely used for comparison. The 3D comparison programs of Sali and Blundell are one of the few exceptions, they use a hierarchy of levels such as atoms, residues, secondary structures and domains [19, 20].

3D structures contain atoms and entities provided with Cartesian coordinates as descriptors, as well a chemical (covalent) connectivity. This description is used by most of the molecular modelling and structure comparison programs. Structural databases contain the entities and their labels; the connectivity maps are included with the analysis programs.

Distance matrices. Distances calculated between the elements of the same structure constitute a distance matrix. In 3D structures, one can use the positional coordinates to define distance vectors, whereas the number of edges between two nodes can be used to define a distance in a graph. Both are extensively used in similarity analysis.

Finite sequences. All graphs can be represented in terms of finite sequences. A protein sequence is a special graph where the residues are the entities and the polypeptide chain connectivities are the edges. 1D plots (such as the hydrophobicity plot) can be derived from an amino acid sequence by representing one single numeric parameter as a function of the residue position. This parameter can be either an experimentally determined value (such as a physicochemical parameter, or a quantity computed from the sequence or from the 3D structure.

Surfaces used for proteins include the Van der Waals surface or the electrostatic surfaces that are computable from the 3D structure. Surface similarity analysis is not included in this review, an excellent review is in [21-23].

Integrable scalar fields. In this representation the molecule is treated as a spatial distribution of a single quantity, such as electron density or mass density [24].

Transforms. There are various methods to calculate topological transforms from graphs. Fourier transforms of 1D sequence plots have been used to identify amphifilic regions in proteins, as well as to compare proteins.

Finite sets are unstructured descriptors that can be obtained e.g. by omitting all relationships from a labeled graph. The resulting set of entities provides description that can be ordered according to kinds. A typical example is the amino acid composition, or other fragment-composition type descriptions (dipeptide, tripeptide etc. compositions). This is a vector-representation, the parameters of the vector corresponds to the number of times a certain entity is present in a structure A subcase of finite set descriptions consists in reducing the set of entities to a set (list) of kinds. This can be achieved by omitting the numbers from a compositional description.

Distributions. A vector consisting of nonnegative numbers that sum to unity constitutes a parameter vector of a multinomial distribution. A typical example is the amino acid composition expressed in percentages, or the distribution of inter-atomic distances within a protein structure, or distribution of connectivity degrees in large networks.

Vectors, product spaces. In addition to the special vectors mentioned in 5 and 6, arbitrary parameters of a given molecules can be assembled into vectorial descriptions. Such complex descriptions are used as input in machine-learning, and are also often used in general pattern-recognition applications.

Real numbers (molecular sizes, molecular weight etc.) are perhaps the simplest descriptors of molecules.

## 2. Mathematical concepts related to similarity

### *2.1 Relations*

#### *2.1.1 Equivalence*

Equivalence relations (denoted here by "≅") are related to the commonly used term of identity. Strictly speaking, a molecule can only be identical with itself; here we are concerned with the cases when two molecules have identical mathematical descriptions, which does not mean that they are identical. For example, two proteins that have an identical description in terms of amino acid sequence may undergo phosphorylation or other posttranslational modifications at different sequence positions).

Equivalence relations in mathematics are defined by three properties: reflexivity, symmetry, and transitivity. A relation is reflexive if $A \cong A$ for all molecular descriptions A. It is symmetric if $A \cong B$ implies $B \cong A$. It is transitive if $A \cong B$, and $B \cong C$ implies $A \cong C$. Let [A] denote the family of those molecules equivalent to A with respect to ≅. If B denotes some other molecule, it can be proven mathematically that either [A] and [B] denote the same set of molecules or the two sets have no members in common. The set [A] is called an equivalence class. For example two proteins are considered identical if and only if their (amino-acid) sequences are the same. It is noted that "identity" refers to a given description; in this example the potential differences in post-translational modifications are disregarded.

#### *2.1.2 Partial ordering*

Partial ordering relations are related to the commonly used terms "to be a substructure of", "to be a part of". A relation ≤ is called a partial order if it is reflexive, antisymmetric, and transitive. The reflexive and transitive properties of a relation were defined earlier. A relation is antisymmetric if A ≤ B and B ≤ A implies that A and B are identical. For example if A ≤ B means that A is a subsequence of B, then ≤ is a partial ordering relation.

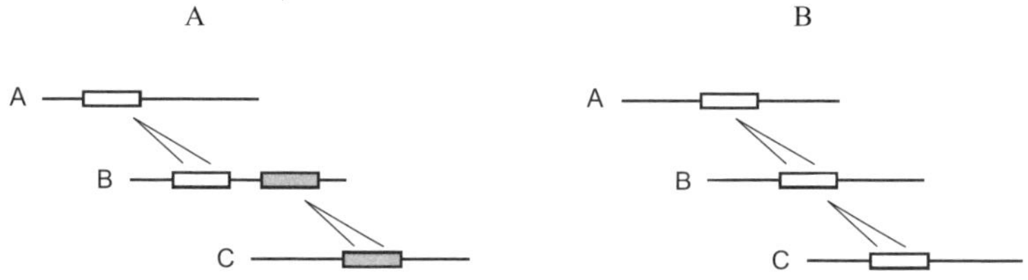


**Figure 5.** Similarity of molecules can be considered either a tolerance relationship (A), or an equivalence relationship (B) depending on whether or not the basis of similarity – the shared substructure – is fixed.

### *2.1.3 Tolerance, general and specific similarity*

Tolerance relations denote the common sense situation in which two things have a common part or feature, or two structures share a common substructure. A relation ~ is called a tolerance if it is reflexive, symmetrical, but – in contrast to equivalence relations – not necessarily transitive. In other words, A~A, A~B implies B~A. Tolerance comes closest to the common sense concept of similarity, however there is an important distinction to be made. Based on the psychological concept of Goldmeier [15, 16], we can call two structures similar if they share some common substructure (see **Figure 3**, above). This general similarity is not transitive, as shown in **Figure 5a**, it is in fact a tolerance relationship. On the contrary, we may use the term specific similarity, if two structures share a well-defined substructure (feature). Fixing the shared substructure renders the relationship transitive, so specific similarity is an equivalence relationship (**Figure 5b**).

If biological sequences are found similar to each other by BLAST, this is a general similarity, i.e. it is not necessarily true that all of them share a subsequence, such as a protein domain. However, those sequences that turn out to share a common subsequence form an equivalence class. It is noted that a "common subsequence" is often defined in an empirical way: biologists usually decide based on their prior knowledge whether or not a subsequence of a protein is a true member of a domain group (like EGF domains), and once a positive decision is made, the protein sequence is accepted as a member of the equivalence class of EGF-containing proteins. We might say that evaluation of BLAST searches consists in distinguishing general and specific similarity.

The use of relations in chemical structure analysis is reviewed in [2, 18].

### *2.2 Proximity measures*

Proximity measures (*PM*) are numeric measures designed to characterize similarity or dissimilarity of two molecular descriptions. Two general types of proximity measures are in use. Similarity measures are high for similar molecules and low for dissimilar ones. The distance measures, on the other hand are zero for identical molecular descriptions and high for dissimilar ones. In the foregoing we will use proximity measures, distance measures and similarity measures.

Proximity measures can be used in vastly different contexts, and it is useful to define two situations that are common in bioinformatics applications. A) *Simple proximity* between two objects is computable by an unequivocal algorithm. A distance of two stars in space is a good example. For instance, such measures are computed between unstructured descriptions like vectors. One can define simple proximity measures also for structured descriptions, provided the equivalences of the entities (residues, atoms) are *a priori* defined. The Hamming distance and *rmsd* are such measures for character strings and 3D structures respectively. They are based on straightforward algorithms for calculating a distance between the two objects. Such distances are often calculated between fragments of larger structures hence they are sometimes called fragment distances. B) *Substructure proximity* measures are computed between parts structured descriptions. They require a simple measure as well as an algorithm to select the "optimal substructure" in the two objects. For instance, the distance of two galaxies can be defined as the distance between their closest stars. In this case, we have to measure the distance between all possible pairs (simple proximity), and then select the smallest one. The two central problems of bioinformatics, – sequence alignment and 3D structural alignment – are substructure similarity problems. So instead of two objects, we need compare "galaxies of substructures" which is a compute intensive task. The complexity of the calculation is different (sequence alignment has a complexity *O(n,m)* while structural alignment is *np*-complete), but the the basic concepts of substructure selection matching– described in the subsequent paragraph 3.3 – is common to both. The present section will concentrate mostly on simple proximity measures. We will follow the classification of Sneath and Sokal [25] who distinguished four classes proximity measures: distance coefficients, association coefficients, correlation coefficients and probabilistic coefficients.

## *2.3 Distance measures*

Vector distance measures are perhaps the simplest class of similarity measures owing to their geometric interpretation. The most common is probably the Euclidean distance, which, for some pair of objects A and B, described by n-dimensional vectors $A_i$ and $B_i$, respectively, is defined as:

[1] $$D_E = \left( \sum_{i=1}^{n} (A_i - B_i)^2 \right)^{\frac{1}{2}}$$

This is a distance defined in the n-dimensional space. A simple variant of this formula is the average distance, which is simply the Euclidean distance divided by the number of dimensions, i.e., by $n$ in this case. The generalization of the Euclidean distance leads to a class of metric distance functions called the Minkowski metrics, defined by the following general formula:

[2] $$D_M(r) = \left( \sum_{i=1}^{n} |A_i - B_i|^r \right)^{\frac{1}{r}}$$

$r = 1$ corresponds to the "city block" distance and $r = 2$ to the Euclidean distance. The so-called sup distance is the Minkowski metric of $r = \infty$ and corresponds to

[3] $$D_M(\infty) = \max_{1 \le i \le n} |A_i - B_i|$$

Distance measures calculated between identical molecular descriptions (vectors) are zero, and may grow without limit for non-identical vectors. It is sometimes desirable to have bounded values, for example the so-called Canberra metric is defined as:

[4]
$$C_M = \frac{\sum_{i=1}^{n} |A_i - B_i|}{\sum_{i=1}^{n} (A_i + B_i)}$$

so it is zero for identical values but remains below unity for non-identical vectors. The metric properties of vector distance measures are important for clustering and for evolutionary studies. For $M$ to be a metric (metric distance), the following criteria have to be fulfilled for all A, B, C from X: *i)* $M(A,B) \geq 0$ the equality holding if and only if $A = B$; *ii)* $M(A,B) = M(B,A)$ (symmetry); *iii)* $M(A,B) + M(B,C) \geq M(A,C)$ (triangular inequality). Metric properties are essential if a distance measure is to be used for clustering. A string similarity measures S (eqn. 5) is applicable to clustering if there is an associated distance measure $M = f(S)$ that has metric properties. $f$ is a monotonous function, and distance measures such as *1-kS* (where *k* is a constant) are routinely used in clustering applications.

## *2.4 String similarity measures for biological sequences*

A special class of proximity measures, sequence similarity scores are used to quantify the matching (alignment) of protein and DNA sequences. The underlying mathematical concept is the string distance. Let us first concentrate on bit strings consisting of zero/one values. (**Figure 6A**). The Hamming distance is the number of (zero to one or one to zero) changes necessary to change string 1 to string 2. This can be used immediately for short character strings of identical length, with the condition that only exchanges are possible, gaps are not allowed. If we keep this condition, we can use a simple lookup table to store the costs of exchanging one character against another. The situation is the same if we use overlapping doublet or triplet words, etc. i.e. the Hamming distance is a simple distance that can be unequivocally computed based on lookup tables, because the matching of the two strings is considered unique.

The situation is quite different if we match strings of arbitrary length and allow gaps (Figure 7). The string edit distance is defined as the minimal number of steps (insertions, deletions and replacements) necessary to transform one word into the other. The proximity measures used for biological sequences are defined as similarity coefficients (high values for similar, low for dissimilar sequences) and contain cost factors for residue substitutions as well as gaps (insertions, deletions).

[5]
$$S_{1,2} = \sum \cos t_{identities,replacementss} - \sum \cos t_{gaps}$$

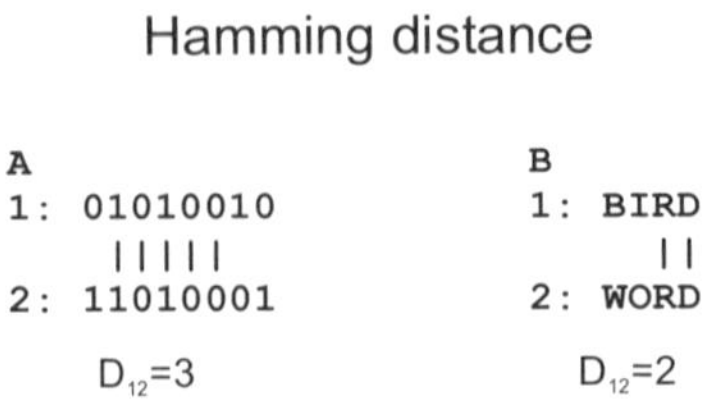


**Figure 6.** The Hamming distance is the number of exchanges necessary to turn one string of bits or characters into another one (the number of positions not connected with a straight line). It is assumed that the two strings are of identical length and that no alignment is necessary. The exchanges in character strings can have different costs, stored in a lookup table. In this case the value of the Hamming distance will be the sum of costs, rather than the number of the exchanges.

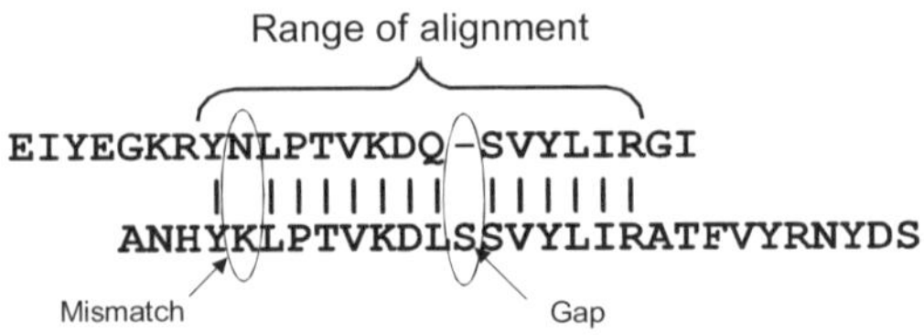


**Figure 7.** A string similarity measure can be defined as a sum of costs assigned to matches, replacements and gaps (insertions and deletions). The two strings do not need to be of the same length. A string similarity measure between biological sequences is a maximum value calculated within a range of alignment. The maximum depends on the scoring system that includes a lookup table of costs, such as the Dayhoff matrix, and the costing of the gaps.

The alignment used here is no longer unique, like in the case of a Hamming distance, and there are different (arbitrary) ways to cost gaps (different cost factors for gap opening and gap extension etc.). Establishing an alignment between two sequences consists in maximizing a similarity measure given in equation [5]. This problem can be solved if in addition to the formula of $S$ we have a cost matrix for replacements and identities, or some other lookup table that contains the similarity/distance values of the elements used in the description. In the case of proteins, the cost factors of amino acid substitutions are included in the well-known Dayhoff and BLOSUM matrices, and there are several established strategies for costing the gaps – for recent reviews see (). The algorithm for finding a maximal similarity between two longer sequences is an optimization problem. The actual algorithms of similarity search are beyond our scope. The basic principle is mentioned in section 3.5, and some examples are given in section x.x. There are number of comprehensive reviews on this subject.

### *2.5 The rmsd distance for protein 3-D structures*

A very popular quantity used to express the structural similarity of 3-D structures is the root-mean-square distance (*rmsd*) calculated between equivalent atoms, defined as

[6]
$$rmsd = \sqrt{\frac{\sum_i d_i^2}{N}}$$

where $d$ is the distance between each of the $N$ pairs of equivalent atoms in two optimally superposed structures. For the calculation of *rmsd* a range of alignment has to be defined within which the matching of atoms (establishment or equivalent atoms within the two structures) is determined which is a computationally much harder problem than the alignment of sequences in one dimension. Once the equivalence of atoms is established, the optimal superposition has to be found which is carried out by such straightforward algorithms as that of Kabsch [26].

If the equivalences are fixed, then rmsd can be considered as a simple distance that can be computed with a straightforward algorithm. This is the case for instance when one compares different conformations of the same protein such as produced by NMR methods. In this case the equivalences of the atoms are *a priori* known, since each conformation consists of the same atoms. The *rmsd* is 0 for identical structures (identical conformations) while its value increases as the two structures become more divergent. In fact *rmsd* values are considered as reliable indicators structural variability when applied to very similar proteins (say *rmsd* < 5-6 A). But even in this case, the *rmsd* value obviously depends on the number of residues $N$ included in the structural alignment. A statistical analysis of a large number of structures showed that the dependence can be described as:

[7] $$rmsd = rmsd_{100}(1 + \ln\sqrt{\frac{N}{100}})$$

where $\text{rmsd}_{100}$ is a constant, an *rmsd* value standardized to 100 residues [27]. The *rmsd* values also depend on the crystallographic resolution, which is more difficult to take into consideration (Carugo, 2002). As a result, *rmsd* does not behave as a metric distance for divergent structures so it cannot be used in itself for automated clustering. Clearly, an *rmsd* value of, say 3 Å has a different significance for proteins of 500 residues and for those of 50 residues, so e.g. the structural variability of fold types can not be easily compared ($rmsd_{100}$ on the other hand may be useful for such comparisons[27]). In other terms, *rmsd* is a good indicator for structural identity, but less so for structural divergence.

The algorithms for calculating *rmsd* are beyond our scope, the reader is referred to recent reviews [28]. The philosophy of the calculation depends on whether or not the alignment, i.e. the equivalences between residues (represented as $C_\alpha$ atoms) are known. If yes, the very popular algorithm of [26] and McLachlan (1978) can be used. If this is not the case, and when the two 3-D models that are compared are too different, there are two alternatives. Either a partial alignment is available or no a priori assumptions can be made. In the first case, few equivalences between atom pairs are assumed and they are extended (and some time rejected) through dynamic programming techniques [29]. In the other case an exhaustive search is performed by rotating and translating a 3-D model over the other in a six-dimensional way Diedrichs, 1995).

It has to be noted that superposition of divergent protein 3-D structures is often a quite arbitrary exercise and various superposition algorithms may lead to completely different results. An effective, recently proposed procedure to reconcile different structural alignment procedures consists in an iterative reduction of the number of aligned $C_\alpha$ atom pairs [30]. After each superposition, the worse pair is eliminated and a new superposition is performed leading, eventually, to the identification of the protein core that shows a significant degree of similarity.

Finally we mention that the *rmsd* distance does not allow the costing of gaps. For this reason, it can not be used directly for finding an optimum alignment between two arbitrary proteins.

### 2.6 Association measures

For the comparison of chemical graphs of small molecules, association measures are used almost as widely as distance measures [31]. The majority of these coefficients are intended for use with simple two-state, i.e., binary, variables which are conventionally coded as 0 or 1 depending upon their presence or absence within an object description. Although these coefficients can be described in terms of a vector it is conceptually simpler to formulate the coefficients as follows. For two objects A and B let $ab$ be the number of attributes in common and $\overline{ab}$ the number of attributes in neither in A nor in B. Let $a\bar{b}$ and $\bar{a}b$ be the number of attributes occurring in only in $A$ or $B$, respectively. Let $a$ and $b$ the total number of attributes in $A$ and $B$ respectively. Let $n$ be the total number of attributes, i.e. $a$+$b$. Two frequently used association coefficients are the Jaccard (also called Tanimoto) coefficient:

[8]
$$J = \frac{ab}{a\bar{b} + \bar{a}b + ab}$$

(which can also be written as $a \cap b / a \cup b$, the ratio of common attribute types to all attribute types) and the Dice coefficient

[9]
$$D = \frac{2 \cdot ab}{a + b}$$

The coefficients may readily be generalised to non-binary data. For instance, if the data vectors contain the actual frequencies of occurrence of each fragment type, rather than their mere presence or absence, the Jaccard coefficient can be rewritten as

[10]
$$J' = \frac{\sum_{i=1}^{n} A_i B_i}{\sum_{i=1}^{n} A_i^2 + \sum_{i=1}^{n} B_i^2 - \sum_{i=1}^{n} A_i B_i}$$

A related association measure is the so-called cosine coefficient that corresponds to the cosine of the angle between the vectors $A$ and $B$:

[11]
$$C = \frac{\sum_{i=1}^{n} A_i B_i}{\left( \sum_{i=1}^{n} A_i^2 \sum_{i=1}^{n} B_i^2 \right)^{\frac{1}{2}}}$$

### 2.7 Correlation measures

Another widely used coefficient of similarity in cluster analysis has been the Pearson product-moment or correlation coefficient. Given two structures $A$ and $B$, let $\overline{A}$ be the mean value for all of the variables in the vector $A$(and similarly $\overline{B}$ for). Then the coefficient is defined as

[12] $$r = \frac{\sum_{i=1}^{n}(A_i - \bar{A})(B_i - \bar{B})}{\left(\sum_{i=1}^{n}(A_i - \bar{A})^2\right)^{\frac{1}{2}}\left(\sum_{i=1}^{n}(B_i - \bar{B})^2\right)^{\frac{1}{2}}}$$

The correlation coefficient is 1 for identical vectors, is around zero of dissimilar vectors and is –1 for anticorrelated vectors ($A_i = -B_i$). A large number of related coefficients are given in [32, 33].

### *2.8 Probability based measures*

The final class of coefficients identified by Sneath and Sokal [25], probability based coefficients, take account of the frequency distribution of variables over the entire data set. Probability-based coefficients are less often used for small molecules, on the other hand they are the most often used method of scoring in biological sequence comparison. Probability-based measures are obtained by first calculating a raw proximity measure PM between a query and all members of a dataset. This is followed by rescaling the raw PM using knowledge on the distribution of scores. This operation places the PM values on a common scale and thus provides an obvious way to set significance threshold for the hits of interest. It is customary to distinguish "biologically meaningful" and "random" similarities. The former are those between evolutionarily (homologs, orthologs, paralogs) or structurally related proteins (molecules with a common fold), the rest of the similarities are usually considered "random".

One approach is based on the distribution of random similarities. If the distribution is known in analytical or numeric form, then the statistical significance of any computed measure – the probability $P$ ($0 \leq P \leq 1$) for finding a given value in the given dataset by chance – can be estimated. Random similarities occur more likely for larger queries and for larger databases, so the description of random distributions usually includes query size and database size as variables. (The product *query size x database size* is sometimes referred to as the *search space*). Current biological databases provide a sufficiently large number of data for modeling the distribution of random similarities, and – at least for sequence data – various random shuffling techniques can be used to generate larger datasets. This approach thus consist in rescaling a proximity measure PM to give a probability P for a given search space. This P is called statistical significance, in other words, if the value of proximity measure lies far outside the distribution of random scores (P is very small), one tends to consider it biologically significant, and conversely, large P values indicate random similarities that are unimportant in the biological sense.

Another approach relies on the distribution of the target similarities, i.e. the distribution of PM within a biologically important group of objects. Often there are not enough reliable data for the analytical modelling of this target distribution, and random shuffling techniques may not be easily applicable same as for random similarities. A compromise solution consists in concentrating on the distribution of biologically significant as well as random similarities in the neighbourhood of a target group[34, 35]. This approach relies on the fact that space defined by existing macromolecules is sparsely and unevenly populated (as compared to the hypothetical space of all possible molecules), and the neighbourhoods of existing similarity groups may be quite different.

Further kinds of probabilistic coefficients can be obtained if one represents the objects themselves by some kind of a distribution, and then compares two distributions so as to obtain a probabilistic estimate of their identity [36]. There are established methods for the comparison of distributions, such as the $\chi^2$ test and contingency table analysis, etc. [37] that all yield probability values between 0 and 1.

Probability-based measures are widely used for the evaluation of prediction methods [32, 33]. Similarity measures for chemical structures have been reviewed by Willett [31].

*2.9 Proximity measures for groups of objects*

Proximity measures originally defined to pairs of structural descriptions can be generalized to groups. Given a single description *S* and a group of descriptions $[A]=\{A_1, A_2, ...A_n)$, a proximity measure $P(X,Y)$ between *S* and *[A]* can be defined using the $P(S,A_i)$ values of the pairwise comparisons; for example, one can take the minimal, the maximal or the average of the $P(S,A_i)$ values as the proximity measure between *S* and the group. Another possibility is to calculate from the descriptions $A_i$ a "consensus value" $<A>$, sometimes called the centroid of *[A]*. If the descriptions are simple numeric values or vectors, $<A>$ can be defined as their average. If $A_i$-s are vectors, $<A>$ can be their vectorial average, etc. Then, the proximity measure between *S* and *A* can be calculated as $P(S,<A>)$.
Proximity measures between two groups of objects *[A]* and *[B]* can be defined in a similar way: we can take the minimum, maximum or average of the $P(A_i,B_j)$ proximity measures, or determine the proximity of the two centroids, $P(<A>,<B>)$.
If a single object is compared to group *[A]* in terms of a feature *f* that is supposed to be normally distributed in *[A]*, with mean *m* and standard deviation *sd*, then, instead of the simple difference $|f-m|$ we can use a scaled value $\left|\dfrac{f-m}{sd}\right|$ for calculating a distance between an object and the group. Similarly, one can calculate a distance between two groups (denoted by upper indices 1 and 2, respectively) using the values $\left|\dfrac{m^1-m^2}{\sqrt{(sd^1)^2+(sd^2)^2}}\right|$. The resulting distance values will thus incorporate a natural scaling based on the different variance of the groups. This scaling can be generalized to cases in which the objects to be compared are represented as vectors of features $f_1, f_2 ...f_n$ characterized by a covariance matrix *C*. In this case, the so-called Mahalanobis distance is defined as:

[13] $$MD = (m^1 - m^2)' C^\wedge (m^1 - m^2)$$

where $m^1$ and $m^2$ are average vectors for group 1 and group 2, respectively, $(m^1 - m^2)'$ is the transpose of $(m^1 - m^2)$ and $C^\wedge$ is the inverse of the variance-covariance matrix *C*. *MD* can be viewed as an Euclidean distance scaled by the covariance matrix, the latter being assumed to be identical for both groups.

## 3. Matching (alignment)

For two structures to be similar, one has to find a matching in terms of entities and relationships. Such a matching is shown in **Figure 3**. A matching resembles an analogy. In an analogy, features of one object are paired with features of another object. Which features are paired is often a subjective choice. Matching of molecular structure relies in comparing molecular descriptions finding an optimal match between the features of the mathematical representations.

The simplest example of matching is the alignment of short character strings of equal length described in Figure 6. Another example is to find the exact occurrence of a short character string query within another, longer string.

With these examples one can illustrate, without formal definition, some of the important properties of matching used in bioinformatics. Given two descriptions A, B consisting of $i$ and $j$ elements, respectively, let's define a mapping $m$: $A \rightarrow B$ that assigns certain $k$ elements in $A$ to certain $l$ elements in $B$; it is not necessary that all elements in $A$ have an element assigned in $B$ and vice versa. In the simplest case, the mapping is one-to-one, so certain elements in $A$ will have a pair in $B$ (so $k=l$). In other cases, we may get multiple matching.

*Types of alignments* [2]. From the philosophical point of view, matching (alignments) of structural descriptions can arise from three specific sources. i) Due to prior knowledge, only some parts of the molecules are considered when establishing a matching. For example, backbone atoms of a protein must match backbone atoms of the other protein, when the 3-D structures are compared. ii) If we deal with unstructured descriptions, such as vectors, the elements – the vector components – match by definition, which is called canonical matching; iii) Finally, we might be interested in the maximal matching of two larger structures given in the form of structured descriptions (consisting of both entities and relationships), and then we need an optimizeable similarity measure such as in equation 5, in order to find a maximal alignment. The number of all possible alignments is very high, so finding the optimum is often very compute-intensive and sometimes intractable problem.

*Algorithmic solutions* From the algorithmic point of view, the methods can be subdivided a) according to the structure types (character strings, graphs, 3D structures), b) according to the nature of the matches that are being sought (exact matching, approximate matching), or according to number of partners compared (pairwise alignments or multiple alignments).

The majority of algorithms can cope only with the simplest descriptions, character sequences. Finding exact matches between two sequences of n and m characters respectively, has a complexity of $O(n,m)$. Comparison of graphs is much more difficult, here the majority of the problems are NP complete, i.e. computationally not tractable. Identity of graphs is determined by graph isomorphism algorithms, and similarity of graphs (such as protein structures, metabolic pathways) is a subgraph isomorphism problem, which is more difficult, and is aggravated by the fact that the structures in question are labelled graphs. Rigorous comparison of complex descriptions such as 3D structures is np-hard.

Time raqirements can be decreased if we use unstructured descriptions that do not need an alignment for comparison. These descriptions are simpler, and perform well only if one can find an adequate resolution, such as a multidimensional vector. The calculations are fast, but there is no guarantee that the results will be similar to those produced by alignment methods. On the other hand, such methods are useful as preliminary filters used to screen large databases.

Heuristic solutions provide the second general avenue for diminishing computer times. Most of the alignment methods use some kind of heuristics. There are two important heuristics that are used to simplify the process of alignment, the principle of linearity and the use of higher order descriptions.

The *principle of linearity* is based on the chain-like topology of protein and DNA molecules. All biologically important alignments contain short stretches that are very similar or identical between the two molecules. So instead of testing all possible alignments, one can start identifying the highly similar chain segments and then combining them into larger alignment, which is computationally much less expensive. This philosophy it can be used both for sequences and for 3-D chain. Let $A$ and $B$ be polypeptide chains of $l$ and $k$ residues, respectively, and $A_i^n$ denote a contiguous fragment of $n$ residues of protein $A$ starting at residue $i$. In this case, $[A^n]=A_1^n,\ A_2^n \ldots A_{l-n+1}^n$ will be an ordered list of overlapping fragment descriptions covering the entire chain of protein $A$. Let's provide such a list for both proteins and compare the fragments using a proximity measure $PM_{i,j} = PM(A_i^n, B_j^n)$. $PM$ must be a proximity measure that can be unequivocally determined for any two fragments. In most cases this means that no alignment is needed between fragments compared (alignment and gaps would make the process prohibitively expensive). In some cases the precomputed or *a priori* known values of $PM$ are stored in lookup tables. $PM_{i,j}$ values define a so-called similarity matrix, which is a symmetrical matrix of $k$ x $l$ elements (more accurately (l-n+1)x(k-n+1) elements). If $PM_{i,j}$ is a similarity measure, similar segments within two proteins appear as series of large values parallel to the diagonal of the matrix. The similarity matrix is used – under various names – for the determination of an overall alignment in several algorithms, many of which use dynamic programming techniques. Global alignments that extend from the beginning to the end of both sequences are found via an exhaustive search for the maximal matching, based on such methods as the Needleman-Wunsch algorithm [38]. Local alignments can be found via similar strategies, such as the Smith-Waterman and the Sellers [39] algorithms, as well as by heuristic solutions such as the FASTA and the BLAST algorithms.

All of these algorithms were developed for sequence alignment, where the fragments are overlapping $n$-words of amino acids, the scoring is be based on a sequence similarity score such as in equation 5. Naturally, one can also use 3-D description of the backbone for longer peptide segments of 3D structures, and use the *rmsd* distance for comparison. The actual algorithms are problem dependent, further examples are given in section 4.

The principle of higher order descriptions is based on the simple fact that comparing a smaller number of higher-order elements takes less time. The best example is the comparison of protein structures in terms of secondary structure elements. In addition to decreased computer time, higher order descriptions, such as secondary structure elements incorporate a great deal of human knowledge. As a consequence, the results of comparisons are usually close to human understanding.

Finally we mention that alignment is an optimisation problem, so all optimisation algorithms can be used for aligning structures. The optimum is understood in the context of the chosen representation and scoring scheme and may involve parameters that have to be adjusted on an empirical basis. Most users would therefore agree that alignments produced by computer programs can always be improved upon visual inspection.

## 4. Similarity spaces

In the foregoing, we reviewed the mathematical concepts relevant to the definition of similarity in bioinformatics: equivalence, matching, partial ordering, and proximity. These relationships arise in the context of a mathematical space. A mathematical space suitable for molecular similarity analysis is called a *molecular similarity space* and is defined to consist of a) a set of mathematical representations of molecules and b) one or more similarity relationships defined on this set. For example, one of the possible protein similarity spaces contains the sequences as representations, plus a set of equivalence classes, each containing members of a protein family. It is assumed that a sequence similarity measure is also defined on the set of sequences. Another similarity space used for proteins consists of the structures or protein folds as descriptors, a set of equivalence classes, each containing members of a specific fold group. A distance function, such as *rmsd* is defined on the set of fold structures.

The co-existence of *a priori* known (biologically relevant) classification schemes and computable proximity measures is characteristic of the similarity spaces studied by bioinformatics. In the typical case, the database also contains a large number of unclassified objects (sequences, structures), and much effort is put into either founding new classes for some of these objects, or trying to fit them into one of the existing categories. It is noted that a proximity measure can be used to establish a computable classification using one of the many clustering methods. In a fortunate case the computed clustering is consistent to the *a priori* known classification, and the potential new clusters that have no *a priori* known counterparts are excellent candidates for discovering new, biologically relevant classes.

Methods for representing *a priori* known categories can be grouped according to the nature of description used for the individual categories [40]. Classical summary descriptions are *consensus descriptions* that are valid for all members of a category. Probabilistic summary descriptions are valid only with some probability. Consensus descriptions such as sequence patterns can be pictured as the description of a prototype in the given class. In contrast to consensus descriptions, *exemplar-based descriptions* represent the categories as a database consisting of the members of the category. All of these methods have been used e.g. for protein domain sequences. Domain sequence collections and domain annotations in protein sequence databases are exemplar-based descriptions. Regular expressions are classical summary (consensus) descriptions that are supposed to be valid for all members, and there is a variety of statistical (probabilistic) descriptions [40].

The problem of classification is one of the fundamental exercises in such fields a domain sequence identification, or function prediction. Given a set of classes $A^i$ in a database, the classification of a sequence is often based on minimal distance (or maximum similarity). Oftentimes, the class $A^i$ of the closest object [$\min_{i,j} PM(S, A^i_j)$] is automatically assigned to an unclassified object. In other cases, the closest class is determined from the consensus-representations of the classes, using $\min_i PM(S, <A>^i)$. The use of mathematical spaces in the analysis of chemical structures is reviewed in [2, 18].

## 5. Conclusions

Summarizing we can conclude that the description of structures as entity-relationship networks provides a simple framework to describe the use of similarity in various fields. There are a number of qualitative concepts, such as similarity groups (equivalence classes), patterns as well as quantitative concepts, such as similarity measures that are present in all fields. Mathematical spaces ("similarity spaces") provide a way for describing databases as well as the mathematical tools of analysis in a common framework. The definitions listed in this review are applicable in other fields of bioinformatics not explicitly mentioned in this review, such as the analysis semantic similarities [9] or the analysis of networks [41]. An overview of practical applications will be published in a subsequent chapter in this volume [8].

The description of structures as entity-relationship networks provides a simple framework to describe the use of similarity in various fields. There are a number of qualitative concepts, such as similarity groups (equivalence classes), patterns and quantitative concepts, such as similarity measures that are present in all fields.

## Acknowledgements

This material is partly based on the lectures of the course "Bioinformatics: Computer applications in molecular biology", held in Trieste, Italy, 1992-2003. Special thanks are due to M. Bishop (Hinxton, UK), E. Gasteiger (Geneva, Switzerland), R. Harper (Hinxton, UK). D. Judge (Cambridge, UK), D. Landsman (Bethesda, MD), J. Leunissen (Wageningen, The Netherlands) for advice, as well as to the following individuals for their comments on various topics in the manuscript: Stephen Altschul (Bethesda, US), Steve Bryant (Bethesda, US), Alexandre De Leon, (Calgary, Canada), Jacques Demongeot (Grenoble, France), Mark Gerstein (New Haven, UK), Andrew Harrison (London, UK), Lisa Holm (Hinxton, UK), Jack Leunissen (Wageningen, The Netherlands), Christine Orengo (London, UK), William F. Pearson (US), János Podani (Budapest, Hungary).

## References


[1] Pongor, S., *Novel databases for molecular biology.* Nature, 1988. **332**(6159): p. 24.

[2] Johnson, M.A. and G.M. Maggiora, *Concepts and applications of molecular similarity.* 1990, New York: Wiley-Interscience. 393.

[3] Kanehisa, M., *Post-genome informatics.* 2000, Oxford New York: Oxford University Press. 148.

[4] Baldi, P. and S. Brunak, *Bioinformatics: The Machine Learning Approach, Second Edition (Adaptive Computation and Machine Learning).* 2001, Cambridge, MA: MIT Press. 400.

[5] Durbin, R., et al., *Biological Sequence Analysis : Probabilistic Models of Proteins and Nucleic Acids.* 1999, Cambridge: Cambridge University Press. 368.

[6] Ripley, B.D., *Pattern Recognition and Neural Networks.* 1999, Cambridge: Cambridge Univeristy Press. 403.

[7] Cristianini, N. and J. Shawe-Taylor, *An Introduction to Support Vector Machines.* 2000, Cambridge: Cambridge University Press. 189.

[8] Vlahovicek, K., et al., *Concepts of similarity in bioinformatics: Principles of applications to sequences, protein 3D structures and genomes.*, in *Introduction to Bioinformatics*, S. Jelaska and D.S. Moss, Editors. 2003, Kluwer Academic Publishers: Boston, Dordrecht, London. p. in press.

[9] Lord, P.W., et al., *Investigating semantic similarity measures across the Gene Ontology: the relationship between sequence and annotation.* Bioinformatics, 2003. **19**(10): p. 1275-83.

[10] Carugo, O. and S. Pongor, *The evolution of structural databases.* Trends Biotechnol, 2002. **20**(12): p. 498-501.

[11] Csányi, V., *Evolutionary Systems and Society.* First ed. Vol. 1. 1989, Durham and London: Duke University Press. 257.

[12] Kampis, G., *Self-modifying systems in Biology and Cognitive Science.* First ed. International Series in Systems Science and Engineering, ed. G.J. Klir. Vol. 1. 1991, Oxford, New York: Pergamon Press. 543.

[13] Hátsági, Z., V. Skerl, and S. Pongor, *Motifs in Protein Sequences: Towards a unified view on sequence databases*, in *Biotechnology Computing*, L. Hunter, Editor. 1994, IEEE Computer Society Press: Los Alamos, CA. p. 255-264.

[14] Ashburner, M. and S. Lewis, *On ontologies for biologists: the Gene Ontology--untangling the web.* Novartis Found Symp, 2002. **247**: p. 66-80; discussion 80-3, 84-90, 244-52.

[15] Goldmeier, E., *Über die Ähnlichkeit bei gesehenen Figuren.* Psychol. Forsch., 1936. **21**: p. 146-208.

[16] Goldmeier, E., *Similarity in visually percieved forms.* 1 ed. Psychological Issues, ed. H.J. Schlesinger. Vol. 29. 1972, New Yorik, N.Y.: International Universities Press, Inc. 135.

[17] Gentner, D., *The mechanisms of analogical learning*, in *Similarity and Analogical REasoning*, S. Vosniadou and A. Ortony, Editors. 1989, Cambridge, University Press: Cambridge, U.K. p. 199-241.

[18] Johnson, M.A., *A review and examination of mathematical spaces underlying molecular similarity analysis.* Journal of Mathematical Chemistry, 1989. **3**: p. 117-145.

[19] Sali, A. and T.L. Blundell, *Definition of general topological equivalence in protein structures. A procedure involving comparison of properties and relationships through simulated annealing and dynamic programming.* J Mol Biol, 1990. **212**(2): p. 403-28.
[20] Sali, A., et al., *From comparisons of protein sequences and structures to protein modelling and design.* Trends Biochem Sci, 1990. **15**(6): p. 235-40.
[21] Via, A., et al., *Protein surface similarities: a survey of methods to describe and compare protein surfaces.* Cell Mol Life Sci, 2000. **57**(13-14): p. 1970-7.
[22] Via, A., et al., *Three-dimensional view of the surface motif associated with the P-loop structure: cis and trans cases of convergent evolution.* J Mol Biol, 2000. **303**(4): p. 455-65.
[23] Pawlowski, K. and A. Godzik, *Surface map comparison: studying function diversity of homologous proteins.* J Mol Biol, 2001. **309**(3): p. 793-806.
[24] Ankerst, M., et al., *Nearest neighbor classification in 3D protein databases.* Proc Int Conf Intell Syst Mol Biol, 1999: p. 34-43.
[25] Sneath, P.H. and R.R. Sokal, *Numerical Taxonomy*. 1973, San Fransisco: Freeman. 256.
[26] Kabsch, W., *A solution for the best rotation to relate two sets of vectors.* Acta Crystallogr. A, 1976. **32**: p. 922 –923.
[27] Carugo, O. and S. Pongor, *A normalized root-mean-square distance for comparing protein three-dimensional structures.* Protein Sci, 2001. **10**(7): p. 1470-3.
[28] Johnson, M.S. and J.V. Lehtonen, *Comparison of protein three-dimensional structure*, in *Bioinformatics. Sequence, structure and databanks*, D. Higgins and Taylor, W., Editors. 2000, Oxford University Press: Oxford New York. p. 15-50.
[29] Rossmann, M.G. and P. Argos, *Exploring structural homology of proteins.* J Mol Biol, 1976. **105**(1): p. 75-95.
[30] Irving, J.A., J.C. Whisstock, and A.M. Lesk, *Protein structural alignments and functional genomics.* Proteins, 2001. **42**(3): p. 378-82.
[31] Willett, P., *Similarity and clustering in chemical information systems*. 1987, New York: John Wiley & Sons Inc. 254.
[32] Bajic, V.B., *Comparing the success of different prediction software in sequence analysis: a review.* Brief Bioinform, 2000. **1**(3): p. 214-28.
[33] Baldi, P., et al., *Assessing the accuracy of prediction algorithms for classification: an overview.* Bioinformatics, 2000. **16**(5): p. 412-24.
[34] Murvai, J., K. Vlahovicek, and S. Pongor, *A simple probabilistic scoring method for protein domain identification.* Bioinformatics, 2000. **16**(12): p. 1155-6.
[35] Murvai, J., et al., *Prediction of protein functional domains from sequences using artificial neural networks.* Genome Res, 2001. **11**(8): p. 1410-7.
[36] Carugo, O. and S. Pongor, *Protein fold similarity estimated by a probabilistic approach based on C(alpha)-C(alpha) distance comparison.* J Mol Biol, 2002. **315**(4): p. 887-98.
[37] Evans, M., N. Hastings, and B. Peacock, *Statistical Distributions*. 3nd edition (June 15, 2000) ed. 2000: John Wiley & Sons. 221.
[38] Needleman, S.B. and C.D. Wunsch, *A general method applicable to the search for similarities in the amino acid sequence of two proteins.* J Mol Biol, 1970. **48**(3): p. 443-53.
[39] Sellers, P.H., *The theory and computation of evolutionary distances.* Journal of Algorithms, 1980. **1**: p. 359-373.
[40] Smith, E.E. and D.L. Medin, *Catgories and Concepts*. Cognitive Science Series. 1981, Cambridge, MA: Harvard University Press. 203.
[41] Dorogovtsev, S.N. and J.F.F. Mendes, *Evolution of Networks*. 2003, Oxford: Oxford University Press. 264.

*Essays in Bioinformatics*
*D.S. Moss et al. (Eds.)*
*IOS Press, 2005*


# Comparison of sequences, protein 3D structures and genomes

László KAJÁN[1], Kristian VLAHOVICEK[1,2], Oliviero CARUGO[1,3], Vilmos ÁGOSTON[4], Zoltán HEGEDÜS[4] and Sándor PONGOR[1]

[1]*Protein Structure and Bioinformatics Group, International Centre for Genetic Engineering and Biotechnology, Area Science Park, 34012 Trieste, Italy*

[2]*Molecular Biology Department, Biology Division, Faculty of Science, University of Zagreb, 10000 Zagreb, Croatia*

[3]*Department of General Chemistry, Pavia University, viale Taramelli 12, 27100 Pavia, Italy*

[4]*Bioinformatics Group, Biological Research Center, Hungarian Academy of Sciences, Temesvári krt. 626726 Szeged, Hungary*

**Abstract.** The analysis of similarity is a fundamental task in comparing sequences, three dimensional structures as well as genomes and molecular networks. This chapter reviews the common principles underlying these diverse applications.

## Introduction

The basic concepts of similarity analysis – as presented in the first part of this review – provide a common framework for the classification of newly identified the protein sequence or protein 3D structure. Classification of an object implies placing it into the already existing categories or marking it as "unknown" i.e. as a potential initiator of a new category. This process usually consists of the following steps.

*Recognition of similarity*. This is a qualitative decision that is often based on some approximate quantitative measure. In sequence analysis, if the raw alignment score is above a threshold, the similarity is considered significant and retained for further analysis. In the case of protein 3-D structures the preliminary evaluation is often based on visual inspection.

Next, the basis of similarity, i.e. a common substructure is identified. This is carried out by matching of the equivalent entities and relationships, and sequence alignments as well as structural alignments are the best examples. Determination of matching by computers involves maximization of a similarity measure (or minimization of a distance measure), and the final value of the respective parameters is used as a numeric measure of similarity.

*Evaluation of similarity*. First a decision has to be made whether or not the similarity is biologically important, and the protein is either assigned to a known similarity group or it will be considered as the initiator of a new group. This decision is usually based on one or more similarity scores as well as on the alignment, but human judgment is hard to replace and at this stage.

*Representation of similarity in databases*. Once the similarity is established, it has to be added to the annotation of the protein in the sequence and or 3-D databases. Protein superfamilies, structural domains, orthologous groups etc. are determined by similarity analysis, and there is large number of secondary databases that are dedicated to the curation of the underlying similarity groups. Apart from narrative descriptions there are two general avenues to describe similarity groups. Cladograms are classifications that can be established using proximity measures and represent the internal structure of the similarity group. Common patterns on the other hand are usually derived from alignments and represent common substructures present in the members of the similarity group.

The above steps are not always obvious for the users. For example, sequence similarity search programs present the results corresponding to step II, while some of the 3-D similarity search servers provide only a qualitative suggestion corresponding to step I. What is apparent however that all methods include a preliminary, approximate estimation of similarity, followed by a filtering and finally an alignment step.

This section provides a brief overview of how similarity scoring in used in the comparison of sequences, protein 3-D structures and entire genomes. In these fields, similarity measures are used for database searching, for classification and for phylogenetic analysis. A comprehensive overview of these broad fields would be far beyond the scope of this chapter. Instead, we will attempt to highlight, using the terminology introduced in the previous sections, the common themes underlying these three diverse areas.

## 1. Sequence comparison

Sequences are the simplest descriptions of macromolecules that use residues (amino acids, nucleotides) as entities and sequential vicinity as the only relationship between them. Sequence comparison algorithms use essentially the same principle for similarity scoring. The simple proximity measure is related to the Hamming distance (i.e. no gaps allowed, as shown in Fig. 3.2). The scoring matrices used in DNA as well as protein comparisons are constructed in such a way that similar residues give high scores, so the resulting measure can be called a Hamming similarity measure, rather than a distance. The optimizable substructure similarity is the string similarity measure (equation 5) in which the position and number of the gaps as well as the range of alignment is determined by optimization. The result is a maximal matching, and the alignment score is a *local* or *global* maximum value depending on the algorithm used. Algorithms of global alignment (Needleman-Wunsch, [1]) or local alignment (FASTA [2], BLAST [3],) have been the subject of several excellent, recent reviews [see, e.g. [4,5]], the current section focuses on the principles of scoring, i.e. how a similarity score is transformed into a probabilistic measure.

We will use a simple classification: *General methods* of comparison use a general statistical description of random similarities for calculating the significance value to alignment scores. *Specific methods* use application-specific descriptions of the biologically important target groups, such as protein families, domain sequence groups etc. These groups are often too small for statistics, so specific methods rely instead on additional, *a priori* knowledge.

The most frequently used general methods (BLAST [3], FASTA [2], Smith-Waterman [6]) are based on local sequence alignment. The resulting sequence similarity scores do not preserve the metric properties (can not be converted into metric distances), on the other hand they have the advantage that the distribution of random similarities can be described in an analytical form. This is because scores are maximal values, and the maximum of a large number of independent identically distributed (i.i.d) random variables tends to an extreme value (or Gumbell) distribution, just as the sum of a large number of i.i.d. random variables tends to a normal distribution [7]. The underlying statistics was described in detail by Karlin and Altschul [8,9] for the *BLAST* program. Originally, BLAST used local alignments without gaps called high-scoring segment pair (or *HSP*), in which scores were maximized in the sense that they could not be further improved by extension or trimming. We will use HSPs as an example, adding that the description of gapped BLAST, FASTA and Smith/Waterman scores follows a similar statistics. The random emergence of *HSPs* was studied on random sequences in which the occurrence amino acid residues is independent, with specific background probabilities for the various residues. For two sufficiently long ($m$ and $n$) sequences, the expected number of HSPs with score at least $S$ is given by the formula

[1] $$E = Kmne^{-\lambda S}$$

where K and $\lambda$ are constants that can be considered a can be as natural scales for the search space of size $m \times n$ and the scoring system. The raw score $S$ is defined by a formula given in figure x. The number of random HSPs with score $\geq S$ is described by a Poisson distribution and the probability of finding at least one such HSP is

[2] $$P = 1 - e^{-E}$$

*P* is the statistical significance, the probability of finding a score *S* (or bigger) by chance. It is important to note that this simple statistics is also approximately valid for gapped alignments used by modern alignment programs, and this makes it possible to give a more objective, probabilistic interpretation to similarity scores.

Global alignments are found via an exhaustive search for the maximal matching between two sequences, based on such methods as the Needleman-Wunsch algorithm [1]. Global alignment scores can be transformed to metric distance scores, which is important for clustering. On the other hand, very little is known about the random distribution of optimal global alignment scores, so a rigorous probabilistic interpretation is not possible in this case. A practical approach is based on generating many random sequence pairs of the appropriate length and composition, and calculating the optimal alignment score for each. The average $S_r$ and the standard deviation $\sigma_r$ of the random scores can then be compared with original score $S$ score, and a Z score

[3] $$Z = \frac{|S - S_r|}{\sigma_r}$$

can be used as an approximate measure of significance. Namely, even though Z resembles the Student *t* value, but rigorously speaking it cannot be converted into a *P* value since the underlying distribution is not a normal distribution. Only an approximate interpretation is thus possible, for example if 100 random alignments have scores inferior to the alignment of interest, the *P*-value in question is likely less than 0.01. It is important to note that the meaning of this statistics is different from the one derived from a database of random similarities (equation 16). Namely, for two sequences of similar, but unusual amino acid composition, the Z-score may be a low value, even is the two sequences compared are both very different from the rest of the database.

The general methods of sequence comparison can be used to divide the sequence database into clusters. In principle, a metric distance measure (such as can be derived from global alignment scores) is a prerequisite for statistical clustering. Given the large size of databases, both global alignments and statistical clustering methods are compute-intensive. On the other hand, the protein sequence space is sparsely populated and the existing natural sequences form well-separated clusters, which makes it possible to use efficient, approximate methods for clustering. Krause and Vingron used a threshold-based, iterative procedure based on BLAST for identifying consistent protein clusters [10,11]. The result an objective picture of the sequence space in terms of similarities, but the clusters have to be compared with knowledge based groups, such as protein families etc. With this approach, protein domains that are shared among several protein families lead to the merging of protein family clusters.

A sharp distinction between biologically significant and random similarities is not possible from the scores alone – such decisions still require *a priori* knowledge, namely biological knowledge (e.g. knowledge of the overall domain structure of the protein, the exon-structure of the genes) as well as a knowledge of the previously known similar sequences. In addition to the general methods of sequence comparison mentioned above, there are a number of dedicated specific methods, based on some explicit representation of biologically important similarity groups such as protein domain sequences. A sequence similarity group can be represented by a consensus description that represents e.g. a sequence pattern that is shared by all members of the group. As such patterns can be obtained by multiple sequence alignments, there is a large variety of algorithms that represent multiple alignments in terms of consensus sequences, regular expressions, position-specific scoring matrices or profiles, hidden Markov models (HMMs) or neural networks (for recent reviews see [5,12]). These consensus descriptions can then be used to decide whether or not a new query sequence is member of a given similarity group. The similarity measures used to compare a query with these representations are similar to the ones described in this review, the details can be found in the original publications as well as the reviews cited above.

Another group of specific approaches uses a graph-theoretical representation of similarity groups, which is an exemplar-based description. Sequences within a similarity group are related to each other by specific similarity (Figure 3.1.), for example each member of the group is related to at least one other member with a similarity score greater that a certain threshold [13]. Protein domains are typical examples of well-defined similarity groups. On the other hand, many of the known proteins are composed of modules, so the score determined between two such proteins will express the similarity of the building blocks, rather than that of the two proteins.

The similarities of protein domain groups can be defined on a threshold basis. In the SBASE protein domain sequence library, a sequence is considered as member of a domain group if it is similar to at least $NSD_t$ members of the group, with an average similarity score of $AVS_t$ where $NSD_t$ and $AVS_t$ are threshold values automatically determined from a database vs. database comparison with the BLAST program. A later extension of this scoring system takes into consideration the distribution of similarity scores in the neighborhood of each similarity group and uses a probabilistic score. For each raw scores, four probability values are read from the precomputed distributions shown in **Figure 1**, and the score is derived from the sum of these distributions [14]. From the computational point of view, this approach is similar to the memory-based computing paradigm [15], the memory of the system is a database vs. database comparison [16,17].

The approach underlying the COG (Clusters of Orthologous Sequences) databank is based on grouping sequences together that are mutually the nearest neighbours of each other in terms of sequence similarity score [18]. Such tight groups or cliques can be extended to larger similarity groups, which is the basis of identifying orthologous proteins. This approach is especially successful in prokaryotic genomes in which multidomain proteins are not abundant.

Recent approaches combine many of the previous concepts. The underlying philosophy is that database search results should contain all information necessary to find distant similarities – such as the weak similarities of protein domains – and that these might be found via a clever sorting of the search results. Namely, the alignment scores (an the *P* values) traditionally used to sort the result constitute only one dimension of the sorting.

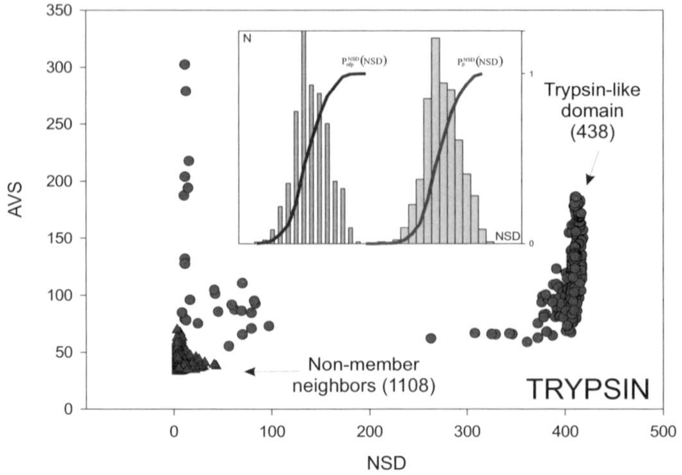


$$C = TP_{NSD} + P_p(NSD), + P_{nfp}(NSD), + TP_{AVS} + P_p(AVS) + P_{nfp}(AVS),$$

**Figure 1.** The principle of classifying domains in SBASE [14] (See text for explanations).

Alignments can be sorted according to their position within the query, as well as according to their common sequence patterns. Recent versions of BLAST, incorporate position specific scoring known from profile methods (PSI-BLAST) as well as pattern-specific searches (PHI-BLAST) [19-21].

Given the ease and speed of current sequence alignment algorithms, approximate methods based on unstructured descriptions are used only in specific applications. In composition-based methods, the sequences are described as vectors, in terms of the amino acid, dipeptide, tripeptide etc. composition, and the comparison is based on simple distances such as the Euclidean distance. Same as with other unstructured descriptors, the calculation is very fast, especially since the database can be stored in the form of pre-calculated vectors. The number of vector components (the resolution of the description) has to be selected with care, and this is done either heuristically, or using an algorithm to automatically select and/or weight those amino acid words that give the best separation between a test group and a control group. In this manner group-specific distance functions can be developed. The resolution of the description can be fine-tuned e.g. by decreasing the amino acid alphabet (to 4,5, etc. letter alphabets instead of 20) and or by increasing the word size (dipeptides, gapped dipeptides, tripeptides etc.). Examples include the composition-based protein sequence search of Hobohm and Sander [22], as well as the promoter-search program of Werner et al. [23-25]. Simple applications include the recognition of coding regions based on codon-usage. Composition-based methods are very useful for building recognizers for any sequence group for which a sufficient number of examples are known. Given a test group and a control group of sequences, one can compare the frequency of arbitrary words (provided as a list) between these two groups. The most characteristic words can be selected based on simple measures such as the Mahalanobis distance, and used for recognizing potential new members of the test group [26]. Similar algorithms are often used in gene prediction systems [27].

Distributions are less frequently used for representing sequences, even though methods of comparing sequence profiles such as hydrophobicity plots, secondary structure propensity plots were developed already in the 1980-es. Fourier transforms of hydrophobicity plots have been used to recognize amphipathic helices as well as to build classifiers to various protein groups. A review on these applications is in [28].

## 2. Comparison of 3D structures

Comparison of 3D structure is used in a variety of fields such as fold recognition, structural evolution studies and drug design, and the protocols are as diverse as the fields themselves. E.g. in the comparison of 3D structures produced on the same protein molecule by NMR methods, all the equivalent atom-pairs are a priori known and can be used in the comparison. In contrast, determination of folds is based on the backbone Cα atoms only and the equivalences have to be determined by the calculation itself. In this section we will briefly summarize the similarity/distance functions used for backbone comparison, concentrating on the similarity/distance measures used rather than the goal and/or implementation of the actual algorithms. In the majority of the cases, the approach used for structural alignments is quite similar to that used in sequence analysis (finding alignment paths in a distance matrix or optimizing the range by successive omission or additions). This is because 3D structures can be compared in terms of their (overlapping) peptide fragments, and a series of peptide fragments is a linear, sequence-like representation. For example, one can compute an *rmsd* between the peptide fragments of two proteins and construct a distance-matrix with the resulting values [29,30]. But there are many ways to represent peptide fragments as vectors, and then one can use any of the vector-distance formulas to produce the values of the distance matrix. For example, vectors of torsional angles [31,32], curvature and torsion parameters of peptide fragments [33,34] have been used by early comparison methods, as reviewed by Orengo [35]. More recent methods include structural alphabets described in terms of dihedral angles [36,37] or on distance geometry [38,39]. In the latter method, the size of the alphabet (the minimum number of fragments necessary to describe the observed data) is 27 derived from statistical optimisation. The similarity search is then carried out by Smith-Waterman alignment.

The similarity measures described in this section can be classified according to the use of atomic (residue-based) descriptions, or higher-order descriptions such as secondary structure elements. Another important difference is that some of the methods can be used to produce structural alignments while others are only preliminary filters indicating similarity without providing a structural alignment.

Methods based on superposition of atoms use the *rmsd* distance (section x, above) Even though the results of atom superposition methods are generally considered superior to most computational alternatives, and very low *rmsd* values are indicative of identical structures – rmsd can be used only with caution as a quantitative indicator of similarity. In addition, there is no accepted and reliable statistical model that would allow to use *rmsd* as a probabilistic score with a statistical significance, moreover *rmsd* does not penalize gaps. Therefore there a number of alternative similarity scores have been developed for obtaining optimal structural alignments even though the final results are always characterized in terms of the rmsd score.

One group of similarity scores is based on vectors or sets of vectors assigned to each position within a protein structure. The parameters of the vector represent various features. Methods developed by Taylor and Orengo [40,41] assigned a set of intramolecular $C_\alpha$–$C_\alpha$ vectors to each residue position, or used various geometric features as parameters of the vector assigned to each residue position. As a result, a protein structure was converted into a series of residue vectors, and two structures could be compared to give a so-called residue matrix in which the elements are calculated as a vectorial difference (city-block distance of vectors, equation [2]). The optimal structural alignment can be determined by a dynamic programming algorithm.

A roughly similar approach was used by Holm and Sander for the very popular DALI server [42]. In the underlying method the $C_\alpha$ atoms are characterized by vectors the parameters of which are the elements of distance matrix. The local vectors are then compared in terms of residue similarity scores such as

[4] $$\phi^R(i,j) = \theta^R - \left|d_{ij}^A - d_{ij}^B\right|$$

or

[5] $$\phi^E(i,j) = \left(\theta^E - \frac{\left|d_{ij}^A - d_{ij}^B\right|}{d_{ij}^*}\right) e^{-(d_{ij}^*)^2/\alpha}$$

The subscript A,B refer to residues in structure $d_{ij}$ are the elements of the hexapeptide distance matrices i.e. elements of the residue vectors. $d_{ij}^*$ denotes the average of $d_{ij}^A$ and $d_{ij}^B$, $\theta^R, \theta^E$ and α are constant. A and B, respectively. Superscript $R$ denotes rigid comparison [eqn. 4], E refers to an elastic comparison dampened by a negative exponential term [eqn.5]. As can be seen, summing the residues similarity measures $\phi^R$ or $\phi^E$ results in quantities related to the city block distance. Comparison of two proteins A and B is then carried out using a distance matrix whose elements are equal to either $\phi^R(i,j)$ or $\phi^E(i,j)$, where i and j refer to two pairs of structurally aligned residues: i(A), i(B), j(A), and j(B). The optimization task is to find the best set of equivalences between A and B that maximize this function and the structural alignment is obtained by an optimization algorithm (Monte Carlo optimization) To improve convergence, various heuristics are used to obtain a reasonable starting point.

The residue similarity score of Levitt and Gerstein [43] has the formula

[6] $$S_{i,j} = M/(1+(d_{ij}/d_0)^2)$$

where $d_{ij}$ is the distance between $C_\alpha$ atoms of the two structures compared, $M$ and $d_0$ are constants. $S_{ij}$ values are elements of a similarity matrix from which an optimizeable substructure similarity measure $S_{str}$ can be calculated by introducing gaps. The $S_{str}$ score is defined as

[7] $$S_{str} = M(\sum_{ij} 1/(1+(d_{ij}/d_0)^2) - N_{gap}/2)$$

The structural alignment is carried out with a dynamic programming method such as the Smith-Waterman algorithm. Levitt and Gerstein found that random structural similarities determined by this method follow the same extreme value distribution as BLAST scores (or Smith-Waterman sequence alignment scores), so the results can be characterized in terms of $P$ values [43].

As superposition methods are compute intensive, a number of simplified representations have been developed. One general strategy is to represent the protein by a set of secondary structure elements (SSEs), characterized by their position within the polypeptide sequence and the position in 3D space and are usually represented as vectors fit to the $C_\alpha$ atoms. This is another kind of entity-relationship description in which SSEs are the nodes and a variety of parameters (such as distances, angles ec) are used to describe relationships. The rationale is that superposition of a few SSEs is less compute intensive than superposing a large number of $C_\alpha$ atoms, so one can use algorithms that could not cope with large atomic detail structures. In addition, SSEs incorporate added knowledge on molecular geometry. The success of the process depends on i) how secondary structures are assigned; ii) how the similarity between two secondary structural elements of two proteins is estimated; iii) how the overall similarity between the two proteins is defined.

Although the SSEs (at least the most common like helices and strands) are clearly defined, different assignment result from different assignment algorithms [44-46]. Consequently, different representations of the protein structures may arise. A further problem is which SSE types are considered. Very often a two-states classification is used: helix, including 3/10 and pi, and stand. There are nevertheless exceptions. Orengo et al. [44-46], for example, adopt a three-states classification: alpha-helix, 3/10-helix, and strand.

The similarity between secondary structural elements in two proteins is usually estimated by comparing each *pair of SSEs* of one protein with each pair of the other. The 3D arrangement of a two secondary structural elements in a protein is usually defined by their distance, their plane angle, and their torsion. A similarity score can then be computed for each pair of two secondary structural elements. The resulting matrix of similarity scores can then be scrutinized with dynamic programming techniques [41,47-49], treated as a maximum clique problem [50], with pseudo-distance matrices [51], or with cluster analysis [52]. The alignment of the secondary structural elements is eventually followed by a superposition of the $C_\alpha$ atoms with an initial structural alignment that depends on the secondary structure alignment. The overall similarity between the two structures can be then estimated on the basis of the *rmsd* values [50] of with more sophisticated figures of merit that considers also the quality of the secondary structure fit.

The fragment-pair approach is also amenable to probabilistic interpretation. The VAST program of Bryant and coworkers [53,54] provides BLAST-like *P* significance values. VAST's elementary unit of comparison is a simplified *rmsd* score resulting from a superposition of the endpoints of SSE pairs "trimmed" to the same length. First *rmsd* values are converted into log-odds scores using precomputed values of comparison of SSE pairs from related and unrelated structures, then a combined score $S_o$ is calculated from the *i* best SSE pairs found to mattch between the query and a database entry. The principle of converting $S_o$ into a *P* value is similar to that used by BLAST, given in equations. 15-17, but relies on tabulated statistics, rather then on analytical formulae. Let the probability of finding a substructure of size *i* with a score $S_i \geq S_o$ be denoted as $P(S_i \geq S_o)$. In VAST, the value of $P(S_i \geq S_o)$ is estimated as a function of *i* and $S_i$, using tabulated values resulting from random comparisons. The expected number *E* of finding at least one score $S_i \geq S_o$ by chance will also depend on the size of the search space which can be defined as the total number of possible common substructures of *i* SSEs between the two proteins, a number denoted by $N_i$. The equation computed by VAST is then

[8] $$E = \sum_i N_i P(S_i \geq S_o)$$

The sum is calculated for all *i* values using the tabulated $P(S_i \geq S_o$ values. Same as with BLAST, if *E* is small (e.g. E<0.01) it is also a *P* value. The method is very fast, due to the precomputed statistics, and accessible at the NCBI web site.

A variety of other procedures that represent the protein 3-D structure as an ensemble of secondary structural elements have also been proposed. In Martin's approach [55], secondary structural elements are given one of the letters of an alphabet that identify the secondary structure type, direction, length, and solvent accessibility. Two proteins can be thus compared with the simple Needlemann-Wunsch algorithm. Murthy [56] used dynamic programming techniques to optimally superpose secondary structural elements.

Mitchel [57] developed a graph-like representation, using secondary structural elements as nodes, and angles and distances as edges, the largest common substructure was then identified by subgraph isomorphism algorithm developed by Grindley et al. [58]. Harrison and associates [59] further developed this approach and introduced a similarity measure $S_{grath}$ based on the number of SSEs and residues in two proteins and in the largest common subgraph:

[9] $$S_{grath} = \frac{5}{8}\left(\sqrt{\frac{CS}{SS1}\frac{CS}{SS2}}\right) + \frac{Min(R1,R2)}{Max(R1,R2)} + 2\left(\sqrt{\frac{CR1}{R1}\frac{CR2}{R2}}\right)$$

where the two proteins compared have *SS1* and *SS2* secondary structures and *R1* and *R2* amino acid residues, respectively, their comparison generates a largest clique of size *CS*. The largest clique is produced from a set of secondary structures that contain a total of *CR1* and *CR2* residues in protein 1 and protein 2, respectively. This similarity measure is reported to be independent of fold size and was used to characterize the fold space represented by the CATH database [59].

Finally, there are methods that do not use superposition but define simple similarity scores instead. PRIDE [60] is based on the distribution of intramolecular $C_\alpha$-$C_\alpha$ distances incorporated into a set of histograms for $C_\alpha$ pairs separated by 3 to 30 residues. The comparison between two proteins is thus reduced to the comparison of 28 distribution pairs, which can be carried out by a standard statistical method of contingency table analysis and yields a probability value. The average value of these 28 single similarity scores was defined as the Probability of identity or PRIDE score [60]. Pride has a value between 0 and 1, and has metric properties which makes it suitable for clustering large datasets. The calculation is extremely fast (perhaps the fastest available today), database search and fold assignment, clustering of structures are possible on line. When used as a simple nearest neighbor classifier, PRIDE reaches 99.5% success in fold recognition, based on the C,A,T, H classes of the CATH database. This method is available via a web server at ICGEB.

Another recent, fast comparison method by [61] uses a vector representation of protein folds which is based on topological invariants called Gauss integrals, each representing a topological property of the backbone space curve [62]. 30 such integrals are calculated for two proteins, which are then compared in terms of a 30 dimensional Euclidean distance. A classifier built on Gauss integrals has a reported accuracy of 96.8% on the C,A,T classes of the CATH database [61].

## 3. Genomes, proteomes, networks

Designing representations for genomes, proteomes and networks is a real challenge and we are only at the first steps of this new era. The representations in current use follow the entity-relationship tradition, for example genomes are represented as linear array of genes and other DNA segments. The entities – genes – are predicted with gene-prediction programs or are determined experimental methods, and this adds a new layer of knowledge to the molecular data. The relationships are manifold but are predominantly binary in nature. Examples of relations include physical vicinity, distance along the chromosome, regulatory links extracted from DNA chip data and so on. The resulting picture is a graph of several ten-thousand nodes and relatively few edges per node denoting various relationships. The description of proteomes is only somewhat different. The proteins are described in functional, biochemical and structural terms, and the relationships between proteins include metabolic relationships (sharing substrates in metabolic pathways) as well as structural relationships (sequence and structural similarities). Even this sketchy introduction implies that we deal with new a kind of complexity that originates, from the numerous and to a large extent, unknown interactions between the molecules. On the other hand, the study of large networks – such as Internet, social- road- and electric networks, etc. – has provided interesting insights that have been successfully applied to genomes, proteomes and bibliographic networks.

From the computational point of view, genomes and proteomes are described as very large graphs in which the nodes (genes, proteins) and the edges (relations) are unknown or unsure. These large and fuzzy descriptions are in sharp contrast with the descriptions developed for well-defined molecular structures, but the methods are not dissimilar to those used in other fields. Given the large and different genome sizes as well as the uncertainties of the data, structured descriptions are not very useful for comparison. Simple, unstructured representations like sets or vectors that can be easily compared in terms of their known components are widely used. The approaches differ how the components are selected and compared. (It is noted that this section concentrates only on those genome-comparison studies that use genome-level descriptions. For phylogenetic approaches to genome comparison see [63,64].

One group of approaches use predefined components, given in the form of a classification. Proteins can be classified into several thousand orthologous groups (COGs) and a genome (proteome) can be described by as a vector with a corresponding number of components, each component denoting the presence or absence of a given protein group [63]. This is an extremely simplified unstructured description, but the selected components of the vector adequately describe the entire universe of protein functions as we know it today. Two such vectors can then be compared using the Jaccard coefficient, and a related distance measure (1 minus the coefficient) can be used as a metric for classifying the genomes. This is a fast procedure that has no adjustable parameters, nevertheless it gave results in good agreement with other, more subjective methods. In a similar way, proteins can be grouped according to their similarity to sequences representing known 3D folds. In a similar manner, the genomes can be classified in terms of the 3D folds[65,66].

Metabolic data are a further example of predefined component classification that can be used for genome comparison [67,68]. A proteome can be described in terms of the constituting enzymes, substrates, intermediate complexes such as given in the WIT database [69]. Organism data can then be converted into vectors representing enzymes or substrates in pathways or pathway-groups. For example, in a vector representing the metabolic pathways of *E. coli* in terms of enzymes, a parameter $e_i$ is an integer denoting the number of times enzyme $i$ occurs in the metabolic pathways of the organism. Such vectors can then be compared using any of the vector similarity/distance measures described above. A classification of genomes based on vectors representing the metabolic and information-processing pathways in terms of enzymes and substrates has shown that the system-level organization of *Archea* and *Eukarya* are similar [70]. This comparison was based partly on presence/absence data and the Jaccard coefficient, and partly on comparing the ranking data of component frequencies.

Another group of representations uses sequence comparison to dynamically define matching components between two genomes. The matching pairs of genes can be selected based on BLAST scores [71], Smith-Waterman scores [72,73]. The intergenomic distances can then be based on the list of shared (as well as total) components present in two genomes, using e.g. the Jaccard coefficient. A particularly interesting version of this method uses vicinal gene-pairs with conserved direction of transcription, identified from Smith-Waterman searches [74-76]. Given the matching vicinal pairs in the two genomes as substructures, one can proceed in the usual way. This method thus preserves the speed of the comparison but uses substructures that are richer in detail i.e. capture a part of the gene order.

Unfortunately, the gene-based substructures cannot be increased without practical limits: at present, quantitative genome comparisons are seemingly limited to gene (protein) pairs. On the other hand, higher-order patterns can be very informative in the qualitative sense. Studies on conserved local gene-order revealed that in addition to the known operons, there are larger units – über-operons or super-operons – that are conserved in terms of functional and regulatory context [77,78]. This takes us to a familiar world of known patterns: operons and metabolic pathways are both higher order patterns (directed graphs) defined in terms of entities and relationships. Comparison of related metabolic pathways is a subgraph isomorphism problem [79], and related techniques underlie the Kyoto encyclopedia of genes and genomes [80].

The study of technological networks such as the Internet, has provided important insights into genetic and metabolic networks. The methods of comparison are qualitative, rather than quantitative. The currently known network types (scale-free, small-world, modular and random networks) all have been observed in various biological systems [81]. Identification of network type is based on graph-measures, such as clustering coefficients, betweenness centrality, etc. [82,83]. In principle, any numeric measure that can be "locally" computed for vertices or edges of a graph, can be used to draw a distribution. In fact, the main network types are currently qualitatively defined by these distributions. For instance, the number of connections (i.e. the degree) in scale-free networks that are characteristic of many biological systems can be described by a power-law type degree distribution [84]. They can be further subdivided into groups based on the distribution of betweenness-centrality [85], or of the local clustering coefficient[85]. Network patterns defined as small directed subgraphs are also used to characterize network classes, statistical studies revealed similar network patterns shared by genetic and electronic networks [86-88].

## 4. Conclusions

The framework of entity-relationship networks provides a simple method to describe similarity groups (equivalence classes), patterns, similarity measures that are used in the comparison of sequences as well as protein 3D structures. The strength of this analysis is shown by the fact that it can be extended to the analysis of large and fuzzy structures, such as genomes and networks.

## Acknowledgements

This review is partly based on the lectures of the course "Bioinformatics: Computer applications in molecular biology", held in Trieste, Italy, 1992-2003. Special thanks are due to M. Bishop (Hinxton, UK), E. Gasteiger (Geneva, Switzerland), R. Harper (Hinxton, UK). D. Judge (Cambridge, UK), D. Landsman (Bethesda, MD), J. Leunissen (Wageningen, The Netherlands) for advice, as well as the following individuals for their comments on various topics in the manuscript: Stephen Altschul, Steve Bryant (Bethesda, US), Alexandre De Leon, (Calgary, Canada) Jacques Demongeot (Grenoble, France), Mark Gerstein (New Haven, UK), Andrew Harrison (London, UK), Lisa Holm (Hinxton, UK), Christine Orengo (London, UK) and William F. Pearson (US)

## References

[1] Needleman SB, Wunsch CD. A general method applicable to the search for similarities in the amino acid sequence of two proteins. J Mol Biol 1970;48 (3):443-53.

[2] Pearson WR, Lipman DJ. Improved tools for biological sequence comparison. Proc Natl Acad Sci U S A 1988;85 (8):2444-8.
[3] Altschul SF, Gish W, Miller W, Myers EW, Lipman DJ. Basic local alignment search tool. J Mol Biol 1990;215 (3):403-10.
[4] Koonin EV, Galperin MY. Sequence, evolution, function. Boston, Dordrecht, London: Kluver Academic Publishers, 2003.
[5] Higgins D, Taylor WR. Bioinformatics, Sequence, structure, and databanks. Oxford, New York: Oxford University Press, 2000.
[6] Smith TF, Waterman MS. Identification of common molecular subsequences. J. Mol. Biol. 1981;147:195-7.
[7] Gumbel EJ. Statistics of extremes. New York, NY.: Columbia University Press, 1958.
[8] Karlin S, Altschul SF. Applications and statistics for multiple high-scoring segments in molecular sequences. Proc Natl Acad Sci U S A 1993;90 (12):5873-7.
[9] Karlin S, Altschul SF. Methods for assessing the statistical significance of molecular sequence features by using general scoring schemes. Proc Natl Acad Sci U S A 1990;87 (6):2264-8.
[10] Krause A, Vingron M. A set-theoretic approach to database searching and clustering. Bioinformatics 1998;14 (5):430-8.
[11] Krause A, Stoye J, Vingron M. The SYSTERS protein sequence cluster set. Nucleic Acids Res 2000;28 (1):270-2.
[12] Attwood TK. The role of pattern databases in sequence analysis. Brief Bioinform 2000;1 (1):45-59.
[13] Murvai J, Vlahovicek K, Pongor S. A memory-based approach to protein sequence similarity searching. In: Pifat G, editor. Supramolecular Structure and Function. Dordrecht/Plenum Press, New York, USA: Kluwer Scientific, 2001. pp. 167-84.
[14] Murvai J, Vlahovicek K, Pongor S. A simple probabilistic scoring method for protein domain identification. Bioinformatics 2000;16 (12):1155-6.
[15] Stanfill C, Waltz D. Toward memory-based reasoning. Communications of the ACM 1986;29 (12):1213-28.
[16] Murvai J, Vlahovicek K, Szepesvari C, Pongor S. Prediction of protein functional domains from sequences using artificial neural networks. Genome Res 2001;11 (8):1410-7.
[17] Vlahovicek K, Carugo O, Murvai J, Pongor S. Prediction of protein structure and function: Towards a memory-based interpretation of proteome data. In: Gromiha M, Selvaraj S, editors. Recent Research Developments in Protein Folding Stability and Design. Trivandrum, India: Research Signpost, 2001. pp. 141-50.
[18] Tatusov RL, Natale DA, Garkavtsev IV, Tatusova TA, Shankavaram UT, Rao BS, Kiryutin B, Galperin MY, Fedorova ND, Koonin EV. The COG database: new developments in phylogenetic classification of proteins from complete genomes. Nucleic Acids Res 2001;29 (1):22-8.
[19] Altschul SF, Madden TL, Schaffer AA, Zhang J, Zhang Z, Miller W, Lipman DJ. Gapped BLAST and PSI-BLAST: a new generation of protein database search programs. Nucleic Acids Res 1997;25 (17):3389-402.
[20] Schaffer AA, Wolf YI, Ponting CP, Koonin EV, Aravind L, Altschul SF. IMPALA: matching a protein sequence against a collection of PSI-BLAST- constructed position-specific score matrices. Bioinformatics 1999;15 (12):1000-11.
[21] Schaffer AA, Aravind L, Madden TL, Shavirin S, Spouge JL, Wolf YI, Koonin EV, Altschul SF. Improving the accuracy of PSI-BLAST protein database searches with composition-based statistics and other refinements. Nucleic Acids Res 2001;29 (14):2994-3005.
[22] Hobohm U, Sander C. A sequence property approach to searching protein databases. J Mol Biol 1995;251 (3):390-9.
[23] Werner T. The promoter connection. Nat Genet 2001;29 (2):105-6.
[24] Werner T. Promoter analysis. Ernst Schering Res Found Workshop 2002;38:65-82.
[25] Frech K, Danescu-Mayer J, Werner T. A novel method to develop highly specific models for regulatory units detects a new LTR in GenBank which contains a functional promoter. J Mol Biol 1997;270 (5):674-87.
[26] Solovyev VV, Makarova KS. A novel method of protein sequence classification based on oligopeptide frequency analysis and its application to search for functional sites and to domain localization. Comput Appl Biosci 1993;9 (1):17-24.
[27] Solovyev VV, Salamov AA. INFOGENE: a database of known gene structures and predicted genes and proteins in sequences of genome sequencing projects. Nucleic Acids Res 1999;27 (1):248-50.
[28] Pongor S. The use of structural profiles and parametric sequence comparison in the rational design of polypeptides. Methods Enzymol 1987;154:450-73.
[29] Remington SJ, Matthews BW. A systematic approach to the comparison of protein structures. J Mol Biol 1980;140 (1):77-99.

[30] Remington SJ, Matthews BW. A general method to assess similarity of protein structures, with applications to T4 bacteriophage lysozyme. Proc Natl Acad Sci U S A 1978;75 (5):2180-4.

[31] Levine M, Stuart D, Williams J. A method for the systematic comparison of the three-dimensional structures of proteins and some results. Acta Cryst. 1984;A40:600-10.

[32] Karpen ME, de Haseth PL, Neet KE. Comparing short protein substructures by a method based on backbone torsion angles. Proteins 1989;6 (2):155-67.

[33] Rackovsky S. Quantitative organization of the known protein x-ray structures. I. Methods and short-length-scale results. Proteins 1990;7 (4):378-402.

[34] Rackovsky S, Scheraga HA. Influence of ordered backbone structure on protein folding. A study of some simple models. Macromolecules 1978;11 (1):1-8.

[35] Orengo CA. A review of methods for protein structure comparison. In: Taylor WR, editor. Patterns in Protein Sequence and Structure. Heidelberg: Springer-Verlag, 1992. pp. 159-88.

[36] De Brevern AG, Benros C, Gautier R, Valadie H, Hazout S, Etchebest C. Local backbone structure prediction of proteins. In Silico Biol 2004;4 (2):0031.

[37] de Brevern AG, Valadie H, Hazout S, Etchebest C. Extension of a local backbone description using a structural alphabet: a new approach to the sequence-structure relationship. Protein Sci 2002;11 (12):2871-86.

[38] Camproux AC, Gautier R, Tuffery P. A hidden markov model derived structural alphabet for proteins. J Mol Biol 2004;339 (3):591-605.

[39] Guyon F, Camproux AC, Hochez J, Tuffery P. SA-Search: a web tool for protein structure mining based on a Structural Alphabet. Nucleic Acids Res 2004;32 (Web Server issue):W545-8.

[40] Taylor WR, Orengo CA. Protein structure alignment. J Mol Biol 1989;208 (1):1-22.

[41] Orengo CA, Brown NP, Taylor WR. Fast structure alignment for protein databank searching. Proteins 1992;14 (2):139-67.

[42] Holm L, Sander C. Protein structure comparison by alignment of distance matrices. J Mol Biol 1993;233 (1):123-38.

[43] Levitt M, Gerstein M. A unified statistical framework for sequence comparison and structure comparison. Proc Natl Acad Sci U S A 1998;95 (11):5913-20.

[44] Colloc'h N, Etchebest C, Thoreau E, Henrissat B, Mornon JP. Comparison of three algorithms for the assignment of secondary structure in proteins: the advantages of a consensus assignment. Protein Eng 1993;6 (4):377-82.

[45] Andersen CA, Bohr H, Brunak S. Protein secondary structure: category assignment and predictability. FEBS Lett 2001;507 (1):6-10.

[46] Andersen CA, Palmer AG, Brunak S, Rost B. Continuum secondary structure captures protein flexibility. Structure (Camb) 2002;10 (2):175-84.

[47] Yang AS, Honig B. An integrated approach to the analysis and modeling of protein sequences and structures. I. Protein structural alignment and a quantitative measure for protein structural distance. J Mol Biol 2000;301 (3):665-78.

[48] Yang AS, Honig B. An integrated approach to the analysis and modeling of protein sequences and structures. II. On the relationship between sequence and structural similarity for proteins that are not obviously related in sequence. J Mol Biol 2000;301 (3):679-89.

[49] Yang AS, Honig B. An integrated approach to the analysis and modeling of protein sequences and structures. III. A comparative study of sequence conservation in protein structural families using multiple structural alignments. J Mol Biol 2000;301 (3):691-711.

[50] Alexandrov NN, Fischer D. Analysis of topological and nontopological structural similarities in the PDB: new examples with old structures. Proteins 1996;25 (3):354-65.

[51] Richards FM, Kundrot CE. Identification of structural motifs from protein coordinate data: secondary structure and first-level supersecondary structure. Proteins 1988;3 (2):71-84.

[52] Mizuguchi K, Go N. Comparison of spatial arrangements of secondary structural elements in proteins. Protein Eng 1995;8 (4):353-62.

[53] Madej T, Gibrat JF, Bryant SH. Threading a database of protein cores. Proteins 1995;23 (3):356-69.

[54] Gibrat JF, Madej T, Bryant SH. Surprising similarities in structure comparison. Curr Opin Struct Biol 1996;6 (3):377-85.

[55] Martin AC. The ups and downs of protein topology; rapid comparison of protein structure. Protein Eng 2000;13 (12):829-37.

[56] Murthy MR. A fast method of comparing protein structures. FEBS Lett 1984;168 (1):97-102.

[57] Mitchell EM, Artymiuk PJ, Rice DW, Willett P. Use of techniques derived from graph theory to compare secondary structure motifs in proteins. J Mol Biol 1990;212 (1):151-66.

[58] Grindley HM, Artymiuk PJ, Rice DW, Willett P. Identification of tertiary structure resemblance in proteins using a maximal common subgraph isomorphism algorithm. J Mol Biol 1993;229 (3):707-21.

[59] Harrison A, Pearl F, Mott R, Thornton J, Orengo C. Quantifying the similarities within fold space. J Mol Biol 2002;323 (5):909-26.

[60] Carugo O, Pomgor S. Protein fold similarity estimated by a probabilistic approach based on Calpha-Calpha distance comparison. J. Mol. Biol. 2002;315 (4):887-98.

[61] Rogen P, Fain B. Automatic classification of protein structure by using Gauss integrals. Proc Natl Acad Sci U S A 2003;100 (1):119-24.

[62] Rogen P, Bohr H. A new family of global protein shape descriptors. Mathematical Biosciences 2003;182 (2):167-81.

[63] Wolf YI, Rogozin IB, Grishin NV, Koonin EV. Genome trees and the tree of life. Trends Genet 2002;18 (9):472-9.

[64] Mirkin BG, Fenner TI, Galperin MY, Koonin EV. Algorithms for computing parsimonious evolutionary scenarios for genome evolution, the last universal common ancestor and dominance of horizontal gene transfer in the evolution of prokaryotes. BMC Evol Biol 2003;3 (1):2.

[65] Hegyi H, Lin J, Greenbaum D, Gerstein M. Structural genomics analysis: characteristics of atypical, common, and horizontally transferred folds. Proteins 2002;47 (2):126-41.

[66] Gerstein M, Hegyi H. Comparing genomes in terms of protein structure: surveys of a finite parts list. FEMS Microbiol Rev 1998;22 (4):277-304.

[67] Jeong H, Tombor B, Albert R, Oltvai ZN, Barabasi AL. The large-scale organization of metabolic networks. Nature 2000;407 (6804):651-4.

[68] Jeong H, Mason SP, Barabasi AL, Oltvai ZN. Lethality and centrality in protein networks. Nature 2001;411 (6833):41-2.

[69] Overbeek R, Larsen N, Pusch GD, D'Souza M, Selkov E, Jr., Kyrpides N, Fonstein M, Maltsev N, Selkov E. WIT: integrated system for high-throughput genome sequence analysis and metabolic reconstruction. Nucleic Acids Res 2000;28 (1):123-5.

[70] Podani J, Oltvai ZN, Jeong H, Tombor B, Barabasi AL, Szathmary E. Comparable system-level organization of Archaea and Eukaryotes. Nat Genet 2001;29 (1):54-6.

[71] Tekaia F, Lazcano A, Dujon B. The genomic tree as revealed from whole proteome comparisons. Genome Res 1999;9 (6):550-7.

[72] Snel B, Bork P, Huynen MA. Genome phylogeny based on gene content. Nat Genet 1999;21 (1):108-10.

[73] Korbel JO, Snel B, Huynen MA, Bork P. SHOT: a web server for the construction of genome phylogenies. Trends Genet 2002;18 (3):158-62.

[74] Snel B, Lehmann G, Bork P, Huynen MA. STRING: a web-server to retrieve and display the repeatedly occurring neighbourhood of a gene. Nucleic Acids Res 2000;28 (18):3442-4.

[75] Huynen MA, Snel B. Gene and context: integrative approaches to genome analysis. Adv Protein Chem 2000;54:345-79.

[76] Huynen M, Snel B, Lathe W, Bork P. Exploitation of gene context. Curr Opin Struct Biol 2000;10 (3):366-70.

[77] Lathe WC, 3rd, Snel B, Bork P. Gene context conservation of a higher order than operons. Trends Biochem Sci 2000;25 (10):474-9.

[78] Rogozin IB, Makarova KS, Murvai J, Czabarka E, Wolf YI, Tatusov RL, Szekely LA, Koonin EV. Connected gene neighborhoods in prokaryotic genomes. Nucleic Acids Res 2002;30 (10):2212-23.

[79] Kanehisa M. Post-genome informatics. Oxford New York: Oxford University Press, 2000.

[80] Kanehisa M, Goto S. KEGG: kyoto encyclopedia of genes and genomes. Nucleic Acids Res 2000;28 (1):27-30.

[81] Oltvai ZN, Barabasi AL. Systems biology. Life's complexity pyramid. Science 2002;298 (5594):763-4.

[82] Bollobás B. Random Graphs. Cambridge: Cambridge University Press, 2001.

[83] Dorogovtsev SN, Mendes JFF. Evolution of Networks. Oxford: Oxford University Press, 2003.

[84] Barabasi AL, Albert R. Emergence of scaling in random networks. Science 1999;286 (5439):509-12.

[85] Goh KI, Oh E, Jeong H, Kahng B, Kim D. Classification of scale-free networks. Proc Natl Acad Sci U S A 2002;99 (20):12583-8.

[86] Yook SH, Jeong H, Barabasi AL. Modeling the Internet's large-scale topology. Proc Natl Acad Sci U S A 2002;99 (21):13382-6.

[87] Milo R, Shen-Orr S, Itzkovitz S, Kashtan N, Chklovskii D, Alon U. Network motifs: simple building blocks of complex networks. Science 2002;298 (5594):824-7.

[88] Shen-Orr SS, Milo R, Mangan S, Alon U. Network motifs in the transcriptional regulation network of Escherichia coli. Nat Genet 2002;31 (1):64-8.

*Essays in Bioinformatics*
*D.S. Moss et al. (Eds.)*
*IOS Press, 2005*


# GenBank: The NCBI Nucleotide Sequence Database

Ilene MIZRACHI
*National Center for Biotechnology Information, National Library of Medicine, Building 38A, Bethesda, MD 20894, USA*

**Abstract.** The GenBank sequence database is an annotated collection of all publicly available nucleotide sequences and their protein translations. This database is produced at National Center for Biotechnology Information (NCBI) as part of an international collaboration with the European Molecular Biology Laboratory (EMBL) Data Library from the European Bioinformatics Institute (EBI) and the DNA Data Bank of Japan (DDBJ). GenBank and its collaborators receive sequences produced in laboratories throughout the world from more than 115,000 distinct organisms. GenBank continues to grow at an exponential rate, doubling every 10 months. Release 142, produced in June 2004, contained over 40.3 billion nucleotide bases in more than 35.5 million sequences. GenBank is built by direct submissions from individual laboratories, as well as from bulk submissions from large-scale sequencing centers. Direct submissions are made to GenBank using BankIt [http://www.ncbi.nlm.nih.gov/BankIt/], which is a Web-based form, or the stand-alone submission program, Sequin[1]. Upon receipt of a sequence submission, the GenBank staff assigns an Accession number to the sequence and performs quality assurance checks. The submissions are then released to the public database, where the entries are retrievable by Entrez or downloadable by FTP. Bulk submissions of Expressed Sequence Tag (EST), Sequence Tagged Site (STS), Genome Survey Sequence (GSS), and High-Throughput Genome Sequence (HTGS) data are most often submitted by large-scale sequencing centers. The GenBank direct submissions group also processes complete microbial genome sequences.

## 1. History

Initially, GenBank was built and maintained at Los Alamos National Laboratory (LANL). In the early 1990s, this responsibility was awarded to NCBI through congressional mandate. NCBI undertook the task of scanning the literature for sequences and manually typing the sequences into the database. Staff then added annotation to these records, based upon information in the published article. Scanning sequences from the literature and placing them into GenBank is now a rare occurrence. Nearly all of the sequences are now deposited directly by the labs that generate the sequences. This is attributable to, in part, a requirement by most journal publishers that nucleotide sequences are first deposited into publicly available databases (DDBJ/EMBL/GenBank) so that the Accession number can be cited and the sequence can be retrieved when the article is published. NCBI began accepting direct submissions to GenBank in 1993 and received data from LANL until 1996. Currently, NCBI receives and processes about 25,000 direct submission sequences per month, in addition to the approximately 700,000 bulk submissions that are processed automatically.

---

[1] [http://www.ncbi.nlm. nih.gov/Sequin/index.html]

## 2. International Collaboration

In the mid-1990s, the GenBank database became part of the International Nucleotide Sequence Database Collaboration with the EMBL database ( European Bioinformatics Institute [http://www. ebi.ac.uk/], Hinxton, United Kingdom) and the Genome Sequence Database (GSDB; LANL, Los Alamos, NM). Subsequently, the GSDB was removed from the Collaboration (by the National Center for Genome Resources, Santa Fe, NM), and DDBJ [http://www.ddbj.nig.ac.jp/] (Mishima, Japan) joined the group. Each database has its own set of submission and retrieval tools, but the three databases exchange data daily so that all three databases should contain the same set of sequences. Members of the DDBJ, EMBL, and GenBank staff meet annually to discuss technical issues, and an international advisory board meets with the database staff to provide additional guidance. An entry can only be updated by the database that initially prepared it to avoid conflicting data at the three sites. The Collaboration created a Feature Table Definition[2] that outlines legal features and syntax for the DDBJ, EMBL, and GenBank feature tables. The purpose of this document is to standardize annotation across the databases. The presentation and format of the data are different in the three databases, however, the underlying biological information is the same.

## 3. Confidentiality of Data

When scientists submit data to GenBank, they have the opportunity to keep their data confidential for a specified period of time. This helps to allay concerns that the availability of their data in GenBank before publication may compromise their work. When the article containing the citation of the sequence or its Accession number is published, the sequence record is released. The database staff request that submitters notify GenBank of the date of publication so that the sequence can be released without delay. The request to release should be sent to gb-admin@ ncbi.nlm.nih.gov.

## 4. Direct Submissions

The typical GenBank submission consists of a single, contiguous stretch of DNA or RNA sequence with annotations. The annotations are meant to provide an adequate representation of the biological information in the record. The GenBank Feature Table Definition [http://www.ncbi. nlm.nih.gov/collab/FT/index.html] describes the various features and subsequent qualifiers agreed upon by the International Nucleotide Sequence Database Collaboration.

Currently, only nucleotide sequences are accepted for direct submission to GenBank. These include mRNA sequences with coding regions, fragments of genomic DNA with a single gene or multiple genes, and ribosomal RNA gene clusters. If part of the nucleotide sequence encodes a protein, a conceptual translation, called a CDS (coding sequence), is annotated. The span of the CDS feature is mapped to the nucleotide sequence encoding the protein. A protein accession number (/protein_id) is assigned to the translation product, which will subsequently be added to the protein databases. Multiple sequences can be submitted together. Such batch submissions of non-related sequences may be processed

[2] [http://www.ncbi.nlm.nih.gov/collab/FT/ index.html]

together but will be displayed in Entrez [http://www.ncbi.nlm.nih.gov/Entrez/] as single records. Alternatively, by using the Sequin submission tool[3], a submitter can specify that several sequences are biologically related. Such sequences are classified as environmental sample sets, population sets, phylogenetic sets, mutation sets, or segmented sets. Each sequence within a set is assigned its own Accession number and can be viewed independently in Entrez. However, with the exception of segmented sets, each set is also indexed within the PopSet division of Entrez, thus allowing scientists to view the relationship between the sequences. What defines a set? Environmental sample, population, phylogenetic, and mutation sets all contain a group of sequences that spans the same gene or region of the genome. Environmental samples are derived from a group of unclassified or unknown organisms. A population set contains sequences from different isolates of the same organism. A phylogenetic set contains sequences from different organisms that are used to determine the phylogenetic relationship between them. Sequencing multiple mutations within a single gene gives rise to a mutation set. All sets, except segmented sets, may contain an alignment of the sequences within them and might include external sequences already present in the database. In fact, the submitter can begin with an existing alignment to create a submission to the database using the Sequin submission tool. Currently, Sequin accepts FASTA+GAP, PHYLIP, MACAW, NEXUS Interleaved, and NEXUS Contiguous alignments. Submitted alignments will be displayed in the PopSet section of Entrez. Segmented sets are a collection of noncontiguous sequences that cover a specified genetic region. The most common example is a set of genomic sequences containing exons from a single gene where part or all of the intervening regions have not been sequenced. Each member record within the set contains the appropriate annotation, exon features in this case. However, the mRNA and CDS will be annotated as joined features across the individual records. Segmented sets themselves can be part of an environmental sample, population, phylogenetic, or mutation set.

## 5. Bulk Submissions: High-Throughput Genomic Sequence (HTGS)

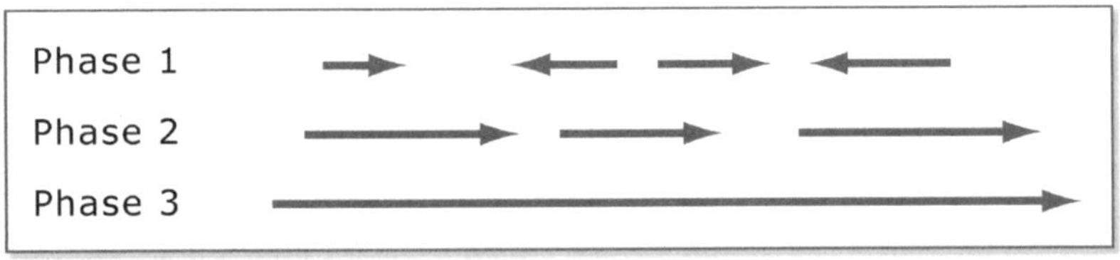


HTGS entries are submitted in bulk by genome centers, processed by an automated system, and then released to GenBank. Currently, about 30 genome centers are submitting data for a number of organisms, including human, mouse, rat, rice, and *Plasmodium falciparum*, the malaria parasite. HTGS [http://www.ncbi.nlm.nih.gov/HTGS/] data are submitted in four phases of completion: 0, 1, 2, and 3. Phase 0 sequences are one-to-few reads of a single clone and are not usually assembled into contigs. They are low-quality sequences that are often used to check whether another center is already sequencing a particular clone. Phase 1 entries are assembled into contigs that are separated by sequence gaps, the relative order and orientation of which are not known . Phase 2 entries are also unfinished sequences that may or may not contain sequence gaps. If there are gaps, then the contigs are in the correct order and orientation. Phase 3 sequences are of finished quality and have no gaps. For each

[3] [http://www.ncbi.nlm.nih.gov/Sequin/index.html]

organism, the group overseeing the sequencing effort determines the definition of finished quality.

Phase 0, 1, and 2 records are in the HTG division of GenBank, whereas phase 3 entries go into the taxonomic division of the organism, for example, PRI (primate) for human. An entry keeps its Accession number as it progresses from one phase to another but receives a new Accession. Version number and a new gi number each time there is a sequence change.

## 6. Submitting Data to the HTG Division

To submit sequences in bulk to the HTG processing system, a center or group must set up an FTP account by writing to htgs-admin@ncbi.nlm.nih.gov. Submitters frequently use two tools to create HTG submissions, Sequin [http://www.ncbi.nlm.nih.gov/HTGS/sequininfo.html] or fa2htgs [http://www.ncbi.nlm.nih.gov/HTGS/fa2htgsinfo.html]. Both of these tools require FASTA-formatted sequence, i.e., a definition line beginning with a "greater than" sign (">") followed by a unique identifier for the sequence. The raw sequence appears on the lines after the definition line. For sequences composed of contigs separated by gaps, a modified FASTA format [http://www. ncbi.nlm.nih.gov/HTGS/sequininfo.html] is used. In addition, Sequin users must modify the Sequin configuration file so that the HTG genome center features are enabled.

fa2htgs is a command-line program that is downloaded to the user's computer. The submitter invokes a script with a series of parameters (arguments) to create a submission. It has an advantage over Sequin in that it can be set up by the user to create submissions in bulk from multiple files. Submissions to HTG must contain three identifiers that are used to track each HTG record: the genome center tag, the sequence name, and the Accession number. The genome center tag is assigned by NCBI and is generally the FTP account login name. The sequence name is a unique identifier that is assigned by the submitter to a particular clone or entry and must be unique within the group's submissions. When a sequence is first submitted, it has only a sequence name and genome center tag; the Accession number is assigned during processing. All updates to that entry must include the center tag, sequence name, and Accession number, or processing will fail.

## 7. The HTG Processing Pathway

Submitters deposit HTGS sequences in the form of Seq-submit files generated by Sequin, fa2htgs, or their own ASN.1 dumper tool into the SEQSUBMIT directory of their FTP account. Every morning, scripts automatically pick up the files from the FTP site and copy them to the processing [http://www.ncbi.nlm.nih.gov/HTGS/processing.html] pathway, as well as to an archive. Once processing is complete and if there are no errors in the submission, the files are automatically loaded into GenBank. The processing time is related to the number of submissions that day; therefore, processing can take from one to many hours. Entries can fail HTG processing because of three types of problems: 1. Formatting: submissions are not in the proper Seq-submit format. 2. Identification: submissions may be missing the genome center tag, sequence name, or Accession number, or this information is incorrect. 3. Data: submissions have problems with the data and therefore fail the validator checks. When submissions fail HTG processing, a GenBank annotator sends email to the sequencing center, describing the problem and asking the center to submit a corrected entry. Annotators do not fix incorrect submissions; this ensures that the staff of the submitting genome center fixes the problems in their database as well. The processing pathway also generates reports. For successful submissions, two files are generated: one contains the submission in GenBank flat file format (without the sequence); and another is a status report file. The status report file, ac4htgs, contains the genome center, sequence name, Accession number, phase, create date, and update date for the submission. Submissions that fail processing receive an error file with a short description of the error(s) that prevented processing. The GenBank annotator also sends email to the submitter, explaining the errors in further detail.

## 8. Additional Quality Assurance

When successful submissions are loaded into GenBank, they undergo additional validation checks. If GenBank annotators find errors, they write to the submitters, asking them to fix these errors and submit an update.

## 9. Whole Genome Shotgun Sequences (WGS)

Genome centers are taking multiple approaches to sequencing complete genomes from a number of organisms. In addition to the traditional clone-based sequencing whose data are being submitted to HTGS, these centers are also using a WGS[4] approach to sequence the genome. The shotgun sequencing reads are assembled into contigs, which are now being accepted for inclusion in GenBank. WGS contig assemblies may be updated as the sequencing project progresses and new assemblies are computed. WGS sequence records may also contain annotation, similar to other GenBank records. Each sequencing project is assigned a stable project ID, which is made up of four letters. The Accession number for a WGS sequence contains the project ID, a two-digit version number, and six digits for the contig ID. For instance, a project would be assigned an Accession number AAAX00000000. The first assembly version would be AAAX01000000. The last six digits of this ID identify individual contigs. A master record for each assembly is created. This master record contains information that is common among all records of the sequencing project, such as the biological source, submitter, and publication information. There is also a link to the range of Accession numbers for the individual contigs in this assembly. WGS submissions can be created using tbl12asn[5], a utility that is packaged with the Sequin submission software. Information on submitting these sequences can be found at Whole Genome Shotgun Submissions [http://www.ncbi.nlm.nih.gov/Genbank/wgs.html].

## 10. Bulk Submissions: EST, STS, and GSS

Expressed Sequence Tags (EST), Sequence Tagged Sites (STSs), and Genome Survey Sequences (GSSs) sequences are generally submitted in a batch and are usually part of a large sequencing project devoted to a particular genome. These entries have a streamlined submission process and undergo minimal processing before being loaded to GenBank. EST [http://www.ncbi.nlm.nih.gov/dbEST/]s are generally short (<1 kb), single-pass cDNA sequences from a particular tissue and/or developmental stage. However, they can also be longer sequences that are obtained by differential display or Rapid Amplification of cDNA Ends (RACE) experiments. The common feature of all ESTs is that little is known about

---

[4] [http://www.ncbi.nlm.nih.gov/ GenBank/ wgs.html]

[5] [http://intranet.ncbi.nlm.nih.gov:6224/ieb/DIRSUB/tbl2asn2.html]

them; therefore, they lack feature annotation. STSs [http://www.ncbi.nlm.nih.gov/dbSTS/] are short genomic landmark sequences (1). They are operationally unique in that they are specifically amplified from the genome by PCR amplification. In addition, they define a specific location on the genome and are, therefore, useful for mapping.
GSS [http://www.ncbi.nlm.nih.gov/dbGSS/]s are also short sequences but are derived from genomic DNA, about which little is known. They include, but are not limited to, single-pass GSSs, BAC ends, exon-trapped genomic sequences, and AluPCR sequences. EST, STS, and GSS sequences reside in their respective divisions within GenBank, rather than in the taxonomic division of the organism. The sequences are maintained within GenBank in the dbEST, dbSTS, and dbGSS databases.

## 11. Submitting Data to dbEST, dbSTS, or dbGSS

Because of the large numbers of sequences that are submitted at once, dbEST, dbSTS, and dbGSS entries are stored in relational databases where information that is common to all sequences can be shared. Submissions consist of several files containing the common informa-tion, plus a file of the sequences themselves. The three types of submissions have different requirements, but all include a Publication file and a Contact file. See the dbEST [http://www.ncbi. nlm.nih.gov/dbEST/], dbSTS [http://www.ncbi.nlm.nih.gov/dbSTS/], and dbGSS [http://www.ncbi. nlm.nih.gov/dbGSS/] pages for the specific requirements for each type of submission. In general, users generate the appropriate files for the submission type and then email the files to batch-sub@ncbi.nlm.nih.gov. If the files are too big for email, they can be deposited into a FTP account. Upon receipt, the files are examined by a GenBank annotator, who fixes any errors when possible or contacts the submitter to request corrected files. Once the files are satisfactory, they are loaded into the appropriate database and assigned Accession numbers. Additional formatting errors may be detected at this step by the data-loading software, such as double quotes anywhere in the file or invalid characters in the sequences. Again, if the annotator cannot fix the errors, a request for a corrected submission is sent to the user. After all problems are resolved, the entries are loaded into GenBank.

## 12. Bulk Submissions: HTC and FLIC

HTC records are High-Throughput cDNA/mRNA submissions that are similar to ESTs but often contain more information. For example, HTC entries often have a systematic gene name (not necessarily an official gene name) that is related to the lab or center that submitted them, and the longest open reading frame is often annotated as a coding region. FLIC records, Full-Length Insert cDNA, contain the entire sequence of a cloned cDNA/mRNA. Therefore, FLICs are generally longer, and sometimes even full-length, mRNAs. They are usually annotated with genes and coding regions, although these may be lab systematic names rather than functional names.

## 13. HTC Submissions

HTC entries are usually generated with Sequin [http://www.ncbi.nlm.nih.gov/ Sequin/index.html] or tbl2asn [http://www.ncbi.nlm.nih.gov/Sequin/table.html], and the files are emailed to gb-sub@ncbi.nlm.nih.gov. Larger files may be submitted by SequinMacrosend [www.ncbi.nlm.nih. gov/LargeDirSubs/ dir_submit.cgi].. HTC entries undergo the same validation and processing as non-bulk submissions. Once processing is complete, the records are loaded into GenBank and are available in Entrez and other retrieval systems.

## 14. FLIC Submissions

FLICs are processed via an automated FLIC processing system that is based on the HTG automated processing system. Submitters use the program tbl2asn to generate their submissions. As with HTG submissions, submissions to the automated FLIC processing system must contain three identifiers: the genome center tag, the sequence name (SeqId), and the Accession number. The genome center tag is assigned by NCBI and is generally the FTP account login name. The sequence name is a unique identifier that is assigned by the submitter to a particular clone or entry and must be unique within the group's FLIC submissions. When a sequence is first submitted, it has only a sequence name and genome center tag; the Accession number is assigned during processing. All updates to that entry include the center tag, sequence name, and Accession number, or processing will fail.

## 15. The FLIC Processing Pathway

The FLIC processing system is analogous to the HTG processing system. Submitters deposit their submissions in the FLICSEQSUBMIT directory of their FTP account and notify us that the submissions are there. We then run the scripts to pick up the files from the FTP site and copy them to the processing pathway, as well as to an archive. Once processing is complete and if there are no errors in the submission, the files are automatically loaded into GenBank. As with HTG submissions, FLIC entries can fail for three reasons: problems with the format, problems with the identification of the record (the genome center, the SeqId, or the Accession number), or problems with the data itself. When submissions fail FLIC processing, a GenBank annotator sends email to the sequencing center, describing the problem and asking the center to submit a corrected entry. Annotators do not fix incorrect submissions; this ensures that the staff of the submitting genome center fixes the problems in their database as well. At the completion of processing, reports are generated and deposited in the submitter's FTP account, as described for HTG submissions.

## 16. Submission Tools

Direct submissions to GenBank are prepared using one of two submission tools, BankIt or Sequin. BankIt BankIt [http://www.ncbi.nlm.nih.gov/BankIt/] is a Web-based form that is a convenient and easy way to submit a small number of sequences with minimal annotation to GenBank. To complete the form, a user is prompted to enter submitter information, the nucleotide sequence, biological source information, and features and annotation pertinent to the submission. BankIt has extensive Help [http://www.ncbi.nlm.nih.gov/BankIt/help.html] documentation to guide the submitter. Included with the Help document is a set of annotation examples that detail the types of information that are required for each type of submission. After the information is entered into the form, BankIt transforms this information into a GenBank flatfile for review. In addition, a number of quality assurance and validation checks ensure that the sequence submitted to GenBank is of the highest quality. The submitter is asked to include spans (sequence coordinates) for the coding regions and other features and to include amino acid sequence for the proteins that derive from these coding regions. The BankIt validator compares the amino acid sequence provided by the submitter with the conceptual translation of the coding region based on the provided spans. If there is a discrepancy, the submitter is requested to fix the problem, and the process is halted until the error is resolved. To prevent the deposit of sequences that contain cloning vector sequence, a BLAST similarity search is performed on the sequence, comparing it to the Vec-Screen [http://www.ncbi.nlm.nih.gov/VecScreen/VecScreen.html] database. If there is a match to this database, the user is asked to remove the contaminating vector sequence from their submis-sion or provide an explanation as to why the screen was positive. Completed forms are saved in ASN.1 format, and the entry is submitted to the GenBank processing queue. The submitter receives confirmation by email, indicating that the submission process was successful. Sequin Sequin[6] is more appropriate for complicated submissions containing a significant amount of annotation or many sequences. It is a stand-alone application available on NCBI's FTP [ftp://ftp.ncbi.nih.gov/sequin/] site. Sequin creates submissions from nucleotide and amino acid sequences in FASTA format with tagged biological source information in the FASTA definition line. As in BankIt, Sequin has the ability to predict the spans of coding regions. Alternatively, a submitter can specify the spans of their coding regions in a five-column, tab-delimited table [http://www.ncbi.nlm.nih. gov/Sequin/table.html] and import that table into Sequin. For submitting multiple, related sequences, e.g., those in a phylogenetic or population study, Sequin accepts the output of many popular multiple sequence-alignment packages, including FASTA+GAP, PHYLIP, MACAW, NEXUS Interleaved, and NEXUS Contiguous. It also allows users to annotate features in a single record or a set of records globally. For more information on Sequin, see Chapter 12.

Completed Sequin submissions should be emailed to GenBank at gb-sub@ncbi. nlm.nih.gov. Larger files may be submitted by SequinMacrosend [www.ncbi.nlm.nih. gov/LargeDirSubs/ dir_submit.cgi].

## 17. Sequence Data Flow and Processing: From Laboratory to GenBank

### *17.1 Triage*

All direct submissions to GenBank, created either by Sequin or BankIt, are processed by the GenBank annotation staff. The first step in processing submissions is called triage. Within two working days of receipt, the database staff reviews the submission to determine whether it meets the minimal criteria for incorporation into GenBank and then assigns an Accession number to each sequence. All sequences must be >50 bp in length and be sequenced by, or on behalf of, the group submitting the sequence. GenBank will not accept sequences constructed *in silico*; noncontiguous sequences containing internal, unsequenced spacers; or sequences for which there is not a physical counterpart, such as those derived from a mix of genomic DNA and mRNA. Submissions are also checked to determine whether they are new sequences or updates to sequences submitted previously. After receiving Accession numbers, the sequences are put into a queue for more extensive processing and review by the annotation staff.

---

[6] [http://www.ncbi.nlm.nih.gov/Sequin/index.html]

### *17.2 Indexing*

Triaged submissions are subjected to a thorough examination, referred to as the indexing phase. Here, entries are checked for:
1. Biological validity. For example, does the conceptual translation of a coding region match the amino acid sequence provided by the submitter? Annotators also ensure that the source organism name and lineage are present, and that they are represented in NCBI's taxonomy database. If either of these is not true, the submitter is asked to correct the problem. Entries are also subjected to a series of BLAST similarity searches to compare the annotation with existing sequences in GenBank.
2. Vector contamination. Entries are screened against NCBI's UniVec[7] database to detect contaminating cloning vector.
3. Publication status. If there is a published citation, PubMed and MEDLINE identifiers are added to the entry so that the sequence and publication records can be linked in Entrez.
4. Formatting and spelling. If there are problems with the sequence or annotation, the annotator works with the submitter to correct them.

Completed entries are sent to the submitter for a final review before release into the public database. If the submitters requested that their sequences be released after processing, they have 5 days to make changes prior to release. The submitter may also request that GenBank hold their sequence until a future date. The sequence must become publicly available once the Accession number or the sequence has been published. The GenBank annotation staff currently processes about 2200 submissions per month, corresponding to approximately 26,000 sequences. GenBank annotation staff must also respond to email inquiries that arrive at the rate of approximately 300 per day. These exchanges address a range of topics including:

- updates to existing GenBank records, such as new annotation or sequence changes
- problem resolution during the indexing phase
- requests for release of the submitter's sequence data or an extension of the hold date
- requests for release of sequences that have been published but are not yet available in GenBank
- lists of Accession numbers that are due to appear in upcoming issues of a publisher's journals
- reports of potential annotation problems with entries in the public database
- requests for information on how to submit data to GenBank

One annotator is responsible for handling all email received in a 24-hour period, and all messages must be acted upon and replied to in a timely fashion. Replies to previous emails are forwarded to the appropriate annotator.

### *17.3 Processing Tools*

The annotation staff uses a variety of tools to process and update sequence submissions. Sequence records are edited with Sequin, which allows staff to annotate large sets of records by global editing rather than changing each record individually. This is truly a time saver because more than 100 entries can be edited in a single step. Records are stored in a database that is accessed through a queue management tool that automates some of the processing steps, such as looking up taxonomy and PubMed data, starting BLAST jobs, and running automatic validation checks. Hence, when an annotator is ready to start working on an entry, all of this information is ready to view. In addition, all of the correspondence

---

[7] [http://www.ncbi.nlm. nih.gov/VecScreen/UniVec.html]

between GenBank staff and the submitter is stored with the entry. For updates to entries already present in the public database, the live version of the entry is retrieved from ID, and after making changes, the annotator loads the entry back into the public database. This entry is available to the public immediately after loading.

## 18. Microbial Genomes

The GenBank direct submissions group has processed more than 200 complete microbial genomes since 1996. These genomes are relatively small in size compared with their eukaryotic counterparts, ranging from five hundred thousand to five million bases. Nonetheless, these genomes can contain thousands of genes, coding regions, and structural RNAs; therefore, processing and presenting them correctly is a challenge.

Submitters of complete genomes are encouraged to contact us at genomes@ncbi.nlm.nih.gov before preparing their entries. A FTP account is required to submit large files, and the submission should be deposited at least 1 month before publication to allow for processing time and coordinated release before publication. In addition, submitters are required to follow certain guidelines, such as providing unique identifiers for proteins and systematic names for all genes. Entries should be prepared with the submission tool tbl2asn[8], a utility that is part of the Sequin package .This utility creates an ASN.1 submission file from a five-column, tab-delimited file containing feature annotation, a FASTA-formatted nucleotide sequence, and an optional FASTA-formatted protein sequence. For more information about using tbl2asn to submt microbial see http://www.ncbi.nlm.nih.gov/Genbank/genomesubmit.html

Complete genome submissions are reviewed by a member of the GenBank annotation staff to ensure that the annotation and gene and protein identifiers are correct, and that the entry is in proper GenBank format. Any problems with the entry are resolved through communication with the submitter. The microbial genome records in GenBank are the building blocks for the Microbial Genome Resources in Entrez Genomes.

## 19. Third Party Annotation (TPA) Sequence Database

The vast amount of publicly available data from the human genome project and other genome sequencing efforts is a valuable resource for scientists throughout the world. A laboratory studying a particular gene or gene family may have sequenced numerous cDNAs but has neither the resources nor inclination to sequence large genomic regions containing the genes, especially when the sequence is available in public databases. The researcher might choose then to download genomic sequences from GenBank and perform experimental analyses on these sequences. However, because this researcher did not perform the sequencing, the sequence, with its new annotations, cannot be submitted to DDBJ/EMBL/GenBank. This is unfortunate because important scientific information is being excluded from the public databases. To address this problem, the International Nucleotide Sequence Database Collaboration established a separate section of the database for such TPA (see Third Party Annotation Sequence Database [www.ncbi.nlm.nih.gov/Genbank/tpa.html]).

All sequences in the TPA database are derived from the publicly available collection of sequences in DDBJ/EMBL/GenBank. Researchers can submit both new and alternative annotations of genomic sequence to GenBank. TPA entries can be also created

---

[8] [http://intranet.ncbi.nlm.nih.gov:6224/ieb/DIRSUB/tbl2asn2.html]

by combining the exon sequences from genomic sequences or by making contigs of EST sequences to make mRNA sequences. TPA submissions must use sequence data that are already represented in DDBJ/EMBL/GenBank, have annotation that is experimentally supported, and appear in a peer-reviewed scientific journal. TPA sequences will be released to the public database only when their Accession numbers and/or sequence data appear in a peer-reviewed publication in a biological journal.

## References

[1] Olson M, Hood L, Cantor C, Botstein D. A common language for physical mapping of the human genome. Science 245(4925):1434–1435; 1989. (PubMed).

*Essays in Bioinformatics*
*D.S. Moss et al. (Eds.)*
*IOS Press, 2005*



# Swiss-Prot: juggling between evolution and stability

Amos BAIROCH, Brigitte BOECKMANN, Serenella FERRO ROJAS, and Elisabeth GASTEIGER
*Swiss Institute of Bioinformatics, Centre Médical Universitaire, 1 Rue Michel Servet, 1211 Geneva 4, Switzerland*

**Abstract.** We describe some of the aspects of Swiss-Prot that make it unique, explain what are the developments we believe to be necessary for the database to continue to play its role as a focal point of protein knowledge, and provide advice pertinent to the development of high quality knowledge resources on one aspect or the other of the life sciences.

## Introduction

The goal of this article is not to depict the history of Swiss-Prot [1], this has already been done elsewhere [2], but rather to explore some of the consequences of decisions taken about 20 years ago, to discuss how the database has constantly evolved and to describe the challenges that it currently faces. To say that the last twenty years have been exciting would be a major understatement. Most young scientists that are now starting a career in the Life Science fields are not aware of how much the combined technological revolutions that led to high throughput sequencing and the WWW have quantitatively and qualitatively changed the universe of knowledge on proteins. Yet, while we now have to cater in the Swiss-Prot and TrEMBL sections of the UniProt knowledgebase [3] for more than 1 million protein sequences, there is a continuously widening chasm between truly characterised proteins and those which have been solely predicted by genome-sequencing projects. For us, in Swiss-Prot, the ultimate in terms of a well-characterised protein is one for which not only the exact sequence, post-translational modifications, sub-cellular location, tissue specificity, interaction partners and 3D structure are known, but more crucially for which a functional role can be assigned.

What we hope to convey in this article are the particular aspects of Swiss-Prot that make it unique, and hopefully derive some advice that would be pertinent to someone embarking on the development of a high quality knowledge resource on one aspect or the other of the life sciences. But before we do so, we want to enumerate six observations that we believe are important to communicate to any would-be developers of such databases:

- Your task will be much more complex and far bigger that you ever thought it could be;
- If your database is successful and useful to the user community, then you will have to dedicate all your efforts to develop it for a much longer period of time than you would have thought possible;
- You will always wonder why life scientists abhor complying with nomenclature guidelines or standardization efforts that would simplify your and their life;
- You will have to continually fight to obtain a minimal amount of funding;

- As with any service efforts, you will be told far more what you do wrong rather than what you do right;
- But when you will see how useful your efforts are to your users, all the above drawbacks will loose their importance!!

## 1. A small bit of historical introspection

### *1.1 How Swiss-Prot started and how it institutionally evolved*

In 1965, the late Margaret Dayhoff published the first edition of the Atlas of Protein Sequence and Structure [4]. It contained information on 65 protein sequences. In the introduction she expressed the mission of the Atlas as "*locating all of the relevant publications; critically reviewing the data and resolving conflicting reports; transforming the data into a uniform format to reflect those aspects of the structure that have been experimentally determined and those that could reasonably be inferred by homology; identifying the material with regard to chemical function, biological source, genetic control, and evolutionary origin...*". This ambitious and still highly pertinent mission statement is a tribute to the vision shown by Margaret Dayhoff. She pursued her task until her untimely death in 1983. At that time the Atlas had evolved into a protein sequence data bank known as the Protein Identification Resource (PIR) of the National Biomedical Research Foundation (NBRF). When in 1985, one of us (Amos Bairoch) was, in the context of a PhD thesis, developing a software package (PC/Gene [5]) to analyse protein sequences, he was faced with some deficiencies and omissions in the PIR database. As he did not receive satisfactory feedback from PIR, he resolved to develop a version of PIR in the format of the European Molecular Biology Laboratory (EMBL) nucleotide sequence database that would contain additional sequences and, more crucially, additional annotations on various aspects of the protein universe.

In mid-1986, the first release of Swiss-Prot came out. Almost immediately we approached the EMBL to see if they were interested in distributing and helping with the maintenance of the database. With foresight they immediately accepted. The collaboration that grew from this early decision gave rise to the current situation: Swiss-Prot is a fully collaborative endeavour of what has become the Swiss-Prot group at the Swiss Institute of Bioinformatics (SIB) and the European Bioinformatics Institute (EBI), an outstation of EMBL. The last institutional development was the decision, in late 2003, of the NIH to award a major grant to a consortium composed of the EBI, the SIB and PIR to produce a universal resource on proteins, known as UniProt.

Today, in 2004, there are more than 120 people that directly work on Swiss-Prot and TrEMBL (see 1.2) or on resources that evolved out of Swiss-Prot. While the first reaction to this figure can be "that's a lot of people", it pales when compared to the amount of work to be carried out. In fact this is a major issue shared by all life sciences information resources: long-term high-quality curation of information is not cheap. It is not as glamorous as whole genome sequencing projects or any such well-defined scientific and technological efforts, yet it needs to be adequately and stably funded. Sadly, this is not yet widely recognised by funding bodies.

### *1.2 Why TrEMBL was developed*

In the mid-90s it was already clear that the increased data flow from genome projects was going to be a major challenge for Swiss-Prot. As it will be explained further on, maintaining the high quality of the database requires careful sequence analysis and detailed annotation of every entry. This was, and still is, a major rate-limiting step. We did not wish to relax the editorial standards of Swiss-Prot and there was a limit to how much the annotation procedures could be accelerated. Yet it was vital to make new sequences available as quickly as possible. To address this concern, we introduced in 1996 TrEMBL (Translation of EMBL). TrEMBL consists of computer-annotated entries derived from the translation of all coding sequences in the EMBL database, except for those already included in Swiss-Prot. TrEMBL is therefore a complement to Swiss-Prot and sequence entries only move out from TrEMBL and enter Swiss-Prot after having been manually curated by an annotator.

From 1996 to the end of 2003, Swiss-Prot grew by 83,000 sequences to reach a total of 140,000 entries. In this period of time, TrEMBL grew from the 86,000 entries in its first release to about 1.1 million entries!

## 2. What makes Swiss-Prot special

### *2.1 Aiming for the perfect sequence*

Even if it may be obvious to many of its users, it is important to restate that Swiss-Prot is a corpus of knowledge centred on protein sequences. As it will be apparent in the following sections of this article, we add many layers of information around the sequence data, yet most of that information is in one way or another dependent on the sequence. It is therefore important to capture and to represent the most correct sequence. This is an important aspect of the work of Swiss-Prot that escapes the notice of most of its users.

The overwhelming majority (>99%) of the sequence data represented in Swiss-Prot originates from the translation of nucleotide sequences submitted to the EMBL/Genbank/DDBJ database. Only a very small proportion of the sequences are obtained directly at the amino-acid level using Edman degradation or mass spectrometry. This situation already existed in 1986. What happened since was obviously an enormous quantitative increase in the amount of nucleotide sequence data, but also, more relevant to our quest toward quality, a significant increase in nucleotide sequence quality and a sociological change in the breakdown of the originators of sequence data. The increase in sequence quality is mainly due to the growing use of very sophisticated automated sequencing machines. In 1986, most nucleotide sequences which were submitted to the DNA databases originated from individual laboratories that were sequencing a single gene or a small region of a genome. Today, the biggest (in terms of quantity) contributors are major sequencing centres that either provide complete genomic sequences or massive amounts of data from full-length cDNAs.

As we depend on primary sequence data that has been submitted to the nucleotide sequence databases, it would seem at first glance that there is not really anything we can do to improve the quality of the derived protein sequences. This is far from being true, and in fact there are many things we can do by comparing sequences. Sequence comparison is essential to the process of creating or updating a Swiss-Prot entry. One needs to remember that Swiss-Prot is a non-redundant database. What this means is that we took the decision from the very beginning to merge the protein sequences from the same organism originating from the same gene. Thus we are often faced with many complete or partial sequences that need to be merged together and whose discrepancies have to be taken into account. Sequence discrepancies are annotated with the feature (FT) keys CONFLICT, VARIANT, MUTAGEN or VARSPLIC. The FT key VARIANT is used to describe polymorphisms and disease mutations, MUTAGEN for experimentally altered sites and

CONFLICT for sequence differences of any other reason. Insertions or gaps within alignments of otherwise identical sequences are usually due to alternative splicing events, which are annotated using the FT key VARSPLIC.

Thus sequence comparisons can already help us in determining what is the most correct sequence. This is especially true in organisms that are the focus of many sequencing efforts. For example, we currently have an average of 3.7 independent sequence reports (cDNA or genomic DNA) for each human protein. Such a redundancy in the nucleotide sequence database helps flagging potential sequencing errors. Further errors can be found when comparing orthologous and paralogous sequences across species. The relevance of such approaches is increasing as more and more full genome sequences are becoming available.

One of the advantages of comparing many sequences is the detection of probable frameshift errors. They stand up in multiple protein sequence alignments as locally divergent regions. If the divergence can be explained at the nucleotide level by the insertion or deletion of a single nucleotide, it is likely (but not certain) that it is due to a sequencing error. The total number of potential frameshift errors that were corrected by Swiss-Prot annotators is difficult to estimate as it often happens that incorrect DNA sequences are later resubmitted by the original authors, correcting sequencing errors, generally by taking into account the correction made in the corresponding Swiss-Prot entries. In the current release we have 1% of the entries that are flagged with at least one potential frameshift error in one of the cross-referenced nucleotide sequence entries.

In many cases, the N-terminal initiation sites of bacterial or archaeal genes or the exon/intron boundaries of eukaryotic genes are incorrectly predicted. It is important to note that these predictions are of a very heterogeneous quality and to recognise that not all sequencing centres produce the same level of quality in terms of both sequences and of protein-coding gene predictions. Swiss-Prot annotators are aware of this heterogeneity and know what data can be more or less trusted. We currently observe that in 7.1% of our entries we disagree with the translation provided by the submitter.

It often happens that annotators have to translate, from a nucleotide entry, protein sequences that have been overlooked by the original submitters. Currently we have 2.5% of our entries that contain such translations.

Finally, the work of the Swiss-Prot annotators is also to reject putative protein sequences which are obviously bogus, either because they originate from a pseudogene or because they were incorrectly predicted either from non-coding DNA or a wrong open reading frame.

If you take all the above factors and tasks into consideration, you can see why we believe that the correction of amino-acid sequences is an important part of the annotation process, and that it is far from being trivial to achieve. This is not necessarily apparent to the user, but it is one of the reasons why Swiss-Prot has always been considered as the reference database for protein sequences. Of course the drawback of such an approach is that it is time-consuming and can only be applied to manually annotated entries. Such an approach can consequently not be applied to TrEMBL, where the represented protein sequences are those that have been indicated by the submitters of the original nucleotide sequence entry. It would therefore be important to develop semi-automatic systems that allow some aspects of sequence correction to be applied to TrEMBL.

### *2.2 Extracting information from the literature*

Fifteen years ago, Swiss-Prot annotators typically went through the following process: they photocopied all relevant papers from the reference list of the entry they were annotating.

The publications were read and important information was marked in the paper copy. Information was then added to the entry in either free text (comments lines) or structured feature lines. Access to reference databases and computing tools considerably facilitated the above procedures, but also brought along a higher level of complexity. Being an annotator in the early 90s was already not a trivial job, but it has since become a much more demanding task.

When Medline became available at the work place first on CD-ROMs, and later via the Internet, most journal abstracts could immediately be read – or discarded if not relevant - and information was retrieved directly from here, which was particularly helpful when the journal was not available from local libraries. But it is online access to full text articles that has completely changed the life of annotators. They can look at many more relevant papers than they used to do when they needed to go to the library. This is particularly useful nowadays as information on a given protein is generally spread between many different reports in a wide variety of journals. Such a trend is exemplified by the journal citation statistics of Swiss-Prot: in 1993, 461 different journals were cited in the database, while today the number has risen to about 1'400. While some journals (such as J. Biol. Chem. and PNAS) were and still are major sources of articles useful for the annotation process, there has been a clear trend toward a "decentralisation" of the sources of protein-related publications. Of course, journal articles are not the only source of information, and we also make use of electronic journals, book articles, theses, patent applications and external information resources, but overwhelmingly the primary source of experimental information remains published journal articles.

We are often asked whether annotators are 'really sitting there and reading publications'. Yes, they are. Knowledge extracted from the articles is mostly added to the appropriate topics of the comment (CC) lines, and to the feature table (FT), whenever a description concerns a defined region or site within the sequence. But we also add new synonyms for protein names (DE line), gene names (GN line), compare or complete author names with the ones given in a reference block (RA line), annotate a reference block (RP and RC lines), add additional relevant references to an entry, and much more. All experimental findings and authors' conclusions are compared with the knowledge available on related proteins and the results from various protein sequence analysis tools. When contradictory results have been published and there is not enough information to prefer one hypothesis to the others, the annotation is performed in a way that draws the user's attention to the contradictory conclusions. Finally the content of an entry is summarised in form of a list of keywords (KW line) from a controlled vocabulary.

Both abstracts and full text articles are the target of text mining tools, which will soon become an indispensable help for annotators to quickly find the publications of interest from the wealth of information available. We believe that efforts to build efficient software tools allowing the semi-automated extraction of information from repositories of full text articles will be essential to anyone trying to build comprehensive information resources for life scientists. The fact that we will rely on such tools to hunt and extract information is paradoxical. Any one outside of the life sciences field would believe that such important information would be immediately made available in a structured way by the experimentalists to the relevant databases. As we will see in the next section, this is unfortunately not the case.

### *2.3 User submissions and updates*

We have always strongly encouraged user feedback, as well as the submission of updates and corrections, initially by asking people to contact us by email. Also, very early on, a list of "on-line experts" was compiled, i.e. a list of email addresses of scientists working with specific protein families or domains, who agreed to review protein sequences in Swiss-Prot relevant to their field of research. This list is regularly updated and the ~150 experts' email addresses, grouped by fields of expertise, are listed in the document http://www.expasy.org/cgi-bin/experts.

However, it does not seem clear to most users - who have grown accustomed to the repository nature of the nucleotide sequence databases, where only the original authors are allowed to correct and update existing entries - that Swiss-Prot is extremely different in that respect, and that we do have an ongoing editorial policy. We do indeed highly value our users' expertise, and we believe that it is only with the assistance of our user community that we can do our job of being comprehensive and up-to-date. We are therefore actively seeking any type of updates and/or corrections, whether they have been published or not, and would like to be notified about annotations to be updated, e.g. if the function of a protein has been clarified, or if new post-translational modification information has become available. In order to increase the visibility of these aspects, and to encourage our users to let us know about outdated protein entries or errors, we have implemented update forms on the ExPASy server (see 3.). The forms, accessible from the bottom of every Swiss-Prot entry, prompt users to provide their corrections and updates in any format. Update requests are treated with a very high priority by annotators. We are currently receiving about 300 update requests for Swiss-Prot entries per year, a number that we would very much like to see growing in the future!

On the other hand, annotators send newly annotated entries to the original authors of reports cited in these entries so as to check the validity of the annotations. We generally get useful feedback, but not as much as we would like!

Another point of interaction with users is sequence submission directly to Swiss-Prot and TrEMBL. We accept submission of sequences that have been obtained only as amino acid sequence. A web submission tool (SPIN) has just been made available, which guides the submitter through the process, and prompts for all required pieces of information. There are about 300 such sequence submissions per year. It is interesting to note that 10% of the proteins originate from venomous animals. This is explained by the fact that toxins can easily be purified in large quantity from venom and are generally quite small, thus they are easily sequenced at protein level.

We have to admit that we are disappointed by the low level of input from users in the updating of the database. We may have been insufficiently efficient in publicising our willingness and eagerness to welcome any type of help. Yet, after years of discussions with researchers, we believe that the root of the project is of a sociological nature. The career of life scientists is driven by the famous 'publish or perish' injunction and submitting data to a database does not get any credit points on a CV. So we have to rely on the altruism of some individuals. We are indeed indebted to those persons who take the time to make sure that we adequately represent the results of their research in our database. However we believe it is time that the community as a whole addresses this issue and initiates a process of responsibilisation toward the biomolecular databases.

### *2.4 Tools for annotation*

#### *2.4.1 The basic data organization, the editor and the syntax checker*

The working copy of Swiss-Prot is arranged in flat files, grouping proteins by family or other functional criteria. Although it was apparent from the beginning that the complexity of protein relationships could not be simulated simply by grouping entries one-dimensionally into separate files, this system allows curators to immediately find orthologs, which can all be updated when new findings become available for at least one protein, or when a review article summarises relevant knowledge on a protein family or subfamily and comes to new conclusions. The quick availability of all related entries (all in the same file) also ensures consistent annotation of all relevant entries. The ~140,000 entries in the current release are thus split into ~3,000 files.

Most of the annotation is done manually with the help of a continuously growing number of tools. We currently use a text editor, Crisp (from Vital, Inc.), that is easy to use and comes with a powerful C-like macro language that we extensively use both for literature-driven textual annotation and as a platform to launch sequence analysis programs (see 2.4.2). An extensive series of macro-commands have been developed to reformat references, comment lines, feature lines or sequences, to check controlled vocabulary or syntax, and to retrieve entries from other databases. Analysis tools are also run directly from the editor with the help of macro-commands that send the sequence and other relevant information to the analysis program, and then retrieve the result and format it in the annotation platform. All commands are available both from keyboard shortcuts (which are preferred by experienced annotators) and from menus and dialog boxes that are fully integrated in the editor's GUI environment.

Swiss-Prot annotation has always been subjected to very strict rules and guidelines. All entries are reviewed before they enter the database, which guarantees the homogeneity of the annotation. We developed a "syntax checker" so as to make sure that our annotation and format rules are enforced. This syntax checker, implemented in Perl, is much more than a program that verifies the basic syntax of a Swiss-Prot entry. It also enforces the use of controlled vocabularies (see 2.5) and checks for dependencies and consistencies between different portions of an entry. In December 2003, the syntax checker contained almost 1'100 different rules, each of which can lead to the detection of errors or inconsistencies.

Many people are surprised to hear that Swiss-Prot annotation is done from within a text editor. However, those same people are usually even more surprised once they see how powerful the annotation platform developed around that text editor is, and that almost every command can be launched, and its results treated, from within the editor, in a remarkable speed. One major disadvantage of this environment is that it relies heavily on the flat file format. We are now developing a Swiss-Prot specific editor, which will work with the XML-formatted version of the databases, and will include many consistency checks and context-specific menus. The new annotation platform will also include many graphical features, e.g. visualization of domain and site predictions along the sequence. We believe that such a development is highly desirable, as it will allow the implementation of consistency checks directly at the level of the annotation platform while we now have to rely on a regular post-processing check of the data, using the syntax checker to enforce consistency.

### *2.4.2 Sequence analysis tools*

The task of annotating Swiss-Prot entries has always relied on the use of the most appropriate sequence analysis programs so as to predict important sequence features. Over the years we have implemented many different methods and programs in our annotation platform. We have also spent a considerable amount of time testing new methods and selecting the most appropriate ones. In some cases, when no existing program could satisfy our needs, we have developed our own set of predictive methods [6, 7]. All these activities are carried out by a small research component within the Swiss-Prot group whose missions are to carry out technological watch and to develop new methodologies for protein sequence analysis.

Currently we use software tools (a full list with references is available in the Swiss-Prot document annbioch.txt) to predict the following sequence features:

- Signal sequences of type 1, type 2 (lipoprotein) and type 3;
- Mitochondrial and plastid targeting sequences;
- Transmembrane domains;
- Coiled coil domains;
- Specific repeats (LRR, TPR, WD, etc.);
- Statistically significant runs of amino acids and regions enriched in particular amino acids;
- N-glycosylation sites;
- GPI-anchors;
- Sulfation sites;
- N-terminal myristoylation sites.

In addition to the above list, we make extensive use of domain/family databases to annotate specific domains. In fact the development of the PROSITE [8] database, which was first released in 1990, was specifically driven by the need to detect and annotate protein domains. The combined usage of profiles and patterns allows the detection of domains (profile) and the functional sites within domains (pattern). As mentioned in the section on cross-references (3.7), there are now many other protein domain databases and we occasionally make use of most of them to annotate specific domains not yet covered by PROSITE. The reasons of our preference for PROSITE over other similar databases are very pragmatic: PROSITE domain descriptors are specifically tailored for their use in the context of protein sequence annotation in order not to predict overlapping domains. Cut-off values are selected conservatively to minimise the number of false positives: we prefer to miss the occurrence of a domain rather than to over-predict its existence.

We believe that the use of the most up-to-date sequence analysis tools is essential to any protein sequence annotation effort. In addition anyone considering applying such methods on a large scale needs to develop internal benchmarks so as to objectively judge the validity and the scope of the methods. In many instances we have observed that the claims of developers of sequence analysis methods are slightly overblown and that one obtains unexpected results when using such methods on large and highly heterogeneous sets of sequences.

### *2.4.3 Automation: trying to simulate the expertise of annotators*

Thanks to genome sequencing efforts, there has been a tremendous rise in the number of available protein sequences. Yet clearly this is only the beginning and what exists now will only represent a drop in an ocean of uncharacterised sequences. And there lies both the problem and a possible solution: on one hand the overwhelming majority of genome-derived sequences are currently not the target of experimental characterisation and are probably not going to be so in the next decade. On the other hand we have encapsulated in Swiss-Prot a tremendous amount of knowledge, some of which is specific to a given protein, while the majority can be carefully propagated to well defined orthologous sequences. Automatic annotation is far from being a novel concept. But what we want to achieve in Swiss-Prot differs from what others expect from such systems. Their aim is to analyse new genomic sequences and predict a maximum of potential information items so as to be able to infer hypotheses on the potential biological processes present in the organism. Our aim is to make sure that we produce high quality annotation with a minimal amount of incorrect inferences.

Our first automatic annotation project is called HAMAP [9], which stands for High-quality Automated and Manual Annotation of microbial Proteomes. In the context of this project, proteins from complete bacterial and archaeal proteomes, together with the related plastid proteins, are automatically annotated based on manually created family rules for complete protein annotation, with template-based feature propagation. Proteins with no similarity to other proteins in Swiss-Prot, which we call ORFans, undergo an automated protein sequence analysis procedure that looks for many of the sequence features described in the preceding section. These features are then automatically annotated according to rules of consistency and dependency.

We have just developed a second system called Anabelle that strives to annotate not only ORFans and well-defined proteins, but also any protein with one or more conserved or functional domains or sites detected by one of the methods carefully selected for their accuracy by the Swiss-Prot team. The information retrieved from all results is logically combined according to selection rules and logical rules, thus coming to more trustworthy conclusions than possible when just looking at one result at a time. Anabelle is integrated in the annotator's workbench: The automatically pre-selected analysis results are visualized in a graphical system, from which the annotator can choose the true positive results and easily generate annotation based on sequence similarity and sequence analysis. Not only does this speeds up annotation, but it also promotes the consistent transfer of entire information blocks that logically group together, ensuring the usage of standardised vocabulary and minimising the probability of errors and typos.

We believe that careful application of rules to produce automatically or semi-automatically annotated protein entries brings about many advantages for users of Swiss-Prot. We know that many are apprehensive of the word "automation" and are afraid that we will drown high-quality manually annotated entries with lower quality "automated" entries. We are very aware of this danger and are almost paranoid in our effort to ensure that automatic annotation will produce data of a quality up to that of manual curation. Finally it must be noted that one of the important changes planned in the Swiss-Prot format (see 2.6) is very pertinent to this issue, the introduction of "evidence tags" which should allow to unambiguously flag if an information item has been manually or automatically derived.

### *2.5 Standardisation and controlled vocabularies*

#### *2.5.1 A long tradition of using controlled vocabularies in Swiss-Prot*

To allow effective and precise database retrieval and searches, the same concepts need to be described with the same terms everywhere in the database. Controlled vocabularies or indexing terms can serve this purpose. A controlled vocabulary is defined as "an organised list of words and phrases, or notation system, that is used to initially tag content, and then to find it through navigation or search" (Amy Warner[1]).

Since its creation, Swiss-Prot has stored information under specific line types many of which are structured in such a way as to facilitate text searches in the database. Even the fields that appear to contain unstructured text are often written according to strict guidelines to ensure consistency. In some cases, lists are made where "preferred" terms are associated with synonyms, spelling differences, abbreviations, or yet other terms considered as equivalents.

---

[1]http://www.lexonomy.com/publications/aTaxonomyPrimer.html

Table 1 provides a partial description of where and how Swiss-Prot either makes use of existing controlled vocabularies or has developed such corpora.

| | |
|---|---|
| Protein names (DE line) | We use as primary name the ones that seem to be the most appropriate according to the function of a protein, to the nomenclature adopted by the specialists in that field or to the gene name, etc. We keep all synonyms used in publications and authors' submissions except if they are misleading. Furthermore we transfer the same name to the orthologs of related organisms. |
| Gene names (GN line) | Whenever a nomenclature committee (for example HUGO, FlyBase, etc.) provides "official" gene names for a given organism, we try to enforce their choice of gene names, yet keeping what authors originally provided as synonyms. |
| Species names (OS line) | The species names used in Swiss-Prot are listed in a document (speclist.txt). From the very beginning, care has been taken to store not only the official (scientific) name, but also the most useful common names and synonyms. |
| Species taxonomy (OC and OX lines) | We make use of the taxonomy compiled by NCBI which is used by most major biomolecular sequence databases. |
| Organelle (OG line) | We standardize plasmid name usage and list them in a Swiss-Prot document (plasmid.txt). |
| Reference comments (RC line) | Among other uses, the RC line allows to indicate the tissue from which a protein originates (TISSUE), or the strain (STRAIN). The tissues are reported in the file tisslist.txt and the strains in strains.txt. Both lists contain indications on synonyms. |
| Reference authors (RA line) | As far as possible, the names of authors are stored according to consistent rules. For example the German Umlaut is replaced by an 'e' following the vowel on which the Umlaut was perched, the hyphen is retained between two initials (which is removed in Medline/PubMed), we keep all the initials (even where PubMed only keeps two) and we often correct misspelling in author names! |
| Reference location (RL line) | Journal abbreviations in Swiss-Prot follow whenever possible those used by the National Library of Medicine (NLM). We provide a journal list (jourlist.txt) that, in addition to the journal names and abbreviations, also provides ISSN (International Standard Serial Number), CODEN number, publishers and journal home page web addresses. |
| Comments (CC line) | The CC lines mainly contain free text comments classified under 24 different topics. If a piece of information cannot be classified under a specific topic, it is put under 'MISCELLANEOUS'.<br>However, with time, the information in the CC lines is becoming less 'free' so to speak, and more and more CC line topics are subjected to controlled vocabularies. For example, this is the case of the 'CATALYTIC ACTIVITY' topic whose text is taken from the ENZYME database [10] for all known enzymes, referred to by their EC (Enzyme Classification) numbers in the DE lines. We are currently standardizing the use of the 'COFACTOR', 'PATHWAY' and 'SUBCELLULAR LOCATION' topics. |
| Keywords (KW line) | Keywords were one of the first sets of controlled vocabulary in Swiss-Prot. They were introduced to summarize the content of an entry and to group entries according to different aspects related to biological processes, molecular function, subcellular location, domains, ligands, sequence modifications and diseases. We provide a keyword list (keywlist.txt) that is being superseded by a dictionary that provides the precise definition of the usage of a keyword in the context of Swiss-Prot. The dictionary also includes synonyms, groups keywords into categories and provides a mapping between Swiss-Prot keywords and GO terms (see 3.5.2). |
| Feature table (FT line) | We are currently establishing a controlled vocabulary for the features describing posttranslational modifications (PTMs) [11]. We are also building a PTM database to store, for each type of modification, information such as the general description, target(s), chemical formula, subcellular localization of modified site, enzyme(s) carrying out the PTM, etc. Domain-type (DOMAIN, REPEAT, DNA_BIND, ZN_FING, etc.) feature descriptions are also standardized across all of Swiss-Prot. |
| Sequence | The sequences are stored in the one-letter code adopted by the commission on Biochemical Nomenclature of the IUPAC-IUBMB. |

Table 1: Standardization efforts and use of existing or in-house controlled vocabularies in Swiss-Prot, listed by line type.

This list, even if incomplete, is impressive; yet it does not capture the whole complexity of issues surrounding the use of nomenclature and controlled vocabularies in the life sciences. We need to state here that if physicists or chemists behaved like biologists do, we would probably live in a world without computers or plastic (this may sound like an attractive proposition to some!). Life scientists do not receive, during their training, the perception of the importance of following nomenclature rules. Yet, they are the first to complain when they look for specific information across one or many databases and fail to obtain a comprehensive answer because that information is heterogeneously described. Therefore we always felt that Swiss-Prot had a mission to fulfil in enforcing existing rules and more and more, as time passed by, to actively participate in the development of new nomenclature and controlled vocabularies. Anecdotally such an active role can have some unexpected consequence: we were once threatened with a lawsuit because we did not accept to use as a valid gene symbol the one proposed by an author.

All of this leads us to give the following advice to would-be developers of databases:

- Try to follow as much as possible existing controlled vocabularies and nomenclatures;
- Do not hesitate to contact the groups maintaining these resources and to point out inconsistencies and/or errors;
- Do not be afraid to take a firm stand toward your users when they request the representation in your database of terms that do not follow a specific guideline. You can always (and you should!) store this information as a synonym.

### *2.5.2 Going ahead with GO in Swiss-Prot*

If we assume, as mentioned above, that "users and database should agree on the meaning of the term being used", given the large number of biomolecular databases available, this indirectly implies that all databases should agree on the meaning of a term! In an attempt to achieve this ambitious goal, maintainers of FlyBase, MGD and SGD joined forces and formed the GeneOntology (GO) Consortium [12]. They established three ontologies, gathering key terms for cellular components, biological process and molecular function, thus catering for a large need for standardisation that could be observed all across the scientific community.

From the beginning of the GO activities, we were repeatedly approached by users wondering when we would introduce GO terms to Swiss-Prot and TrEMBL. However, while clearly welcoming the effort made by the GO consortium, we were reluctant to add links to GO at that time: Given the initially small scope (GO specialised in three major organism groups, whereas Swiss-Prot has to deal with thousand of different species), and the fact that many mappings had been created automatically and were thus likely to assign GO terms to unrelated proteins, we considered it dangerous to mislead users into incorrect assumptions. We did not want to risk the situation where someone would happily accept a GO assignment indicating a function for an otherwise uncharacterised protein, without further questioning the assignment because they trust the judgement of Swiss-Prot annotators and the high quality of the manual annotations.

It was only in 2003 that we felt what it became "safe" to start introducing GO terms in Swiss-Prot. We felt that GO had indeed considerably matured and had increased its coverage. What's more, several species-specific databases have established manually curated mappings between GO terms and their gene catalogues. The EBI GO team has mapped Swiss-Prot keywords to GO terms. Evidence tags are available in GO to indicate whether an assignment has been done automatically or by manual curation. The time had come to follow the demands, and to introduce cross-references (see 2.7.1) from Swiss-Prot to GO. We added them in all cases where they originated from manual annotation efforts. We also are in the process of introducing GO terms for all members of microbial protein families that fall under the scope of the HAMAP annotation project.

*2.6 Evolution of entry structure and format*

Since its creation in 1986, the basic structure of a Swiss-Prot entry has not changed significantly. The distinct line types defined by a 2-letter code are generally relevant to all entries and cover the core data, while the actual protein information is given in the comment (CC) lines and in the feature table (FT). While the general framework has been very stable, we have carried out many changes over the years. New line types were introduced, the structure of existing line types was constantly refined and new sub-fields (comments topics, feature keys) were added. Such changes are always documented (in release notes and other documents) and users are warned in advance of pending changes so that they can adapt their software tools. While the general stability of the Swiss-Prot flat file format may be seen as a proof of foresight, careful planning and experience, one can also say that in some respect Swiss-Prot had become a victim of its own success: even the smallest modification to the flat file format, or the introduction of new fields, needs to be considered carefully, and it happens that ideas are discarded for the sole reason that "this will cause the crash of thousands of programs out there...".

Swiss-Prot and TrEMBL have traditionally been maintained and distributed as flat files. An inherent problem of flat file databanks is that their maintenance becomes increasingly difficult when they grow in size and many people are involved in the production of the data. Since 2002, Swiss-Prot and TrEMBL are also distributed in XML (http://www.ebi.uniprot.org/support/documents.shtml), the extensible markup language that makes it possible to define the content of a document separately from its formatting, making it easy to reuse that content in other applications or for other presentation environments. XML allows, in contrast to HTML, the authors of a document to create their own markup tags suiting their needs and allowing to best structure the data. But what is more, XML allows implementing rules that are not limited to formatting, but can be used to formulate dependencies. We are also in the process of porting the production of Swiss-Prot and TrEMBL to a Relational Database Management System. In order to develop the relational and XML schema, we have designed conceptual data models, using the Unified Modelling Language (UML) notation, to represent the structure and constraints present in the data.

In the meantime, until the production copy of Swiss-Prot is managed in a relational database management system, we still need to introduce certain format changes to the flat file in order to accommodate more complex concepts. Such changes can be quite substantial and time-consuming, as they are always introduced in a way that not only new annotation is performed according to the new format, but all existing entries need to be converted. As a consequence, this can involve, in addition to the creation of conversion software, and to the modification of documentation and annotation tools, a lot of manual cleaning. That we need to embark on such manual cleaning steps is not due to the structure or the format of the database, but rather to our pathological urge to make sure that all aspects of Swiss-Prot are self-consistent. Therefore, whenever we introduce a new type of data, we try as much as possible to update all the entries where such data has some relevance.

There are many changes we plan to make to the flat file format. For example, in the near future, we plan to overhaul the format of the GN (gene) line so that it will allow a more structured representation of the information concerning gene names. The new format will allow distinguishing official gene name, synonyms, ordered locus name and ORF names. This change allows a better representation of the complexity of gene and locus naming schemes.

As we described in the section on automatic annotation (see 2.4.3), it is important to provide users with a means to track down the origin of all information items in a Swiss-Prot entry. Such a need was not apparent in the early days of Swiss-Prot as most information was derived from a single paper that both reported the sequence and its characterisation. This is no longer true and some entries contain information originating from up to 110 references as well as the results of many sequence analysis tools. It is therefore necessary to provide 'evidence tags'. These are links between an information item and its source, whether a reference, the judgement of annotator or the result of a program. Such evidence tags already exist in TrEMBL. We have been very slow in the process of providing them in Swiss-Prot, partly because they are difficult to implement in the current annotation platform and because they are very cumbersome in the current flat file format. Evidence tags are therefore probably going to be implemented in the XML and relational versions of Swiss-Prot and will probably not be available in the flat file distribution.

### *2.7 Cross-references*

#### *2.7.1 Cross-references in Swiss-Prot*

Cross-references as a way to access related information in other databases have been an integral part of Swiss-Prot almost since the beginning (they were introduced in release 4 of April 1987). Whilst navigating between databases is much less of a challenge now, thanks to the web, than it was back in the late eighties. The early presence of DR (Database cross-Reference) lines in Swiss-Prot shows how anticipatory we were in conceiving the database in a way that facilitates data integration. One of the first important software applications that made use of Swiss-Prot cross-references was the Sequence Retrieval System (SRS) [13], developed by Thure Etzold at EMBL, from 1990 on. In addition to providing a search interface for multiple databases with a single query, an important feature of SRS is its ability to combine all indexed databanks into a network, where new ways of linking information from different sources can be explored. One of the main reasons why this became possible was the fact that Swiss-Prot, one of the first databases indexed under SRS, was so highly cross-referenced. SRS documentation contained in 1990, and still contains in 2003, an image showing biological databases linked to each other in form of a network, the centre of which is Swiss-Prot, connected with practically all the other databases indexed under SRS.

The first databases cross-referenced in Swiss-Prot were the primary DNA and protein sequence databases EMBL and PIR, and the PDB protein structure database. New links were regularly added at each of the major Swiss-Prot releases. Currently Swiss-Prot is linked to 55 different databases and each entry contains an average of 9.1 links. One would naively assume that an entry does not contain more than a single cross-reference to a given external database. This is not always true for a variety of reasons that generally depend on the structure of the external database. For example, there is an average of 1.92 cross-references to the EMBL DNA sequence database per Swiss-Prot entry. This reflects the redundant archival nature of the nucleotide databases. However, this overall average does not convey the true nature of the situation: 58% of all Swiss-Prot entries only contain one single cross-reference to EMBL, while 6.2% contain more than 5 such cross-references.

A special emphasis should be given to the cross-references to family/domain databases. PROSITE was the first of these databases to be created and accordingly the first to be cross-referenced in Swiss-Prot. When cross-references to PROSITE were introduced in 1990, there was an average of 0.42 per Swiss-Prot entry. In 2003, this number is more than twice as high, an increase that can be explained by improved methods to detect domains, but also by the fact that PROSITE increasingly reacts to the demands from Swiss-Prot annotators: Whenever a newly annotated protein family carries a particular domain that is not yet present in PROSITE, the PROSITE staff creates a discriminator (pattern or profile) for that domain. Many other family/domain databases were created in the last ten years, most of which are cross-referenced in Swiss-Prot and also incorporated in the InterPro [14] resource which unites these databases "under one roof". Today a Swiss-Prot entry contains an average of 5.2 links to family/domain databases. These cross-references can also be seen as a pointer to the existence of a specific domain in a given protein sequence.

As mentioned in 2.5.2, in 2003, we have added cross-references to the three GO ontologies. These cross-references have a dual purpose: they allow navigation toward an external resource (here GO), and they also serve as information items. This may be better explained by the following example:

DR GO; GO:0012501; P:programmed cell death; TAS.

In the above line, the GO accession number "GO:0012501" provides a handle to access the GO database (navigation), the "P:programmed cell death" indicated that the protein is involved in the biological process ("P") of programmed cell death and the "TAS" stands for "Traceable Author Statement".

### *2.7.2 Cross-referencing versus integrating*

Over the years, it became clear that our strategy to "delegate" specialist tasks to the specialists (and establish reciprocal links), while concentrating on the more "generalist" annotation was satisfactory. This was facilitated and influenced by the appearance of more and more databases: the word-wide web made it a lot easier to publish expert knowledge. Existing and well-established databases (e.g FlyBase) took advantage of the increased visibility offered by the world-wide web, and many additional new information resources burgeoned. A number of these databases were constructed around the primary sequence or organism-specific gene nomenclature databases, and used the accession numbers of the sequence databases (or the primary gene names) as their set of unique identifiers. An example is GeneCards, a database of "information cards" on every human protein in Swiss-Prot and TrEMBL. Such databases are usually cross-referenced to Swiss-Prot via "implicit" links, created on the fly by the NiceProt tool (see 3) that displays a Swiss-Prot entry on ExPASy. In addition to the explicit cross-references "hard-coded" in the Swiss-Prot DR lines, the concept of implicit links enforces the role of Swiss-Prot as a central hub for molecular biology information [15].

There may seem to be certain drawbacks related to the strategy of establishing extensive cross-links vs. the idea of integration of all data locally: 1) "Loss of control"; 2) Cross-references create a certain dependency (when free public access to the Yeast Proteome Database (YPD) was discontinued, expectations grew again for Swiss-Prot to provide more extensive annotation for Saccharomyces cerevisiae) 3) Necessity to rely on the willingness to collaborate of providers of the specialised cross-referenced databases (e.g. use of standard nomenclature and common identifiers, provide or at least help with mappings between Swiss-Prot accession numbers and their database) 4) Some foresight and knowledge of the related field is necessary, in order not to make the effort of adding links to a resource which will not be updated or which is likely to loose funding – with the consequence of being forced to remove those links after a short while. However, these disadvantages are easily outweighed by a gain in time and the relief not to "have to be an expert in every field", as well as the reward of fruitful collaborations and exchanges. Procedures have been established to obtain mappings between Swiss-Prot sequences on one side, and relatively heterogeneous information on the other: nucleotide sequences, gene names, modification sites, domain descriptors, ontologies, etc. Many cross-references, in particular those that are based on sequence searches, i.e. domain and family classification, are now already applied to TrEMBL. This means that an entry comes with a certain number of DR lines before manual annotation even starts. Some other DR lines however require careful checking by an annotator, and yet others have to be added completely "manually" as they can only be established after perusal of literature and other sources (e.g. MIM). While the list of cross-referenced databases keeps growing, it does happen that we are obliged to remove links to certain databases. This can have several different reasons, the most frequent ones being a lack of funding and subsequent discontinuation of a database, or the decision of a database maintainer to commercialise a resource and discontinue free web access even for academic users.

### *2.7.3 Some thoughts on unique and stable identifiers*

There are some important observations to make about cross-referencing in general. To implement cross-referencing to a database, that database needs to provide unique and stable identifiers (USI) for each of their entries. These USI are often known as accession numbers. Such a requirement may seem obvious, but it is still often the case that databases do not see the need for stable identifiers. For example, a species-specific database may use gene names as their unique identifiers. The problem is that such identifiers may be unique but are certainly not stable as it is most probable that some of the gene names will change over time. Far more important for future developments is our belief that major objects in a database require their own independent sets of USI. We became aware of this when we saw the need to add USI to a number of objects in Swiss-Prot thus allowing external databases to seamlessly implement cross-references to a specific object in Swiss-Prot rather than at the level of the entire entry. A good example of such developments is the creation of feature identifiers (FTId) for all human protein sequence variants in Swiss-Prot. These identifiers allow specialized databases that report mutations concerning a specific set of genes to make a cross-reference to the representation of that mutation in Swiss-Prot.

## 3. Making Swiss-Prot available to the users

In prehistoric times – i.e. before the Web! -, Swiss-Prot reached its users by a variety of means. It was sent on computer tapes by the EMBL, it was distributed on floppy disks by companies selling sequence analysis software and, in 1989, it became the first major biomolecular database to be distributed on CD-ROM. In parallel to the physical distribution of Swiss-Prot, the database was made available by anonymous FTP and was searchable from a number of on-line resources such as BIONET and the NCBI IRX database retrieval software.

When the World-Wide Web began in 1993, Swiss-Prot became available on the ExPASy [16] server (www.expasy.org), which was born on August 1, 1993. At that date there were less than 150 web servers worldwide. To the best of our knowledge it was the first web server for the life science community. We were very pleased to see that it was accessed 7'295 times during its first month of activity. We never imagined that a few years later it would be accessed at a rate of 8-10 million hits per month. It has now been accessed more than 300 million times by a total of more than three million computer hosts from 200 countries. Seven mirror sites, i.e. exact copies of the main site in Switzerland have been established in Australia, Bolivia, Canada, China, Korea, Taiwan and the USA. It is also noteworthy to mention that ExPASy and the EBI server (www.ebi.ac.uk) are far from being

the only web servers that redistribute Swiss-Prot and TrEMBL, we estimate that there are about 50 such sites world-wide.
ExPASy has constantly evolved in its ten years of existence. It is outside of the scope of this article to describe all of what is available on the server, yet we want to point out two significant developments that reflect our response to the needs of users.

In autumn 1998, we initiated "NiceProt", with the intention to provide scientists with a more user-friendly way of looking at Swiss-Prot and TrEMBL entries. Instead of showing the raw Swiss-Prot data format (with its two-letter line types), we decided to make use of html tables to group certain fields under common headings, to replace the line type by a more explicit key (e.g. "Cross-references" instead of "DR"). This was initially targeted at users who are not familiar with the Swiss-Prot data format, but rapidly caught on in the scientific community. Gradually, more and more functionalities were added, including many implicit cross-references, and links to context-specific documentation. During the first eight months of 2003, ExPASy treated about 1 million requests for individual Swiss-Prot or TrEMBL entries on average per month. An overwhelming majority of these hits (85 %) are for NiceProt, whereas the remaining 15 % account for accesses to the raw text version, or the "htmlised" view that was prevalent prior to September 1998.

The NEWT [17] taxonomy browser (http://www.ebi.ac.uk/newt/) is a service introduced in 2002 that serves as an entry point into Swiss-Prot and TrEMBL using taxonomic search criteria. The core of NEWT consists in the integration of Swiss-Prot specific taxonomy information with the NCBI taxonomy data in a relational database. Taxonomic nodes are stored in a hierarchical tree; this allows easy navigation through the taxonomy lineage from every taxon. The web interface to NEWT allows users to search and browse the daily updated taxonomy data. Users can navigate through the taxonomy tree and access corresponding Swiss-Prot and TrEMBL protein entries. Additionally, a manually curated selection of over 24,000 external links (including more than 13,000 photographs) provides specific information on selected species.

Both UniProt and NEWT are representatives of the trend toward a 'customisation' of the representation of knowledge. We believe that this trend will not abate; there are many specific communities of life scientists that require information on proteins, yet want them to be represented in a style or perspective specific to their field of research. We are in the process of developing new types of views.

We also believe that the ExPASy server access log files are a valuable source of information as to the most frequently consulted TrEMBL entries (i.e. unannotated entries that will greatly benefit from manual annotation) scientists' use of search engines, the context in which certain entries are consulted etc. We therefore plan to mine the ExPASy log files and expect to be able to draw enlightening conclusions!

## 4. Conclusions

Being a well-established database, we can say that the tireless effort of juggling between evolution and stability has been an exhausting but suitable strategy for the development of the Swiss-Prot protein knowledgebase. Early design features of the database such as the detailed structuring of the entry format, the standardisation of nomenclature, the regular review of the annotation of protein families have been shown to be indispensable. The explosive growth in uncharacterised sequence data has led us to the implementation of automatic and semi-automatic processes. They are designed to ensure the same high-quality standards that have always been the hallmark of Swiss-Prot. Automation has to go in parallel with the introduction of evidence tags that will allow distinguishing data sources and inferences. We strongly believe that the future of Swiss-Prot and of any similar curated information resource relies on the active participation of the life sciences community. This will require an increased educational effort on our part. It is also dependent on the commitment of scientific societies, publishers and funding agencies to provide a framework to facilitate community efforts and give due credit to the participating scientists.

As a closing remark, we would like to thank all the persons involved in the development of Swiss-Prot at the SIB and EBI as well as all the funding agencies and companies that have financially contributed to the continuous evolution of the Swiss-Prot knowledgebase.

## Acknowledgements

The work described in this article covers activities funded by various sources including NIH:1 U01 HG02712-01, EU:BioMinT; QLRT-2001-02770, EU:Temblor; QLRT-2001-00015, EU:BioBabel; QLRI-CT-2001-00981, SNF:3100-063879. The above review originally appeared in *Briefings in Bioinformatics*, 5:39-55(2004) and is reproduced here by permission of the Journal.

## References

[1] Boeckmann, B., Bairoch, A., Apweiler, R. *et al.* (2003), 'The SWISS-PROT protein knowledgebase and its supplement TrEMBL in 2003', *Nucleic Acids Res.,* Vol. 31, pp. 354-370.

[2] Bairoch, A. (2000), 'Serendipity in bioinformatics, the tribulations of a Swiss bioinformatician through exciting times!', *Bioinformatics*, Vol. 16, pp. 48-64.

[3] Apweiler, R., Bairoch, A., Wu, C.H. *et al.* (2004), 'UniProt: the universal protein knowledgebase', *Nucleic Acids Res.,* Vol. 32, pp. D115-119.

[4] Dayhoff, M.O., Eck, R.V., Chang, M.A., and Sochard, M.R. (1965), 'Atlas of Protein Sequence and Structure', Vol. 1. National Biomedical Research Foundation, Silver Spring, MD.

[5] Moore, J., Engelberg, A. and Bairoch, A. (1988), 'Using PC/GENE for protein and nucleic acid analysis', *Biotechniques*, Vol. 6, pp. 566-572.

[6] Monigatti, F., Gasteiger, E., Bairoch, A. *et al.* (2002), 'The Sulfinator: predicting tyrosine sulfation sites in protein sequences', *Bioinformatics*, Vol. 18, pp. 769-70.

[7] Bologna, G., Veuthey, A.-L., Yvon, C. *et al.* (2004), 'N-terminal myristoylation predictions by ensembles of neural networks', *Proteomics*, Vol. 4, pp. 1626-1632..

[8] Hulo, N., Sigrist, C., LeSaux, V. *et al.* (2004), 'Recent improvements to the PROSITE database', *Nucleic Acids Res.,* Vol. 32, pp. D134-137.

[9] Gattiker, A., Michoud, K., Rivoire, C. *et al.* (2003), 'Automated annotation of microbial proteomes in Swiss-Prot', *Comput. Biol. Chem.*, Vol. 27, pp. 49-58.

[10] Bairoch, A. (2000), 'The ENZYME database in 2000', *Nucleic Acids Res.*, Vol. 28, pp. 304-305.

[11] Farriol-Mathis, N., Garavelli, J.S., Boeckmann B., *et al.* (2004), 'Annotation of post-translational modifications in the Swiss-Prot knowledgebase', *Proteomics*, Vol. 4, pp. 1537-1550.

[12] Ashburner, M., Ball, C.A., Blake, J.A. et al. (2000), 'Gene ontology: tool for the unification of biology. The Gene Ontology Consortium', *Nat. Genet.*, Vol. 25, pp. 25-29.

[13] Etzold, T., Argos, P. (1993), 'SRS - an indexing and retrieval tool for flat file data libraries', *Comput. Appl. Biosci.*, Vol. 9, pp. 49-57.

[14] Mulder, N.J., Apweiler, R., Attwood, T.K., Bairoch, A., *et al.* (2003), 'The InterPro Database, 2003 brings increased coverage and new features', *Nucleic Acids Res.*, Vol. 31, pp. 315-318.

[15] Gasteiger, E., Jung, E., Bairoch, A. (2001), 'SWISS-PROT: Connecting biological knowledge via a protein database', *Curr. Issues Mol. Biol.*, Vol. 3, pp. 47-55.

[16] Gasteiger, E., Gattiker, A., Hoogland, C. et al. (2003), 'ExPASy – the proteomics server for in-depth protein knowledge and analysis'. *Nucleic Acids Res.,* Vol. 31, pp. 3784-3788.

[17] Phan, I.Q., Pilbout, S.F., Fleischmann, W., Bairoch, A. (2003) 'NEWT, a new taxonomy portal'. *Nucleic Acids Res.,* Vol. 31, pp. 3822-3823.

*Essays in Bioinformatics*
*D.S. Moss et al. (Eds.)*
*IOS Press, 2005*


# EMBOSS – A sequence analysis package

Lisa MULLAN[1] and David P. JUDGE[2]
[1]*European Bioinformatics Institute, Wellcome Trust Genome Campus, Hinxton, Cambridge CB10 1SD England*
[2]*Department of Genetics, University of Cambridge, Tennis Court Road, Cambridge CB2 3EH, England*

**Abstract.** EMBOSS evolved from EGCG, a collection of programs written to extend the GCG package, originally written by the **G**enetics **C**omputer **G**roup of Wisconsin University. EMBOSS follows the general structure of GCG and sets out to reproduce and extend the functionality of GCG in an open source package. Currently, EMBOSS only runs on UNIX computers. The programs of EMBOSS can be run from the UNIX command line or from behind a number of **G**raphical **U**ser **I**nterfaces (GUIs). EMBOSS offers a wide range of programs covering most aspects of sequence analysis. In addition, a number of well established public domain programs have been engineered to follow the conventions of EMBOSS and then incorporated into the package. Software developers from many places across the world have written programs for the EMBOSS package. Such contributions are encouraged from the user community and training is offered to aspiring contributors.

## 1. The origins of EMBOSS

The essential structure of the EMBOSS software package for sequence analysis follows that of the older GCG package. Indeed, EMBOSS evolved directly from the EGCG (Extended GCG) package, which was comprised of programs written by various EMBnet[1] researchers to extend the functionality of GCG.

GCG was originally written by the **G**enetics **C**omputer **G**roup at Wisconsin University, USA as an open source bioinformatics package. GCG was originally available relatively inexpensively. As the source code was accessible, algorithms could be verified and adapted to suit the needs of individual researchers. Many new programs were written by researchers who were not part of the GCG team using the GCG libraries. In 1988, the best of these new "GCG extensions" were collected together and the EGCG package was born. This was achieved by a collaboration of groups within EMBnet and elsewhere. EGCG provided new sequence analysis software and advanced features, which were used at approximately 150 sites, and by more than 10,000 users of EMBnet national services.

A few years ago, the GCG software was purchased by a commercial enterprise and the source code was no longer available to users. Development slowed significantly and GCG sometimes failed to sufficiently meet the demands of biological advancement. The EGCG project was no longer viable and had reached the limits of what could be achieved using the GCG libraries. Consequently, the former EGCG developers, and others, designed a totally new generation of academic sequence analysis software – the suite of programs that is now known as EMBOSS (**E**uropean **M**olecular **B**iology **O**pen **S**oftware **S**uite) [1].

EMBOSS was released in 2000 and has been actively developed ever since. At the core of the EMBOSS package are the programs designed to reproduce and extend the

[1]European Molecular Laboratory Network, http://www.embnet.org

functionality of GCG and EGCG. New EMBOSS programs are added and current programs continuously improved.

Once a fully commercial package, the GCG license became much more expensive. Institutions such as the RFCGR[2] had to raise or introduce academic user fees for the use of the GCG package. In contrast, EMBOSS is a totally free package, downloadable from http://emboss.sourceforge.net/download/

Cost considerations, together with a requirement for energetically developing contemporary analysis software have encouraged many institutions to move towards EMBOSS. As the two packages share an overall structure and comparable **G**raphical **U**ser **I**nterfaces (GUIs), experienced GCG users found little difficulty running EMBOSS from the UNIX command line or from behind its various GUIs.

Again, in similar fashion to GCG, EMBOSS includes a number of well established programs not written especially for EMBOSS. Examples of such programs include **clustalw** for multiple sequence alignment (**emma** in EMBOSS) and **primer3** for primer design (**eprimer3** in EMBOSS). This policy is clearly preferable to trying to rewrite tools that are accepted as "the best" by the user community. Programs imported in this fashion are run behind a software wrapper that gives them a look and feel compatible to the core EMBOSS programs.

One of the major features not yet to have been implemented in EMBOSS is a sequence database similarity search tool such as **blast**. Currently, similarity searches must be conducted elsewhere before the results are analysed using EMBOSS tools.

## 2. EMBOSS - A Free Open Software Suite

### *2.1 Overall structure*

The pre-requisite of designing EMBOSS was that the source code be freely available to all. Software developers should be permitted to access to the code and to manipulate it in whatever fashion they chose. All users should be able to access the extensive (currently more than 200) collection of EMBOSS applications from anywhere in the world at no cost. Access via intuitive web sites (and other forms of GUI) is of particular importance.

EMBOSS, following the structure of GCG, is comprised of many small programs each carrying out a single task. This has proved to be of particular benefit to bioinformatics specialist wishing to create larger applications by "stitching together" the simple EMBOSS applications. However, this structure is not always ideal for the less ambitious user. In particular, determining the name of the program appropriate for a given task is not always easy for the more casual user.

For the user already familiar with GCG, a table of GCG/EMBOSS program equivalencies has been constructed at several sites. Some of them can be found at:

- http://www.sanbi.ac.uk/mrc/GCG_replacement.html
- http://www.biobind.com/faq/gcg-emboss.html
- http://helix.nih.gov/apps/bioinfo/emboss-gcg.html

Not every EMBOSS program will have an exact GCG equivalent, but functionality should be reproduced.

Also, a small utility called **wossname** is also provided to act as a dynamic index of programs within the EMBOSS suite (The Jemboss GUI offers users a more powerful keyword search). The program uses keywords to identify EMBOSS programs pertinent to a

---

[2]Rosalind Franklin Centre for Genomics Research (UK), situated near Cambridge. This used to be the Human Genome Mapping Project Resource Centre (HGMP-RC)

given researcher. For example, if a program for carrying out a protein motif search were desired, then the keyword "motif" might be given to the **wossname** application. The resulting output would look like this:

```
SEARCH FOR 'MOTIF'
helixturnhelix Report nucleic acid binding motifs
meme           Motif detection
patmatdb       Search a protein sequence with a motif
patmatmotifs   Search a PROSITE motif database with a protein sequence
prosextract    Builds the PROSITE motif database for patmatmotifs to search
```

As each EMBOSS program is typically designed to perform a single analytical step, it is often the case that several programs are required to achieve any given user objective. For example, a multiple sequence file can be generated with the **seqret** program and read directly into **emma** (the multiple sequence alignment program **clustalw** customised for inclusion into the EMBOSS package). The output from **emma** is a series of sequences containing gap characters to represent insertions or deletions throughout the alignment. This can be manipulated to appeal more to the human eye by reading it into an alignment viewer such as **prettyplot**.

There is also a more sophisticated multiple sequence alignment viewer and editor called the Jemboss Alignment Editor which can be invoked from the EMBOSS command line or the Jemboss GUI.

### *2.2 Input and Output formats*

Many previous bioinformatics packages and databases have defined their own sequence formats, which have become standards. EMBOSS has been created to recognise all these standard sequence format types. Thus, the input to EMBOSS applications is not restricted to sequences stored in a particular way (in contrast to the GCG package). Files generated by using GCG may be read into EMBOSS applications. The default sequence format output in EMBOSS is **fasta** format, but almost anything may be specified (in total, EMBOSS supports 42 different sequence formats).

**fasta** format is a common, simple format for sequences, and can be recognised by a sequence description line succeeded immediately by the sequence itself. A "greater than" sign starts the description line, thus identifying it. Multiple sequences can be stored in the same file, with the description line separating the individual sequences from each other.

Other sequence formats are also common. **raw**, or **plain** format is composed of, as their names suggest, simply the sequence on its own. Certain applications will only accept this type of format, but it does not hold an ID or sequence name, and cannot, be used in a multiple sequence file as there is no indication of where the sequence starts or ends. GCG format was devised for the GCG package, and has several lines of description, together with sequence numbering. One of the features of a GCG format is a "checksum". This is a number relating directly to the sequence, and was implemented in the days when file transfer was less reliable than today. The intention was to allow researchers to know that they transferred an intact and correct sequence. In the current phase of more reliable networking, this has proved a hindrance in many cases, as the sequence cannot be manually edited before being input into another GCG, or other software application. Standard report formats are used for alignment output. For example **gff** format for protein features and **markx** format for alignments.The versatility of EMBOSS means that any sequence format can be read into an appropriate application, or may be changed to an alternative format.

In addition to allowing access to sequences in many formats stored in local files and sequences stored in locally managed databases, EMBOSS is able to access sequences stored in remote databases managed on any SRS server throughout the world. This is of particular value to users wishing to install the EMBOSS programs locally. Installing the software and keeping it up to date is a relatively minor undertaking. Keeping a realistic range of sequence databases up to date locally is another matter, requiring an enormous amount of disc space and far more time than most users can afford. Remote access to sequence databases is possible using elements of the **S**equence **R**etrieval **S**ystem, SRS, developed at the EBI.

### *2.3 The range of applications*

There is a historical bias towards sequence analysis but a wide variety of areas are now catered for in the EMBOSS suite. There are applications for local, global and multiple sequence alignments and tools for generating and scanning various types of profile including hidden Markov models, thanks to the integration of Sean Eddy's HMMER package. Tools for motif identification in protein and nucleotide sequences are also provided. A variety of other tasks including manipulation and display of feature tables, graphical output, database indexing and other administrative tools are also catered for.

Recently, software has been added for proteomics and protein structure. Included are tools for identifying sequence fragments of a specified molecular weight and software for parsing and processing the PDB, SCOP and CATH databases.

### *2.4 The software*

EMBOSS programs are written in the computer programming language "C" and can, currently, only be run under the UNIX operating system. To run the programs without a GUI require that program names, together with any parameter requirements, are typed in response to a UNIX prompt forming a command line. PCs running Windows and Macintoshes can be used as terminals to connect to a UNIX machine offering the EMBOSS package.

### *2.5 Maintenance and Support*

EMBOSS is maintained by a collection of individuals, most of whom are based at the RFCGR. They use a central repository of code that is managed by CVS (Central Version System). EMBOSS is under very active development and the number of applications has almost doubled in the last 18 months. To ensure quality and stability, access to the repository is restricted to a core of active developers. Those who would like to join this team and get full write access to the repository are encouraged to follow the "Developing Code and EMBOSS internals" link on the EMBOSS home page at http://emboss.sourceforge.net

While the developers do not promise that EMBOSS is absolutely bug free, they do perform nightly compilation checks for the whole of EMBOSS on a variety of platforms and, as part of their quality assurance exercise, run each application on a test data set to ensure everything is working as anticipated.

User support, training and a means for feedback are provided via mailing lists and regular training courses. Details can be found from the EMBOSS homepage.

## 3. User interfaces for EMBOSS

Most EMBOSS programs will run in a text only environment, such as can be achieved using telnet or ssh to connect to an EMBOSS UNIX server such as provided by the (RFCGR). However, some programs generate graphical output which requires a terminal with graphical capabilities. This can be achieved by running an X server (such as Hummingbird eXceed for Windows) on the terminal, or by using a suitable EMBOSS GUI (such as the web GUI **W2H** or the Java GUI **Jemboss**).

### *3.1 EMBOSS from the UNIX command line*

The most common method of organising files and folders on a UNIX machine is to type textual commands in response to a UNIX prompt (**unix%**, say) forming a UNIX command line. It is possible also to run programs (including those of the EMBOSS package) by forming command lines. A command line to invoke an EMBOSS program specifies the program to run, together with all parameters for the specific analysis. The command line starts with the name of the application, followed generally by the input files.

Simply pressing <return> will set the program running. The researcher will be prompted for any other information necessary for the program to function, including the name of a file in which textual output should be saved. The application will then run, and a UNIX prompt returned to the screen. The results are now contained in the output file, and must be investigated separately. If further options on the program need to be accessed, the –**opt** qualifier[3] may be added onto the command line. This indicates to the program to prompt for further options.

There are some options that a program will never prompt for, and need always to be written on the command line. For each program, these are found in the EMBOSS help section under the heading "Advanced Qualifiers".

One of the major computational features of EMBOSS, is how it finds sequence files to read in. The Uniform Sequence Address (USA) is the form that the file address must adhere to in order for the program to function. This is generally of the type **format::database:file entry**. A similar address may be used for input or output files. Generally the format section can be omitted for input sequences, as EMBOSS should automatically recognise this. This is not the case if the file is in raw format, or Intelligenetics format.

If a multiple sequence alignment is required, input on the command line must be a single file, or a set of wild-carded file names. It is not always the case that this file contains multiple sequences, it may also contain database and accession numbers, or files names within a folder. Such a file is known as a "list file" or "file of filenames", and must be pre-pended with an @ sign (or list::) in order to be recognised by the EMBOSS application.

Documentation on each of the programs in EMBOSS may be invoked by running the program **tfm**. The program got its name from rather exasperated developers bombarded with questions from users who had not read The Fine4 Manual. It has been written for both users and developers and so may be biased towards computer jargon.

To access the information, users must type **tfm** on the command line, plus the name of the application they need help on. Alternatively, all this documentation may be found on-line at: http://emboss.sourceforge.net/apps/

---

[3] A qualifier alters the behaviour of the original program. EMBOSS programs have many of these, to cater for the vast number of options needed to perform specific analyses.

### *3.2 EMBOSS GUIs*

Using the UNIX command line can be daunting for many research biologists. In an attempt to make the programs more accessible to users not familiar with command line UNIX, several intuitive GUIs have been developed for EMBOSS. Some of these GUIs are web based and require data to be cut and pasted into appropriate fields. Other interfaces operate more closely with the programs of the package and can thus access data files more directly

**W2H (**http://gcg.rfcgr.mrc.ac.uk/cgi-bin/w2h-emboss/w2h.start)

Of the many web pages that have been designed as an interface for EMBOSS, this is possibly the most comprehensive. It allows upload of files from RFCGR and local disk, together with a minimal file management system. There is a simple interface and an advanced one for the more experienced user. Files and applications are accessed by double clicking on the relevant buttons, and each time a piece of information is requested, a new browser window opens. Operation of the site is not always obvious, and if Java is available on your computer, then you might be advised to spend the induction time on Jemboss instead.

**PBI (**http://bioinfo.pbi.nrc.ca:8090/EMBOSS/index.html)

An easy web site to use, this was developed at the Plant Biotechnology Institute in Canada. The basic site is not connected to any file management system, and is simply a means of running the EMBOSS applications on your data. Simply choose the relevant application, cut and past the data, and run. The results will appear in the same browser window. The web pages for this site can be downloaded and installed locally. The installation is very simple.

**SPIN (**http://staden.sourceforge.net)

SPIN is the program in the Staden package that offers sequence analysis tools. In common with other Staden package programs, it allows third party software to be accessed from its Graphical Interface. In particular, the whole of the EMBOSS package can be accessed from behind the SPIN interface. The Graphical Interface of SPIN, although not fully developed, is currently superior to any other EMBOSS GUI. In particular, SPIN offers graphical outputs that are interactive (i.e. it is possible to invoke information from a graphical output by clicking on features) and can be customised and combined freely. The Staden package is now open source and both source code and binaries for Windows, UNIX and MacOS X are freely available from the Staden package sourceforge home page. The only caveat to using SPIN is that it is currently not actively supported which means that integration of EMBOSS is limited to Version 2.8.0.

#### *3.2.4 Jemboss (http://www.rfcgr.mrc.ac.uk/Software/EMBOSS/Jemboss/)*

Jemboss [2,3] is the new interface written for EMBOSS at the RFCGR. It is written in the programming language Java primarily to enable it to run on almost any computer (in particular, Java for MacOS and earlier versions will not run Jemboss, although Jemboss will run under MacOS X). It is designed as a "point and click" interface and offers an easy route to sequence analysis for the biological researcher. Jemboss was written primarily as an interface to the EMBOSS package installed on the UNIX computers of the RFCGR. However, the server and standalone versions of Jemboss is free and can be installed along with the EMBOSS package on any UNIX server.

## References

[1] Rice, P., Longden,I., and Bleasby, A. **(2000)** EMBOSS: The European Molecular Biology Open Software Suite, *Trends in Genetics* **16** (6) *276-277*

[2] Carver T.J., Mullan L.J. **(2002)** A new graphical user interface to EMBOSS, *Comparative and Functional Genomics* **3** (1) *75-78*

[3] Carver T.J., Bleasby A.J., **(2003)** The design of Jemboss: a graphical user interface to EMBOSS, *Bioinformatics* **19** (14) *1837-1843*

*Essays in Bioinformatics*
*D.S. Moss et al. (Eds.)*
*IOS Press, 2005*



# Prediction and visualization of DNA structural properties from sequence

Kristian VLAHOVICEK, László KAJÁN Sándor PONGOR
*International Centre for Genetic Engineering and Biotechnology (ICGEB), Area Science Park, Padriciano 99, 34012 Trieste, Italy*
*Correspondence: S. Pongor, Tel.: +39-040-3757300; Fax: +39-040-226555; E-mail: pongor@icgeb.org*

**Abstract.** Visualisation of local DNA conformation is a useful tool in interpreting and designing experiments at the molecular level. There are a number of methods whereby local curvature as well as other conformational parameters can be predicted. Calculation of these parameters on a genomic scale may help to clarify the role of these elements in genomic architecture.

## Introduction

Simple methods that can guide experimenters to find conspicuous regions in DNA are of considerable interest in view of the genomic sequence generated. Structural properties, such as flexibility or intrinsic curvature that are not consequences of explicit sequence motifs are of particular interest since these cannot be identified from sequence similarity searches. Over the past years our group has been developing and testing simple mechanic models that can describe the local behaviour of DNA in such short segments, in a sequence dependent fashion [1-9]. These methods have been extended to the calculation and visualization of various parameters other than curvature [10], and included into WWW-based server programs located on the ICGEB web site.

Parametric visualization of DNA characteristics consists in mapping of numerical data to visually presentable models. The simplest form of parametric visualization is the sequence plot i.e. a graph in which numeric values are assigned to positions along the DNA sequence. The advantage of comparing sequence plots rather than primary sequences originates from the simple fact that plots, unlike primary sequences, can be subjected to arithmetic operations (averaging, subtraction, etc.) and their similarities can be characterized in quantitative terms such as correlation coefficients and standard deviations. This is essentially a parametric approach of sequence comparison which makes it possible, e.g., to compare groups of sequences, to carry out a semi-quantitative comparison (ranking) of sequences in structural terms, etc. using simple programs. The parametric visualization of DNA sequences uses the same properties on a qualitative basis, and the conspicuous segments can be identified by 1D, 2D or 3D plots of various parameters.

In this chapter we first describe DNA curvature as the paradigmatic concept, followed by a short description of the server algorithms. The last part of this chapter gives examples of applications.

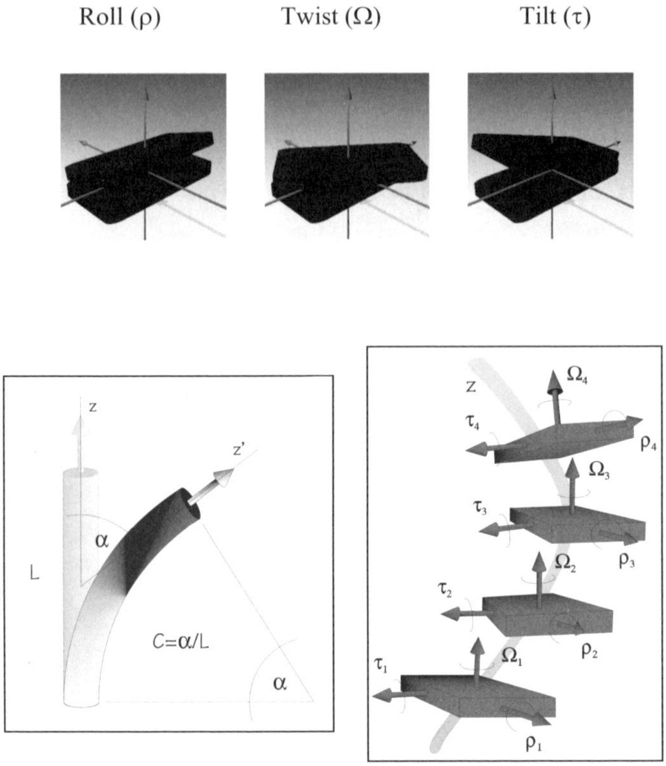


**Figure 1 A**. The molecular parameters describing DNA curvature are assigned to the relative orientation of two successive dinucleotides: roll angle (ρ), tilt angle (τ) and twist (Ω). In the ideal, Watson-Crick model, $\rho=\tau=0$ and $\Omega=36^\circ$ (10 basepairs per helical turn), B DNA in solution has a twist angle $\Omega=34.3^\circ$ (10.5 basepairs per helical turn); for a detailed description of these and other parameters see [54]. **B**. Macroscopic curvature of an elastic rod is characterized by a deflection angle α, in the case of DNA this is sometimes expressed in degrees per helical turn. **C**. The experimentally determined conformation of DNA can be characterized by local roll, tilt and twist angles, and these values can be used to reconstruct the trajectory of the Z-axis.

## 1. Calculation of DNA curvature

The thinking of biologists has been profoundly influenced by the idea of local structural polymorphism in DNA. DNA is no longer considered as a featureless polymer but rather as a series of individual domains differing in flexibility and curvature. Unlike in the case of helical polymorphism (e.g. B, A or Z structures), here we often deal with a localised micropolymorphism in which the original B-DNA structure is only distorted but is not extensively modified [9]. The deviations from ideal, straight DNA are usually expressed as angles of deflection between adjacent base pairs (Figure 1A).

The terms "curved DNA" or "DNA curvature" are used in various contexts. For instance, asymmetrical binding of proteins can induce both kinks and smooth bends in the DNA trajectory. In this review we attempt to summarize another phenomenon, an inherent structural micro-heterogeneity of DNA that occurs in the absence of bound proteins, and depends only on the DNA sequence. In contrast to alternative DNA conformations (such as A and Z-DNA), curvature can be viewed as a slight distortion of the B-DNA geometry that is manifested in the bending of the DNA-trajectory. Such a curvature can be quantitatively described using an analogy of a smoothly bent rod, and in the case of a DNA model, it can be expressed in terms of degree per base pair, or degree per helical turn. In the latter case, the repeat of the helical turn has to be specified. (Figure 1B).

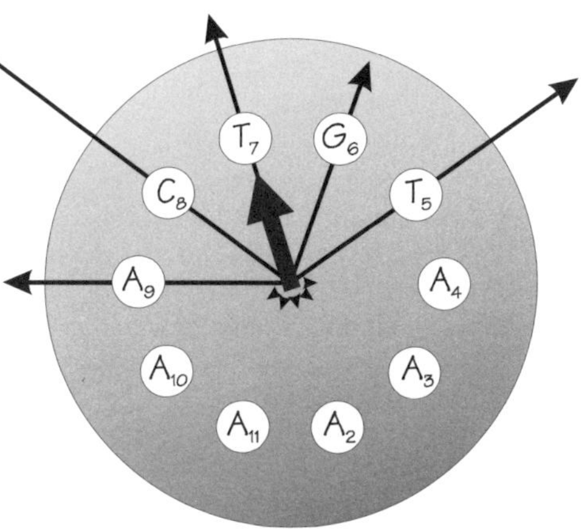


**Figure 2.** DNA curvature as asymmetric bendability. The diagram is a top-view of the DNA helix with the Z-axis perpendicular to the plain of the paper. DNA bendability of subsequent trinucleotides is represented as an arrow perpendicular to the Z-axis. In curved segments, such as the one in the figure, the distribution of the bendability vectors is asymmetrical and the vector-sum (red arrow) is non-zero. In most parts of the genomes the vector sum is small [5].

The discovery of DNA curvature was a slow process. The first evidence that there is an influence of base composition on the average twist between adjacent base pairs came from DNA X-ray fiber diagrams, 20 years after the double-helix paper of Watson and Crick [11]. Subsequent studies by gel-electrophoresis [12], nucleotide/digestions [13] and finally the first X-ray structure of DNA [14] confirmed this view. In 1980, Trifonov and Sussman suggested a correlation between the helical repeat of the DNA and spacing of certain dinucleotides (especially AA and TT) along the sequence which indicated that a substantial part of eukaryotic DNA may in fact be curved [15]. Subsequent experimental data by Marini et al. [16] indicated that periodic A-tracts repeating in phase with the helical repeats cause curvature, which was confirmed both by electron microscopy [17] and by enzymatic circularisation experiments [18]. By the mid nineties, the concept of DNA curvature became generally accepted, and even the apparent controversy between X-ray crystallography and solution experiments could be reconciled by the discovery that divalent cations induce a sequence dependent curvature in DNA [3].

A "curvature model" is a way to derive sequence-dependent DNA geometry parameters from experimental data. The models are different both in terms of the experimental data and the method of calculation. For example, it is common to fix some of the base-pair parameters at the values corresponding to straight B-DNA while letting others vary in a sequence-dependent fashion. In addition, the angles can be assigned to dinucleotides or to trinucleotides; these datasets are referred to as dinucleotide or trinucleotide scales. (All the models described here refer to double-stranded DNA molecules with "classic" phosphate orientations.)

### *1.1 The Wedge Model*

The wedge model is called a "nearest neighbour model" since the geometry of a stack of two base pairs is considered to be defined by the two constituent nucleotides, and the influence of more distant neighbours is ignored [19]. The model is based on gel-electrophoresis data, described in terms of dinucleotide parameters, roll and tilt angles.

### *1.2 The Junction Model*

The junction model was proposed based on gel-mobility experiments using oligonucleotides with "phased" (suitably spaced) adenine tracts [20, 21]. According to this model, curvature is caused by a deflection at each junction between the axes of the normal B-DNA and the B'-DNA of the poly dA, poly dT. The model assumes that the deflection at junction is a result of negative base-pair inclination in adenine tracts and zero inclination in the intervening B-DNA segments, and that this difference generates the bend [21]. According to Haran et al. [22] the wedge and the junction models are not necessarily incompatible. It appears, however, that there are events of curvature that neither the junction model nor the wedge model can sufficiently explain. For example, some GC-rich motifs, such as GGGCCC and CCCGGG have been showed opposite direction of bending [3] to those predicted from both models.

### *1.3 The Elastic Rod Model*

The elastic rod model is based on DNAseI digestion data [1]. This enzyme bends the substrate towards the major grove, so the resulting model allows only one direction of bending, towards the roll angle. The original method described DNA bending in terms of a dimensionless parameter, "relative bending propensity" determined for trinucleotides [1, 4]. Subsequently, a physical model of sequence-dependent anisotropic-bendability (SDAB) was developed [9]. SDAB considers DNA to be an elastic rod, in which the flexibility of each segment (di- or trinucleotide) is anisotropic, namely, greater towards the major groove than it is in other directions. As DNAseI cannot distinguish between *a priori* bent and dynamically "bendable" sites, curvature according to this model is both static as well as dynamic in nature and can be recognized by the phased distribution of bent/bendable sites along the sequences. This can be visualized as a vectorial property along the sequence (Figure 2) which is conceptually analogous to the hydrophobic moment calculations in protein sequences.

There are a number of computer programs that can predict curvature from sequence. One of the first algorithms available for curvature calculations was BEND written by Goodsell and Dickerson [23]. The algorithm can handle both dinucleotide and trinucleotide descriptions, and uses a simplified procedure wherein the successive deflection angles (roll, tilt) are summed up as vectors. This is a well-known approximation that is acceptable however only for low angle values. The BEND algorithm calculates curvature for segments of 11 nucleotides, and outputs a plot of curvature versus sequence position. The algorithm was incorporated into the EMBOSS suite of sequence analysis programs [24] under the name BANANA (which is a reference to curved B-DNA of A and non-A tracts), and is also available on-line [25]. The Haifa University server [26] for DNA structure calculation is built around the program Curvature [27]. The DIAMOD program was written by Mensur Dlakic for PC [28] and handles most curvature models. Finally, several precomputed parametric genome maps are available in the Genome Atlas of the Technical University of Denmark [29, 30].

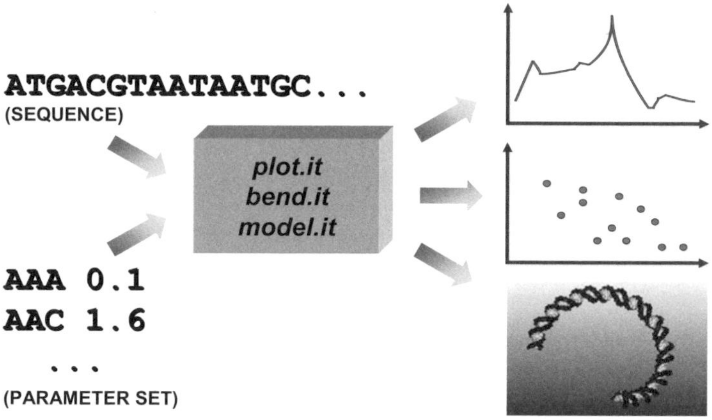


**Figure 3.:** Data flow of the bend.it, plot.it and model.it servers. Each overlapping triplet (or dinucleotide) in a DNA sequence is assigned a corresponding parameter value in a "sliding window" fashion. The resulting numerical vector can then be averaged within a given window (default value is 31bp or approximately three helical turns) and displayed either as a 1C parameter vs. sequence plot, or as a 2D correlation plot from two different parameter sets. (Three-dimensional DNA trajectories are built from basepair geometry parameters without averaging).

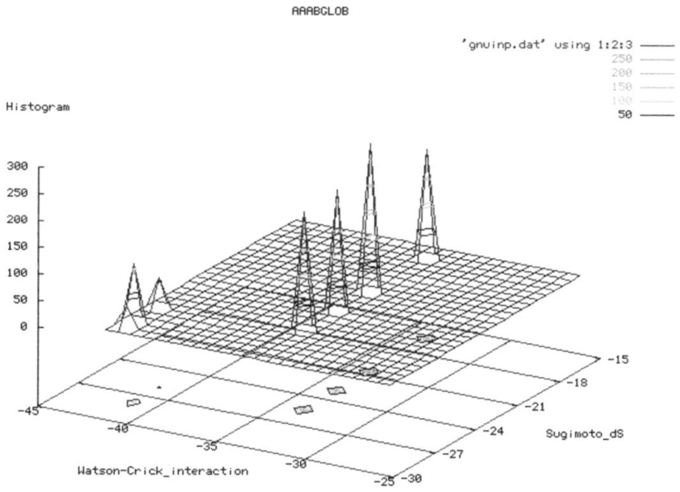


**Figure 4.** Output examples of the plot.it server. A three-dimensional correlation plot of the *Anadara trapezia* (ark clam) beta globin gene (complete cds. genbank:L16978). The vertical Z-axis denotes the number of actual segments represented by data on the XY plane. This type of correlation plot is useful in situations where analysis is performed on a long DNA sequence

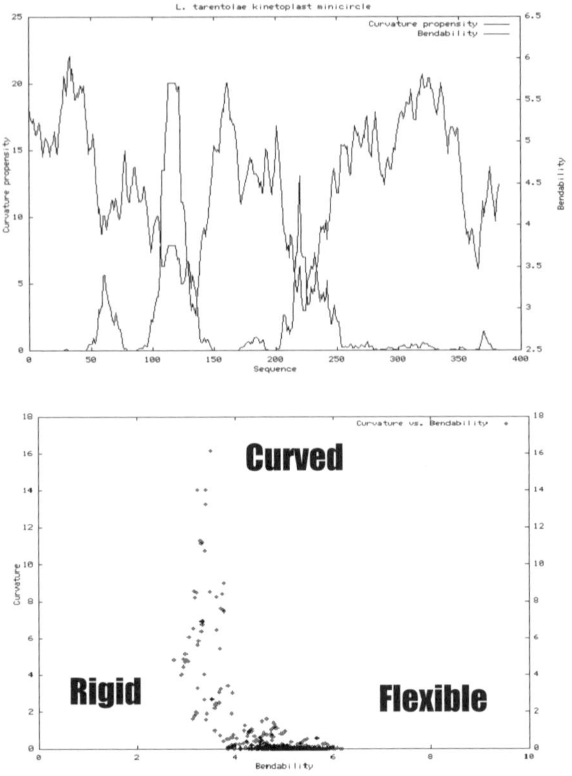


**Figure 5.** Output examples of the bend.it server. **A.** profile plots of bendability (blue) and curvature (red) along the 350bp *L. tarantolae* kinetoplast sequence. Profile plots provide a visual aid to locate "interesting" regions along a DNA sequence. **B.** correlation (2D) plot of curvature vs. bendability of the same sequence.

## 2. Prediction of DNA properties other than curvature

From the computational point of view curvature is a local property of DNA that can be represented by numeric values assigned to each position of a DNA sequence. The same philosophy can be extended to a large number of other DNA properties that can be assigned to a short segment of DNA. There are a few common approximations underlying many parametric descriptions: a) The property is local, i.e. a given n-mer in DNA will have the same property irrespective of its sequence environment ("context"). This may be true for molecular properties depending only on the nucleobases, but is a very rough approximation for complex, statistically derived properties like conformational preferences since, for instance, even dinucleotides are known to adopt a few different conformations depending on their neighbours. b) Segments within DNA (nucleotides, dinucleotides) contribute independently to a given property. This makes it possible to use simple linear or log/linear models to experimental data.

As an example, bending propensity parameters for trinucleotides were deduced from DNAseI digestibility vs. sequence data based on the following principles [1]. (i) Locality: DNase I interacts with the window of 6 nucleotides around the cleaved bond and its cutting efficiency depends only on this window. (ii) This window is represented as four overlapping trinucleotides, and one single structural parameter *p(a)* of the trinucleotides, constituting the enzyme-DNA contact surface, will influence the cutting rate (this is an obvious simplification, since local effects, such as specific residue contacts between the enzyme and the DNA molecule, are not considered); (iii) the bending propensity *p(a)* of each trinucleotide contributes independently to the probability of DNase I cutting, $P_W$. The model thus assumes that the contribution of one element (trinucleotide) does not depend on any other element being present or absent in the window around the cut. So $P_W$ for the 6 nt window can be written as the product of the n different and assumedly independent *p(a)* probabilities:

$$P_w = \prod_{1}^{4} p(a)_i \qquad (1)$$

Equating $P_W$ with the experimentally determined frequencies of cleavage, $F_W$, leads to a linear system of equations

$$F_w = \sum_{1}^{4} \ln p(a)_i \qquad (2)$$

Similar approaches have been used to extract numeric parameters from a wide variety of different experimental data. As an extreme case, DNAseI digestibility data can be obtained on large, continuous DNA fragments, other parameters, such as stability etc. were derived from measurements on short oligonucleotides. Regarding the origins of the data, parameters can be obtained either from measurement or from database statistics, such as evaluation of 3D structures or sequence data. From the computational point of view, the parameters are represented either as tabulated values, or they are computed "on the fly", based on the sequence information itself.

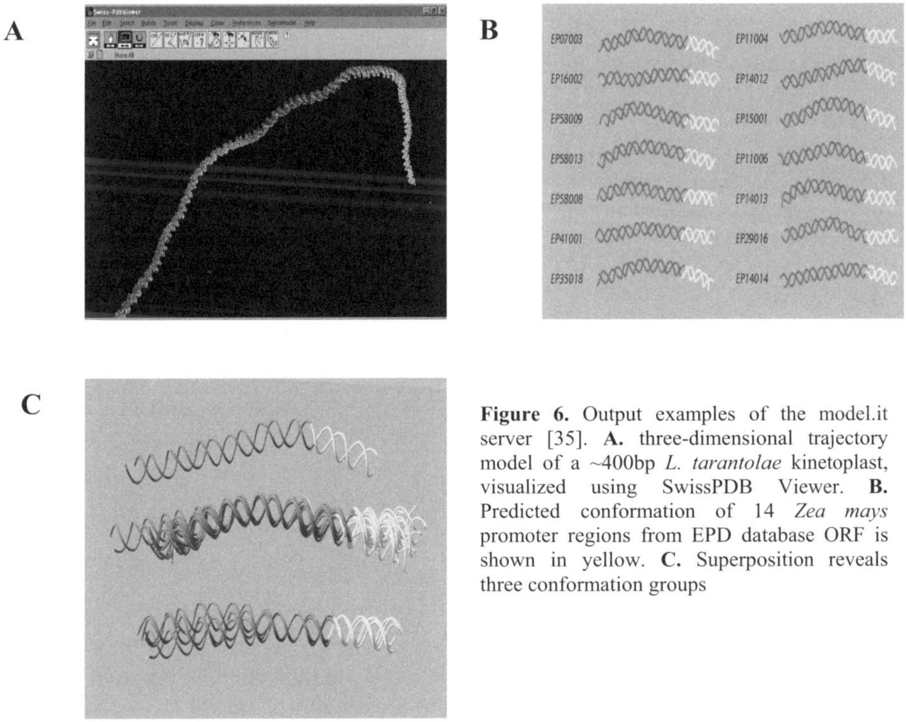


**Figure 6.** Output examples of the model.it server [35]. **A.** three-dimensional trajectory model of a ~400bp *L. tarantolae* kinetoplast, visualized using SwissPDB Viewer. **B.** Predicted conformation of 14 *Zea mays* promoter regions from EPD database ORF is shown in yellow. **C.** Superposition reveals three conformation groups

A

**Observed and expected number of intron phase associations**

| N of exons* | Observed / expected number 0..0 | 1..1 | 2..2 | 0..1 | 0..2 | 1..0 | 1..2 | 2..0 | 2..1 | Sample size | Chi²** |
|---|---|---|---|---|---|---|---|---|---|---|---|
| 1: | 998/923 | 161/163 | 173/163 | 344/388 | 380/388 | 401/388 | 141/163 | 370/388 | 158/163 | 3126 | 16.3 + |
| 2: | 920/839 | 138/122 | 138/127 | 255/302 | 318/325 | 316/340 | 109/131 | 314/327 | 123/118 | 2631 | 24.5 +++ |
| 3: | 805/730 | 95/107 | 105/110 | 248/263 | 253/282 | 253/296 | 128/115 | 295/285 | 107/103 | 2289 | 21.3 ++ |
| 4: | 715/637 | 91/93 | 96/96 | 202/229 | 225/247 | 217/258 | 105/100 | 261/248 | 86/89 | 1998 | 22.3 +++ |
| 5: | 613/554 | 79/81 | 86/84 | 170/200 | 218/215 | 197/225 | 73/87 | 235/216 | 68/78 | 1739 | 19.6 + |
| 6: | 551/478 | 65/70 | 68/72 | 130/172 | 192/185 | 158/194 | 77/75 | 198/186 | 61/67 | 1500 | 30.3 +++ |
| 7: | 502/414 + | 57/61 | 78/63 | 107/149 | 143/160 | 133/168 | 70/65 | 151/162 | 59/58 | 1300 | 44.6 +++ |

*Number of exons incorporated in the intron phase association.
**Confidence levels: + = 95%, ++ = 99%, +++ = 99.5%.

B

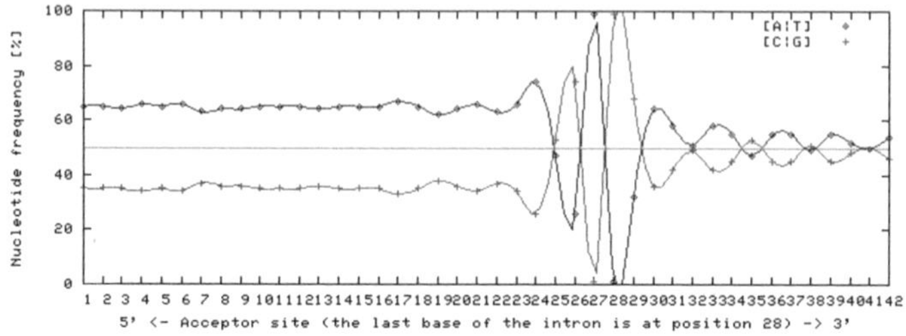


**Figure 7**: Region of 74 kb around the switching point of chromosome 21 of *L. major*. L7171 and L3640 are two cosmids containing the overlapping fragments of chromosome 21. The two strands and the encoded genes are represented in different colors. The curvature analysis of the 40 kb around the switch region is shown in the lower window.

## 3. The DNA-analysis tools developed at ICGEB

***The plot.it server*** produces parametric plots using various statistical physicochemical parameters [31]. A query sequence is divided into overlapping n-mers, and the average value of a given parameter is calculated using tabulated values. The server uses 45 structural parameters (a full list of references is available at the site), the general scheme of calculations is shown in **Figure 3**. The results appear either as simple sequence plots or as 2-D plots in which two parameters are plotted against each other. Examples are shown in **Figure 4**.

***The bend.it server*** calculates the curvature of DNA molecules as predicted from the DNA sequences. The calculation is based on values tabulated for dinucleotides and trinucleotides, and the curvature (degree per helical turn) is calculated using standard algorithms [9]. This calculation was originally based on DNA bendability parameters derived from DNAseI digestion that characterize the (static or dynamic) bending of trinucleotides towards the major groove [2]. Today a number of other dinucleotide [32-34] and trinucleotide models [2, 4] are included, and the results can be visualized as 1D or 2D plots on the screen. Both the bend.it and the plot.it servers are based on C programs provided with GnuPlot graphic routines. Their output appears on the screen and is optionally sent by e-mail to the user (**Figure 5**).

***The model.it server*** was designed to provide 3D models of DNA in response to DNA sequence queries [35]. The results are presented as a standard PDB file that can be viewed directly using any of the widely available molecule manipulation programs such as Swiss-PDBviewer [36] or Rasmol [37]. In addition to straight A and B DNA models, the server is capable of building curved DNA models using the parameter sets mentioned above. The server program was written using "NAB" - a high level molecule manipulation language [38]. Coordinates of the sugar-phosphate backbone are optionally optimised with constrained molecular dynamics using energy parameters from the AMBER package [39]. At present, the server can produce models of 700 bp in length, but models longer than 50 bp will not be optimised. Modelling of canonical, straight B or A DNA structures proceeds in a similar way, but without the need for backbone geometry optimisation (**Figure 6**).

***The IS introns server*** was designed to provide statistical overviews on intron groups [40]. Simple questions, like comparison of introns between various taxonomic groups in terms of intron phases or size-distributions as well as the analysis of splice sites, requires a carefully selected dataset as well as meticulous work that has to be repeated as new data become available. The goal of the introns server was to establish an automatically updated intron resource that allows the evaluation of experimentally validated and statistically balanced intron datasets, as well as a flexible comparison of groups according to various criteria. In addition to sequence retrieval and BLAST similarity search, there are options to compare taxonomic groups based on the NCBI Taxonomy Database, and to perform on the fly statistics. The analysis capabilities of the IS server include statistical evaluation (minimum, maximum, average, standard deviation, etc.) of intron and exon length, of the number of introns per gene, base composition, intron phases, as well as a graphic comparison of two or more groups in terms of the above variables. In addition, the analysis of splice sites and testing of the exon shuffling hypothesis [41, 42] are explicitly included (**Figure 7**).

All the servers are provided with help files that describe the detailed instructions, the theory, the literature citations as well as the instructions for installing the accessory programs such as Swiss-PDBviewer [36] or Rasmol [37].

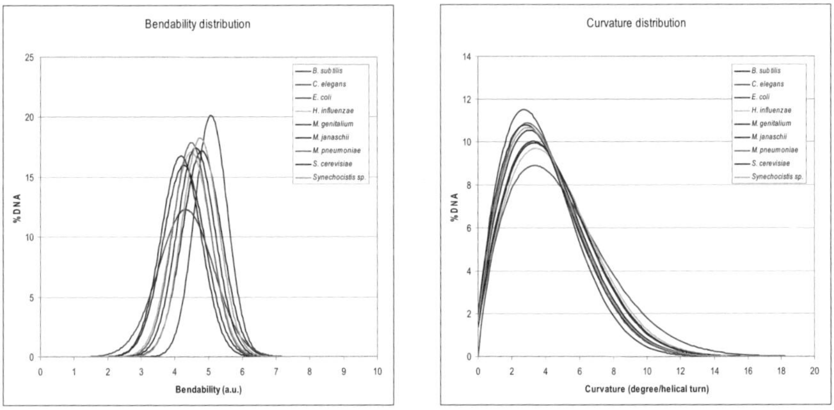


**Figure 8.** Distribution of bendability and curvature in various prokaryotic genomes

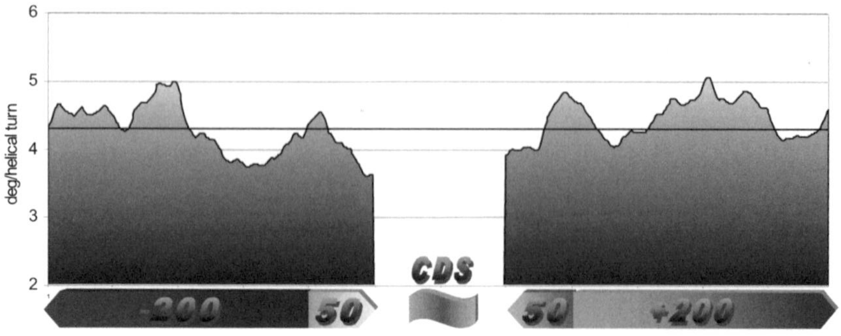


**Figure 9.** Distribution of curvature around open reading frames in yeast chromosome III. Curvature profiles of all fragments 200 bp outside and 50 bp inside the open reading frame were averaged and the result is displayed in terms of a positional preference for curved regions with respect to start and stop codons. The black line represents the average curvature of yeast chromosome III.

## Application examples

One of the obvious applications is to compare the distribution of curvature and other parameters in genomic sequences. Figure 7 shows that bendability has a smooth, symmetrical distribution in genomic DNA, similar to a bell shape. The distribution of curvature is apparently non-symmetrical reminiscent of a gamma function which is often found with randomly distributed variables whose value cannot be negative - curvature is actually such a case. Another possibility is to analyse curved segments along the entire genome. A circular plot is a convenient way to show such distributions even though the graphic resolution is often a limiting factor. Another possibility is to analyse the vicinity of annotated features in genomes, as shown in Figure 9. A comprehensive analysis of the curvature of the *B. subtilis* genomic DNA revealed the percentage of curved motifs within the genome and how many ORFs contain curved segments [43]. As reported in Figure 10, less than 1% of the *B. subtilis* genome contains curved motifs with values above 14° per helical turn. Using this as a cut-off, the majority of the curved DNA is found within the ORFs while using 16°, 64% of the curved segments are within the intergenic regions

(Figure 10, inset), a tendency that continues as the cut-off is raised. In other words, the majority of the most curved segments are concentrated in the intergenic regions.Figure 10 shows the number of ORFs with at least one curved motif. Only 6.2% of all the ORFs in *B. subtilis* shows a curvature with a cut-off value of 14°. These ORFs with at least one curved motif, encode functionally unrelated proteins since their percentage distribution is consistent with the distribution of the known proteins among the different classes, following the standard functional classification reported by SubtiList (cellular processes and cell envelope, intermediary metabolism, information pathways, other functions, [44]). Therefore, only a small percentage of all curved motifs fall inside the coding regions, leading to the hypothesis that a straight DNA is more efficiently transcribed. On the other hand, it could be that intergenic regions have been selected with an intrinsic high curvature to act as genomic signals. Indeed, it is known that, at least in lower eukaryotes such as *Saccharomyces cerevisiae* and *Leishmania major*, promoters and terminators are constituted by flexible DNA stretches [45]. Several coding strand switching points are present within the chromosomes of *L. major* [46]. For example in a region of 74 kb of

A

B

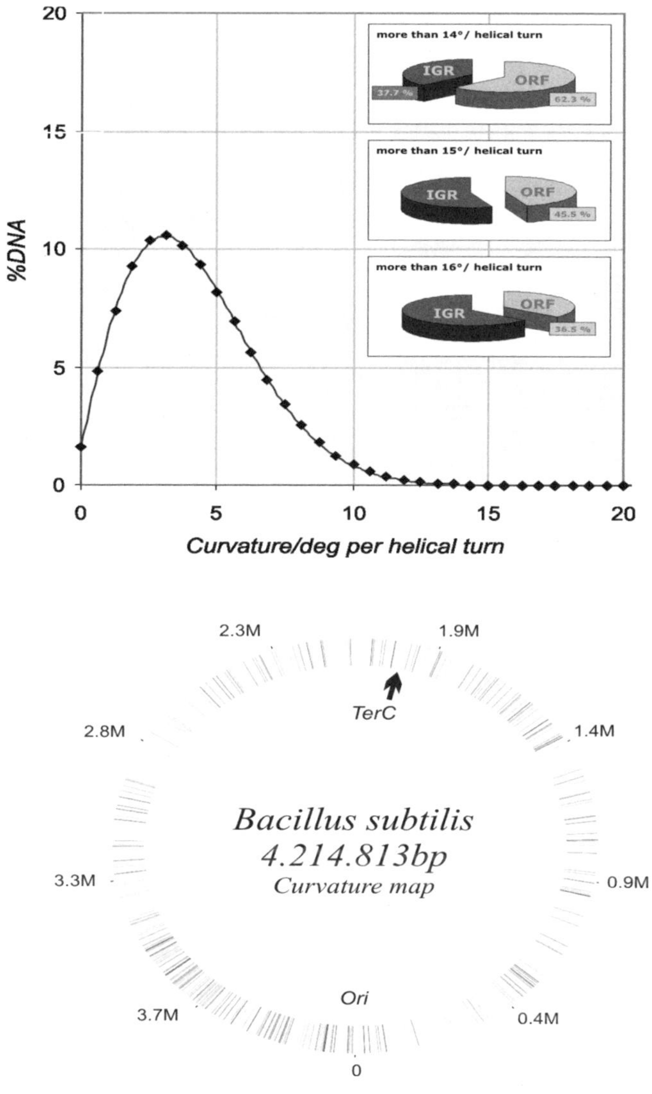


**Figure 10.** Distribution of the curved segments within the *B. subtilis* genome. The distribution of the curved motifs inside Open Reading Frames (ORFs) or inside InterGenic Regions (IGR) obtained with different cut-off degree values is represented in the inset. The graphic representation of the curved motifs is reported at the bottom.

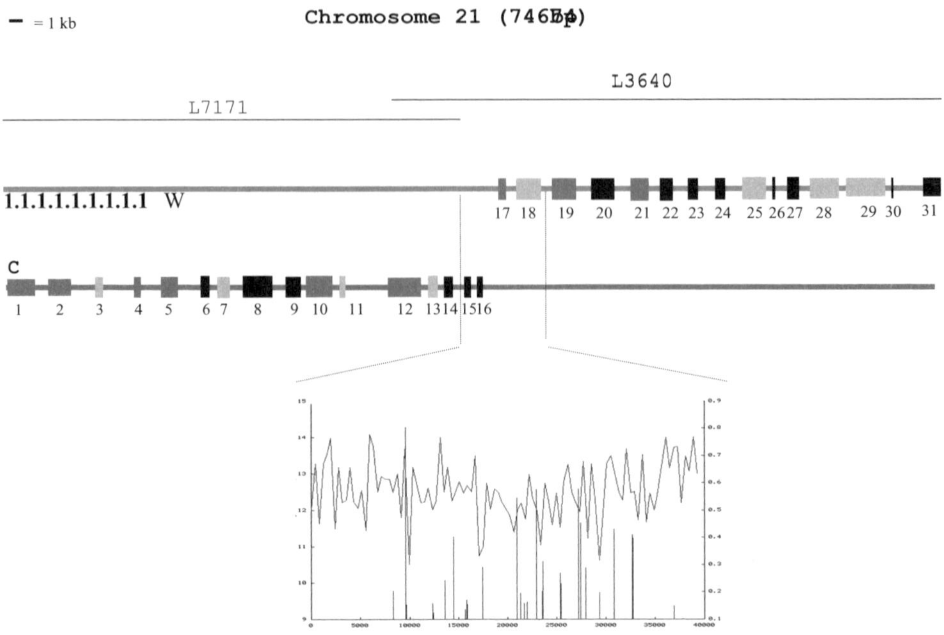


**Figure 11**. Region of 74 kb around the switching point of chromosome 21 of *L. major*. L7171 and L3640 are two cosmids containing the overlapping fragments of chromosome 21. The two strands and the encoded genes are represented in different colors. The curvature analysis of the 40 kb around the switch region is shown in the lower window.

chromosome 21, the first 16 ORFs are encoded on the Crick strand, while the rest of them are localised on the Watson strand (Figure 12). Between the two coding regions there are 1,602 nt, which are part of the so-called switching region, that do not contain neither predicted CDS nor DNA with potential to form hairpin structures. Moreover, this region shows a high DNA curvature with a maximum value of GC skew, as detected by the Bend-it program [46, 47].

The physical features described for the switching point of chromosome 21 characterise also the switching points of other chromosomes of the parasite (chr. 1, 3, 4, 19), suggesting that these regions can be involved in promotion of DNA transcription or can indicate the presence of an origin of replication. In support of the first hypothesis, very recently it was shown, by transfectional studies, that the switching point region seems to drive the expression of the entire chr1 in Leishmania major Friedlin [48].

DNA isolated from normal eukaryotic cells by standard methods exhibit particular fragility resulting in ~50 kb fragments. Breakage at hypersensitive/fragile sites is thought to be due to nucleolytic cleavage and/or localized, non-random release of torsional stress [49-52]. The sequence of several breakpoints of human DNA was recently determined and by multiple alignment, sequence similarities were found among the various breakpoints, both in short and longer stretches of the DNA [53]. An analysis with the plot.it server showed peculiar conformational characteristics (sharp transition or with a centre of symmetry) located exactly at the experimentally determined breakpoints (Figure 13A) [53]. These, however, did not exactly coincide with the position of the short consensus motives. A number of short consensus motives appear to have a curved conformation as predicted by the model.it server (Figure 13B). These instances of correlation between computed and biochemical behaviour imply that the predicted conformations may be useful in the analysis situations where breakage and rearrangements are implicated in pathological scenarios.

## 5. Summary

The WWW servers at ICGEB [10] have been created for the analysis of user-submitted DNA sequences in structural terms. *bend.it* calculates DNA curvature according to various methods, *plot.it* creates parametric plots of 45 physicochemical as well as statistical parameters. Both programs provide 1D as well as 2D plots that allow localisation of peculiar segments within the query. *model.it* creates 3D models of canonical or bent DNA starting from sequence data and presents the results in the form of a standard PDB file, directly viewable on the user's PC using any molecule manipulation program. The *introns* server allows statistical evaluation of introns in various taxonomic groups and the comparison of taxonomic groups in terms of length, base composition, intron type etc. The options include the analysis of splice sites and a probability test for exon-shuffling.

The application examples cited here show that in some cases, genomic segments identified by parametric analysis show interesting correlations even in the absence of sequence similarity. However the correlation is generally weak, so careful analysis and human experts are necessary for the evaluation of the results. On the other hand, parametric plots can be excellent subjects for machine learning studies that might in turn reveal correlations that currently escape the human eye.

## References


[1] Brukner, I., et al., Sequence-dependent bending propensity of DNA as revealed by DNase I: parameters for trinucleotides. Embo J, 1995. **14**(8): p. 1812-8.

[2] Brukner, I., et al., Trinucleotide models for DNA bending propensity: comparison of models based on DNaseI digestion and nucleosome packaging data. J Biomol Struct Dyn, 1995. **13**(2): p. 309-17.

[3] Brukner, I., et al., Physiological concentration of magnesium ions induces a strong macroscopic c urvature in GGGCCC-containing DNA. J Mol Biol, 1994. **236**(1): p. 26-32.

[4] Gabrielian, A. and S. Pongor, Correlation of intrinsic DNA curvature with DNA property periodicity. Febs Lett, 1996. **393**(1): p. 65-8.

[5] Gabrielian, A., A. Simoncsits, and S. Pongor, Distribution of bending propensity in DNA sequences. Febs Lett, 1996. **393**(1): p. 124-30.

[6] Gabrielian, A., K. Vlahovicek, and S. Pongor, Distribution of sequence-dependent curvature in genomic DNA sequences. FEBS Letters, 1997. **406**(1-2): p. 69-74.

[7] Gromiha, M.M., et al., Anisotropic elastic bending models of DNA. J. Biol. Phys., 1996. **22**: p. 227-243.

[8] Gromiha, M.M., et al., The role of DNA bending in Cro protein-DNA interactions. Biophys Chem, 1997. **69**(2-3): p. 153-60.

[9] Munteanu, M.G., et al., Rod models of DNA: sequence-dependent anisotropic elastic modelling of local bending phenomena. Trends Biochem Sci, 1998. **23**(9): p. 341-7.

[10] www.icgeb.org/dna.

[11] Bram, S., Variation of type-B DNA x-ray fiber diagrams with base composition. Proc Natl Acad Sci U S A, 1973. **70**(7): p. 2167-70.

[12] Wang, J.C., Helical repeat of DNA in solution. Proc Natl Acad Sci U S A, 1979. **76**(1): p. 200-3.

[13] Dickerson, R.E. and H.R. Drew, Kinematic model for B-DNA. Proc Natl Acad Sci U S A, 1981. **78**(12): p. 7318-22.

[14] Dickerson, R.E. and H.R. Drew, Structure of a B-DNA dodecamer. II. Influence of base sequence on helix structure. J Mol Biol, 1981. **149**(4): p. 761-86.

[15] Trifonov, E.N. and J.L. Sussman, The pitch of chromatin DNA is reflected in its nucleotide sequence. Proc Natl Acad Sci U S A, 1980. **77**(7): p. 3816-20.

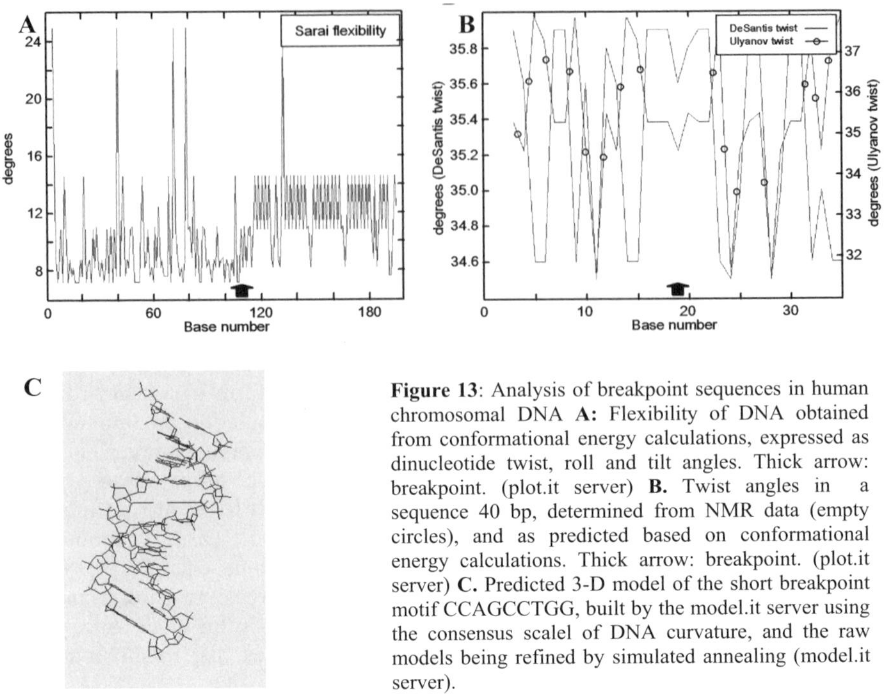


**Figure 13**: Analysis of breakpoint sequences in human chromosomal DNA **A:** Flexibility of DNA obtained from conformational energy calculations, expressed as dinucleotide twist, roll and tilt angles. Thick arrow: breakpoint. (plot.it server) **B.** Twist angles in a sequence 40 bp, determined from NMR data (empty circles), and as predicted based on conformational energy calculations. Thick arrow: breakpoint. (plot.it server) **C.** Predicted 3-D model of the short breakpoint motif CCAGCCTGG, built by the model.it server using the consensus scalel of DNA curvature, and the raw models being refined by simulated annealing (model.it server).


[16] Marini, J.C., et al., A bent helix in kinetoplast DNA. Cold Spring Harb Symp Quant Biol, 1983. **47 Pt 1**: p. 279-83.

[17] Griffith, J., et al., Visualization of the bent helix in kinetoplast DNA by electron microscopy. Cell, 1986. **46**(5): p. 717-24.

[18] Ulanovsky, L., et al., Curved DNA: design, synthesis, and circularization. Proc Natl Acad Sci U S A, 1986. **83**(4): p. 862-6.

[19] Ulanovsky, L.E. and E.N. Trifonov, Estimation of wedge components in curved DNA. Nature, 1987. **326**(6114): p. 720-2.

[20] Diekmann, S., Sequence specificity of curved DNA. Febs Lett, 1986. **195**(1-2): p. 53-6.

[21] Koo, H.S. and D.M. Crothers, Calibration of DNA curvature and a unified description of sequence-directed bending. Proc Natl Acad Sci U S A, 1988. **85**(6): p. 1763-7.

[22] Haran, T.E., J.D. Kahn, and D.M. Crothers, Sequence elements responsible for DNA curvature. J Mol Biol, 1994. **244**(2): p. 135-43.

[23] Goodsell, D.S. and R.E. Dickerson, Bending and curvature calculations in B-DNA. Nucleic Acids Res, 1994. **22**(24): p. 5497-503.

[24] http://www.hgmp.mrc.ac.uk/Software/EMBOSS/.

[25] http://www.hgmp.mrc.ac.uk/Software/EMBOSS/interfaces.html.

[26] http://esti.haifa.ac.il/~leon/cgi-bin/curvatur/.

[27] Shpigelman, E.S., E.N. Trifonov, and A. Bolshoy, CURVATURE: software for the analysis of curved DNA. Comput Appl Biosci, 1993. **9**(4): p. 435-40.

[28] http://www-personal.umich.edu/~mensur/software.html.

[29] Pedersen, A.G., et al., A DNA structural atlas for Escherichia coli. J Mol Biol, 2000. **299**(4): p. 907-30.

[30] Jensen, L.J., C. Friis, and D.W. Ussery, Three views of microbial genomes. Res Microbiol, 1999. **150**(9-10): p. 773-7.

[31] Vlahovicek, K., A. Gabrielian, and S. Pongor, Prediction of bendability and curvature in genomic DNA. J. Mathematical Modelling and Scientific Computing, 1998. **9**: p. 53-57.

[32] Ulyanov, N.B. and T.L. James, Statistical analysis of DNA duplex structural features. Methods Enzymol, 1995. **261**(120): p. 90-120.

[33] Bolshoy, A., et al., Curved DNA without A-A: experimental estimation of all 16 DNA wedge angles. Proc Natl Acad Sci U S A, 1991. **88**(6): p. 2312-6.

[34] Olson, W.K., et al., Influence of fluctuations on DNA curvature. A comparison of flexible and static wedge models of intrinsically bent DNA. J Mol Biol, 1993. **232**(2): p. 530-54.
[35] Vlahovicek, K. and S. Pongor, Model.it: building three dimensional DNA models from sequence data. Bioinformatics, 2000. **16**(11): p. 1044-5.
[36] Guex, N. and M.C. Peitsch, SWISS-MODEL and the Swiss-PdbViewer: an environment for comparative protein modeling. Electrophoresis, 1997. **18**(15): p. 2714-23.
[37] Sayle, R.A. and E.J. Milner-White, RASMOL: biomolecular graphics for all. Trends Biochem Sci, 1995. **20**(9): p. 374.
[38] Macke, T. and D.A. Case, Modeling unusual nucleic acid structures, in Molecular Modeling of Nucleic Acids, N.B. Leontes and J. SantaLucia, Editors. 1998, American Chemical Society: Washington DC. p. 379-393.
[39] Case, D.A., et al., AMBER 5. 1997, University of California: San Francisco.
[40] Barta, E., L. Kajan, and S. Pongor, IS: A web-site for introns statistics. Bioinformatics, 2003. **19**: p. 543.
[41] Long, M., S.J. de Souza, and W. Gilbert, Evolution of the intron-exon structure of eukaryotic genes. Curr Opin Genet Dev, 1995. **5**(6): p. 774-8.
[42] Kriventseva, E.V. and M.S. Gelfand, Statistical analysis of the exon-intron structure of higher and lower eukaryote genes. J Biomol Struct Dyn, 1999. **17**(2): p. 281-8.
[43] Tosato, V., et al., The DNA secondary structure of the Bacillus subtilis genome. FEMS Microbiol Lett, 2003. **218**(1): p. 23-30.
[44] http://bioweb.pasteur.fr/GenoList/SubtiList.
[45] McDonagh, P.D., P.J. Myler, and K. Stuart, The unusual gene organization of Leishmania major chromosome 1 may reflect novel transcription processes. Nucleic Acids Res, 2000. **28**(14): p. 2800-3.
[46] Tosato, V., et al., Secondary DNA structure analysis of the coding strand switch regions of five Leishmania major Friedlin chromosomes. Curr Genet, 2001. **40**(3): p. 186-94.
[47] Myler, P.J., et al., Genomic organization and gene function in Leishmania. Biochem Soc Trans, 2000. **28**(5): p. 527-31.
[48] Martinez-Calvillo, S., et al., Transcription of Leishmania major Friedlin chromosome 1 initiates in both directions within a single region. Mol Cell, 2003. **11**(5): p. 1291-9.
[49] Szabo, G., Jr., F. Boldog, and N. Wikonkal, Disassembly of chromatin into approximately equal to 50 kb units by detergent. Biochem Biophys Res Commun, 1990. **169**(2): p. 706-12.
[50] Szabo, G., Jr., 50-kb chromatin fragmentation in the absence of apoptosis. Exp Cell Res, 1995. **221**(2): p. 320-5.
[51] Gal, I., et al., Protease-elicited TUNEL positivity of non-apoptotic fixed cells. J Histochem Cytochem, 2000. **48**(7): p. 963-70.
[52] Varga, T., I. Szilagyi, and G. Szabo, Jr., Single-strand breaks in agarose-embedded chromatin of nonapoptotic cells. Biochem Biophys Res Commun, 1999. **264**(2): p. 388-94.
[53] Szilagyi, I., et al., Non-random features of loop-size chromatin fragmentation. J Cell Biochem, 2003. **89**(6): p. 1193-205.
[54] Bansal, M., D. Bhattacharyya, and S. Vijaylakshmi, NUVIEW: software for display and interactive manipulation of nucleic acid models. Comput Appl Biosci, 1995. **11**(3): p. 289-92.
[55] Sarai, A., et al., Sequence dependence of DNA conformational flexibility. Biochemistry, 1989. **28**(19): p. 7842-9.
[56] De Santis, P., et al., Validity of the nearest-neighbor approximation in the evaluation of the electrophoretic manifestations of DNA curvature. Biochemistry, 1990. **29**(39): p. 9269-73.

*Essays in Bioinformatics*
*D.S. Moss et al. (Eds.)*
*IOS Press, 2005*


# Protein Structure and its Classification

Andrew J. MILES, Clare E. SANSOM and Bonnie A. WALLACE
*School of Crystallography, Birkbeck College, University of London,, London, UK*

**Abstract**. Description of protein structure is based on a hierarchy ofconcepts, from the peptide bond to secondary structures, motifs and folds. The classification of protein structures is usually achieved by segregating mainly-alpha, mainly-beta, and mixed (alpha/beta and alpha+beta) structures. This chapter gives an overview of structural concepts as well as examples how these are implemented in databases such as CATH, SCOP and FSSP.

## Introduction

There are a vast number of ways to fold a polypeptide chain into a compact structure however the number of possible folds is limited according to the following thermodynamic argument [1]: Protein folding is partly driven by the sequestration of hydrophobic sidechains into the molecule's interior where the backbone polar groups must interact to prevent hydrogen bonding with the solvent that would push the equilibrium towards the unfolded state. Thus short stretches of the chain adopt regular conformations called secondary structure in which internal hydrogen bonding between the backbone amide and carbonyl groups is optimised. The two main secondary structures, α-helices and β-sheets, traverse the molecule from one side to the other where a loop reverses the chain. They pack together to exclude water from the interior and form common motifs that in turn assemble into semi-independent globular regions of the protein called domains. By taking the domain as a fold unit and clustering similar structures at each level (i.e. secondary structures, motifs and folds) it is possible to create a taxonomy of protein families based on structural similarities.

The first X-ray crystal structure of a globular protein was reported by in 1958 by Kendrew [2], and since then thousands of structures have been determined by X-ray crystallography and nuclear magnetic resonance (NMR). The Protein Data Bank (http://www.rcsb.org/pdb) contained over 21000 protein structures with >5500 non-redundant structures in November, 2003 and there are around 3000 additional entries per year. Classification of such a large number of proteins into structural families can best be accomplished using automated methods that require unambiguous definitions at each structural level. This chapter discusses the most common secondary structures, motifs and folds and their classifications.

## 1. The Peptide Unit

From the study of amide and dipeptide crystal structures Pauling *et al.*, [3] determined that the length of the C′-N bond (see figure 1) is 10% shorter than normal whereas the C′-O double bond is more than 1% longer than that seen in ketones and aldehydes. This is due to resonance between the structures shown in figure 2, and corresponds to the C′- N bond having almost 50% double bond character. Consequently the peptide bond is planar and torsion along the polypeptide chain is limited to the $C_\alpha$ - C′ bond and the $C_\alpha$ - N bond of each residue, the angles of rotation referred to as $\phi$ and $\varphi$ respectively (figure 1). By convention $\phi$ and $\varphi$ are set to zero when the $C_\alpha$ -N bond is *trans* to the carbonyl bond and $C_\alpha$-C′ is *trans* to the amide group [4] Looking down the dipeptide from N- to C-terminal, clockwise rotation is defined as positive and anti clockwise defined as negative.

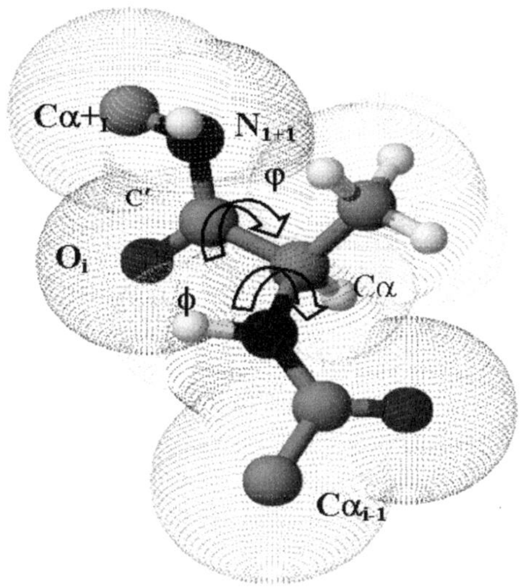


**Figure 1.** Ball and stick model of a peptide bond showing $\phi$ and $\varphi$ angles. The space filling surface shows characteristics mentioned in the text.

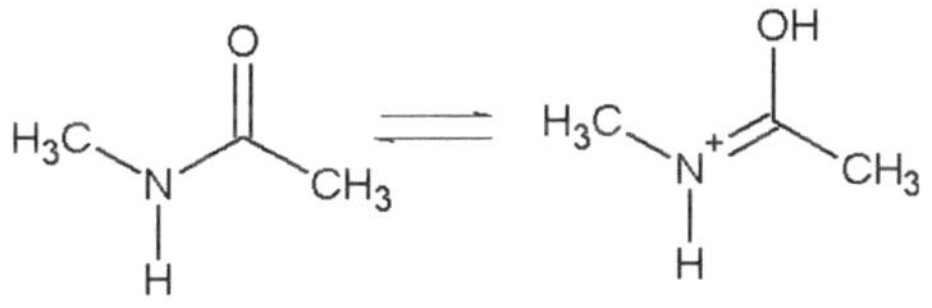

**Figure 2** Amide tautomers

The backbone conformation is also limited by steric constraints illustrated by the spacefilling surface superimposed on the ball and stick model in Figure 1. Ramachandran *et al.* [5] calculated the allowed conformations from Van der Waals contact distances and displayed them on a plot of $\phi$ versus $\varphi$ known as the Ramachndran map (figure 3). The patterned areas represent allowed regions and the grey areas represent outlying regions for which the contact distances were reduced on the basis of empirical data available at the time. The map has been revised a number of times to coincide with accumulated data and quantum mechanical calculations which demonstrate that increased stability due to good hydrogen bond alignment compensates for single steric clashes between hydrogen atoms creating the diagonal distribution shown in figure 4 [6].

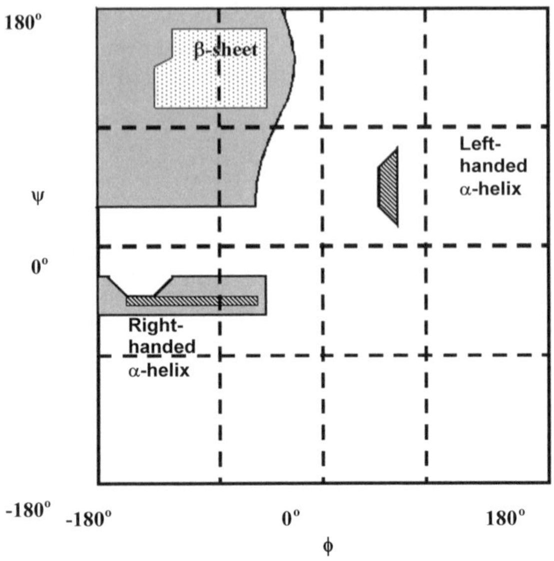


**Figure 3** Ramachandran Plot. Sterically allowed regions are patterned, outlying regions are grey.

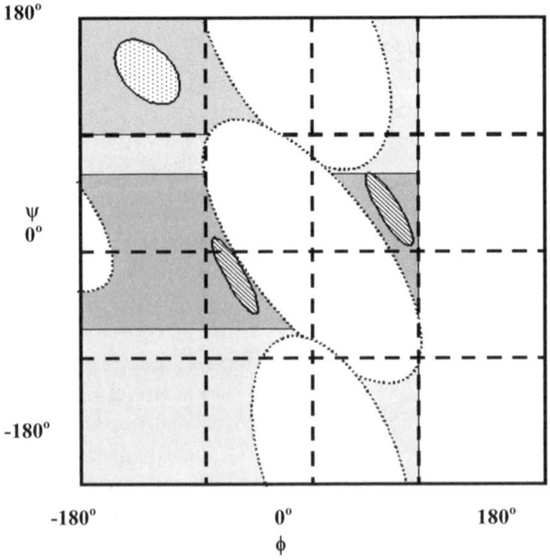


**Figure 4.** Revised steric map (adapted from Ho *et al.*, [6]). Sterically allowed regions are dark grey. Outlying regions that are only excluded by one steric clash are light grey. β, $\alpha_L$ and $\alpha_R$ regions are patterned. Sterically restricted regions are white

## 2. Secondary structure

Secondary structures are stretches of the polypeptide where all the $\phi$ angles and all the $\varphi$ angles are similar so that successive residues have almost identical orientations relative to each other. There are two highly populated regions of $\phi$ $\varphi$ space, one near -60°, -40°, and the other near -120°, +135° that correspond to the conformation of residues in $\alpha$-helices and $\beta$-strands respectively (the patterned areas in the Ramachandran map). These are the most common secondary structures, usually forming the interior of the molecule and spanning its diameter to be joined by turns and loops that are generally exposed at the surface.

### *2.1 Helices*

A regular protein helix can be described by the rise per residue (d), the number of residues per turn (n) and the radius. The helix is stabilized by hydrogen bonding between the amide hydrogen of residue *i* and the carbonyl oxygen of residue *i+n* [7]. Table 1 describes the parameters of the three main types of helix found in proteins.

**Table 1.** Parameters of protein helices

| Structure | $\phi$ and $\varphi$ angles | n | d | r |
|---|---|---|---|---|
| $\alpha$-helix | -60 -40 | +3.6 | 1.5 | 2.3 |
| $3_{10}$ helix | -60 -60 | +3.0 | 2.0 | 1.9 |
| Polyproline II | -75 145 | -3.0 | 2.9 | 1.6 |

*Alpha-Helix* The right-handed $\alpha$-helix, postulated by Pauling and Cory in 1951 [3], was confirmed in the same year in the first crystal structure of haemoglobin [8]. The $\alpha$-helix, with 3.6 residues per turn and a hydrogen bond between the *i* and the *i*+4 residues (figure 5b), is one of the most abundant secondary structure found in proteins, reflecting its high stability due to well-aligned hydrogen bond dipoles and a radius small enough to allow Van der Waals attraction across the helix axis. Alpha-helices in globular proteins can vary in length from four to more than forty residues with an average length of 10 residues in soluble proteins. They generally form straight rods although packing constraints and the incorporation of proline will cause bends and kinks. Figure 5a illustrates the effect of proline 37 on helix B in horse heart myoglobin. The *ab-initio* prediction of $\alpha$-helices from amino acid sequence is difficult since all the sidechains except proline have little effect on the helix backbone. For L-amino acids, left-handed $\alpha$-helices are energetically less favourable than the right-handed variety due to packing constraints. However short sections are found; for example there is one turn of left-handed $\alpha$ helix formed by residues 226-229 in thermolysin [9].

*$3_{10}$ Helix* The $3_{10}$ helix has internal hydrogen bonds between residues *i* and *i*+3 which are not as well aligned as those in the $\alpha$-helix. Moreover its smaller radius leads to more strain due to unfavourable side chain packing and consequently the $3_{10}$ helix is less stable than the $\alpha$-helix. Only short stretches of 1 or 2 turns are found in proteins, usually at the C- or N- terminals of $\alpha$-helices [7].

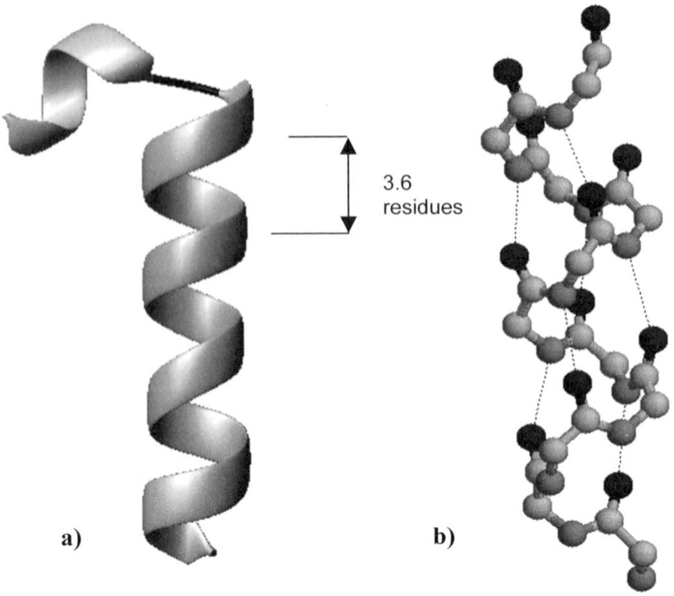


**Figure 5 a)** RASWIN cartoon of a α-Helix from horse heart myoglobin (1ymb), [12]. The kink is caused by a proline residue 37 (dark grey) **b)** RASWIN ball and stick model of an α-helix showing internal hydrogen bonds.

*Polyproline II helix Trans* poly-L-proline forms a left-handed polyproline II (PPII) helix with $\phi$ and $\varphi$ angles of $-75^o$ and $145^o$ and n=-3.0 [10]. In globular proteins short stretches of PPII helix are found on the protein surface. They tend to be mobile generally having few main chain hydrogen bonds with the rest of the protein and are stabilised by hydrogen bonding with the solvent [11]. Proline residues are usually present in the sequence but this is not obligatory.

## *2.2 Beta-Pleated Sheets*

Pauling and Cory [13] also predicted the second major secondary structural element in proteins, the β-strand. Beta-strands have an extended conformation but are technically helices with two residues per turn so that consecutive residues are rotated by $180^o$ with $\phi$ and $\varphi$ angles in the upper left-hand allowed region of the Ramachandran map (figures 3 and 4). Strands tend to be 5 to 10 residues long and are aligned with an adjacent strand so that hydrogen bonding can occur between the C′O of one strand and the NH of the other to form a sheet structure in which all possible main chain hydrogen bonds are formed. Although β-sheets may involve β-strands that are not consecutive in the sequence and are therefore considered to be tertiary structure by some authors (for example Prztycka and co-workers [[14]]) it is usually convenient to classify them as secondary structure. Successive $C_\alpha$ atoms lie above and below the plane of the sheet so that the structure is pleated. Strands can run parallel (in the same biochemical direction) or anti-parallel, each form having a distinctive hydrogen-bonding pattern (figures 6, 7). Sheets can be mixed, parallel and anti parallel but there is some energetic bias against mixed sheets. Anti-parallel chains can pack more closely than parallel sheets resulting in shorter interchain hydrogen bonds; furthermore anti-parallel sheets have well-aligned hydrogen bond dipoles whereas those of parallel sheets are misaligned [15]. The anti-parallel conformation is therefore more favoured, however the presence of bulky or branched sidechains such as valine and isoleucine can favour the formation of parallel β-sheets which accommodate these sidechains more easily [16]. The beta sheet postulated by Pauling and Cory [13] was planar an example of which can be seen in glutathione reductase (figure 8) however this is rarely seen and most sheets have a right-handed twist when viewed in the direction of the polypeptide chain (figure 9) due to intra- and inter-chain interactions involving the sidechains [15] [16]. The twist tends to be greater in anti-parallel sheets, which are more flexible than parallel sheets and can sometimes be exaggerated into a coil. β-sheets can also change directopn by ~90° with the insertion of a residue with the polyproline II conformation (β-bend) or the α- conformation (β-bulge) [17]

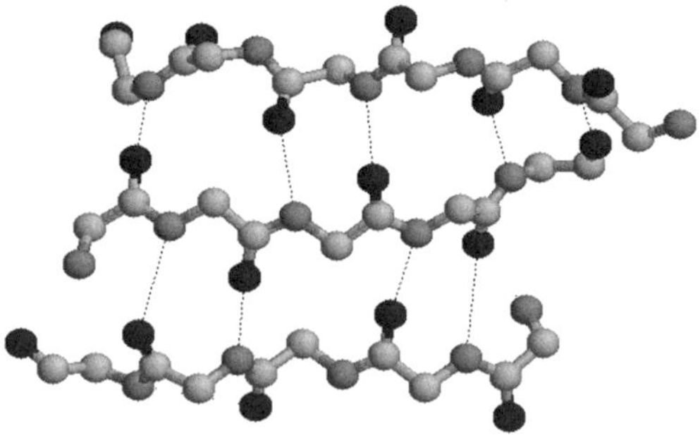

**Figure 6.** RASWIN ball and stick model of an anti-parallel β-sheet showing hydrogen bonds between strands.

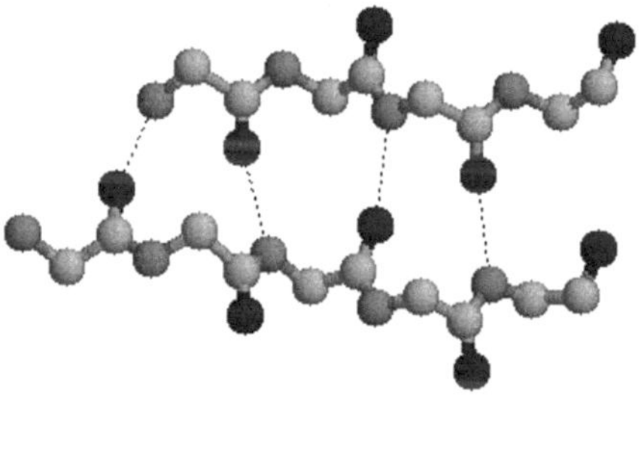

**Figure 7.** RASWIN ball and stick model of a parallel β-sheet showing hydrogen bonds between strands.

### *2.3 Loops and hydrogen-bond stabilised turns*

About one third of the residues in globular proteins are found in turns and loops, which reverse the direction of the polypeptide chain, a prerequisite for the formation of a compact globular structure. Turns are normally located at the surface of a protein, therefore they contain mostly charged or polar residues, are frequently involved in its interactions or in ligand binding (see section 4.3), and are commonly the sites of phosphorylation, glycosylation and other protein modifications.

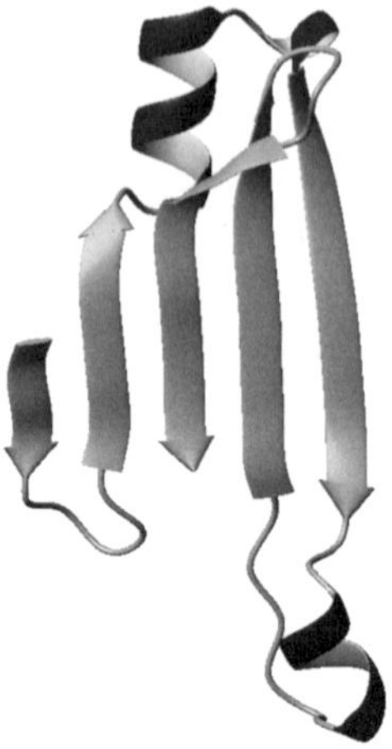

**Figure 8.** Section of glutathione reductase, 1dnc, [18] Example of a planar β-sheet.

**Figure 9.** Section of carboxy peptidase A, 1m4l, [19] showing a right-handed twisted β- sheet

Loops of less than seven residues form predictable structures known as reverse or tight turns, which have been categorised according to the number of residues involved. In each case the turn is not part of a helical structure, the distance between the first and last residue is less than 7Å and there is usually a hydrogen bond between the first and last residues in the turn [20].

The smallest is the δ turn, which contains two amino acids with a hydrogen bond between the backbone NH ($i$) and the backbone CO ($i$+1) while slightly larger is the γ turn involving three residues with a hydrogen bond between the backbone CO($i$) and backbone NH($i$+2). γ-Turns have been classified into two types, classic and inverse, based on the dihedral values of the ($i$+1) residue [20].

The most abundant reverse turn, which is found in most topological environments, is the four-residue β-turn. Beta-turns were identified by Venkatachalam [21] who used model building techniques to characterise three favourable conformations in which a hydrogen bond could form between the backbone CO($i$) and the backbone NH($i$+3). These were designated types I, II and III and their more sterically constrained mirror image conformations were designated types I', II' and III'. Types I and II are identical except that the second residue is rotated by 180$^{o}$ (figure 10) and type III is equivalent to one turn of a $3_{10}$ helix. As the number of known protein structures increased it became apparent that 25% of β-turns were not stabilised by a hydrogen bond [22] and the definition was broadened accordingly. In a more recent classification, nine different types of β-turn were identified based on the ϕ and φ angles of the second and third residues in the turn (Figure 11) [23-25]. These are designated Types I, II, VII, I', II' VIa1, VIa2 VIb and IV. This scheme is used by the authors of the program, PROMOTIF [26], which provides details of protein secondary structure and motifs in the Protein Data Bank and can be accessed at http://www.biochem.ucl.ac.uk bsm/pdbsum/

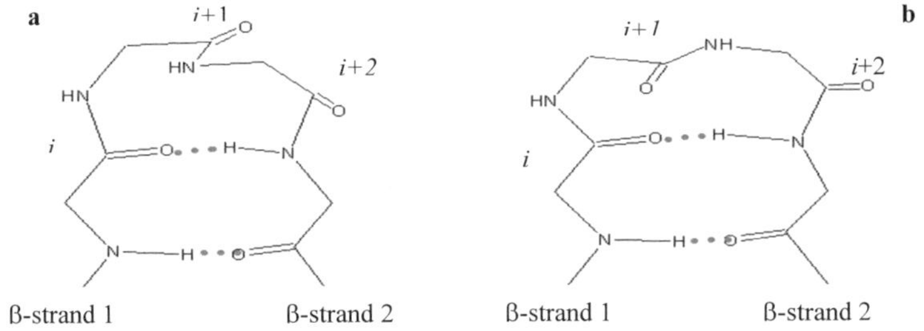


**Figure 10.** Type I (**a**) and type II (**b**) β-Turns. The $i+1$ residue in Type I is reversed in type II.

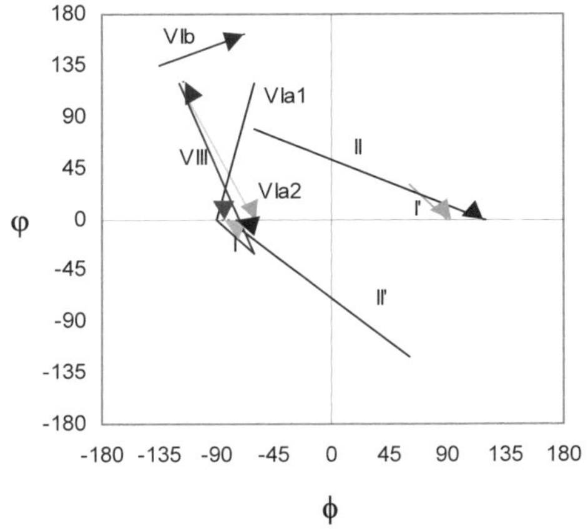


**Figure 11**. Average $\phi$ and $\varphi$ values for residue 2 connecting to average $\phi$ and $\varphi$ values of residue 3 for β-turns. The arrowheads denote the residue 3 $\phi$ and $\varphi$ values (Adapted from Guruprasad & Rajkumar, [27])

The fourth type of turn, the α-turn, contains five residues which may be stabilised by a hydrogen bond between the backbone CO($i$) and the backbone NH($i$+4) [28] although other hydrogen bonding patterns are possible. Nine types have been categorised according to the $\phi$ and $\varphi$ angles of the second, third and fourth residues.

The largest tight-turn is the π-turn with six-residues stabilised by a hydrogen bond between backbone CO ($i$) and the backbone NH ($i$+5). Generally π turns are found at the C-termini of α-helices with the fifth residue adopting left-handed α-helical conformation ($\pi_{\alpha L}$). Three other classes of π turn have been identified [29]. These are the $\pi_{\alpha R}$ turn, and the $\pi_{\beta}$ turn in which the fifth residue $\phi$ and $\varphi$ angles are in the $\alpha_R$ and β regions of the Ramachandran map respectively, and the $\pi'_{\alpha L}$ turn which is the mirror image of the $\pi_{\alpha L}$ turn.

### *2.4 Identifying Secondary Structure*

How are crystal structures analysed to find regions of secondary structure which correlate to the preceding 'text book' descriptions? Before the advent of accessible computer technology secondary structures were identified from visual inspection of atomic models, observing the local conformation of residues relative to those nearby and ascertaining hydrogen bond patterns between closely spaced amides. This method tends to be subjective and is unsatisfactory for deciding where a section of secondary structure ends. For example, in α-helices, the last 3 residues of the carboxyl terminus only contribute NH groups to hydrogen bonding and the N terminus contributes only CO groups. It is also difficult to identify short and irregular sections of secondary structure.

Since the late 1970's a number of pattern recognition algorithms have been developed to determine secondary structure from crystallographic data. These include DEFINE_S, which derives secondary and first level supersecondary structure from the $C_\alpha$ trace [30], Define Secondary Structure of Proteins (DSSP) [31], which uses hydrogen-bonding patterns, STRIDE, which assigns structure from atomic coordinates based on hydrogen-bond patterns and main chain dihedral angles [32], and XTLSSTR [33], uses the same criteria that are used visually.

## 3. Supersecondary Structure (Motifs)

Secondary structures combine to form energetically stable arrangements called motifs, the smallest containing two secondary structural elements connected by a specific turn. In a comparative analysis of 240 proteins, Sun and Jiang [34] were able to classify thirty-four supersecondary motifs of the types: αα, αβ, βα, and ββ in which the connecting peptide consisted of 5 or fewer residues. The classification is based on the hydrogen-bonding pattern and the conformation of the residues in the connecting loop, which in turn determines the relative orientation of the main secondary structure elements. Since the demarcation between secondary structure and simple motifs is somewhat arbitrary, some of the motifs have already been described in section 3.

### *3.1 Hairpins*

The most common basic motifs include, α and β-hairpins (figures 12b and 12a). α-Hairpins consist of two α-helices packed anti-parallel to each other and joined by a δ or χ turn. More abundant are β-hairpins in which two adjacent strands forming part of an anti-parallel β-sheet are connected by either δ or χ turns but most commonly by types I′ or II′ β-turns [35]. Figure 13a shows a Raswin diagram of the snake venom erabutoxin (PDB code1era), which consists of two β-hairpins plus one β-strand.

### *3.2 Mixed αβ motifs*

The segment of peptide chain connecting two parallel β-strands often forms a α-helix giving rise to the βαβ motif (figure 12c). The helix is packed anti-parallel to the sheet and in the most common configuration it lays above the plane of the β-sheet forming a right-handed loop between the β-strands. Other recurrent themes include ααβ, and ββα motifs [36].

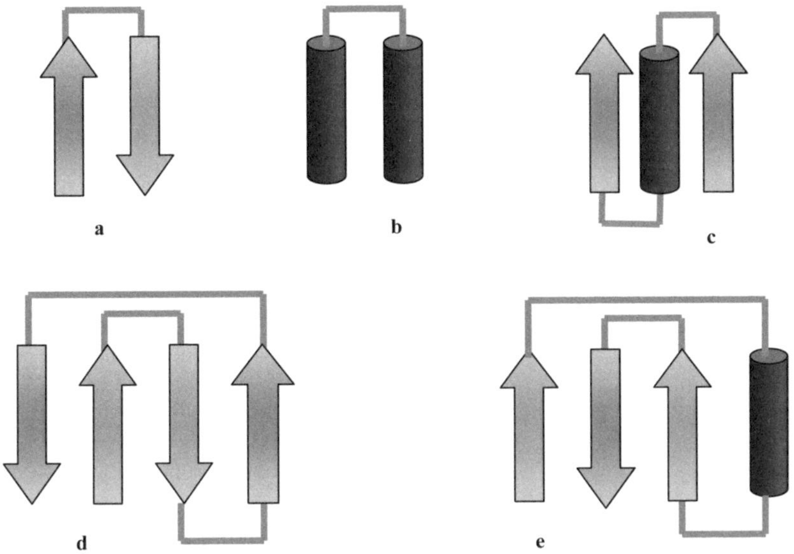


**Figure 12.** Topology diagrams of common supersecondary structures described in the text. **a)** β -hairpin. **b)** α-hairpin. **c)** βαβ-unit **d)** Greek-key **e)** αβ-Greek-Key

*3.3 Greek Key*

Two common extended supersecondary structures have been classified: the $\beta_4$ Greek key [37] and the αβ Greek key [38]. The classic $\beta_4$ Greek key motif illustrated in figure 12d consists of four adjacent anti-parallel β-strands comprising two β-hairpins that need not belong to the same sheet. There are 24 ways in which to arrange a four stranded β-sheet however only eight of these were found in a survey of known structures in 1991 [39], five of which were arranged with all the strands anti-parallel. The Greek key motif is the topological signature of β-barrels and β sandwiches, which are the two most prevalent β-folds (section 6). A variation is the αβ-Greek Key (Figure 12e) in which strand 2 is replaced by a α-helix and the three β-strands are part of the same β-sheet.

*3.4 Larger Loops*

Many supersecondary structures have specific functions for example a helix-loop-helix motif specific for DNA binding is found in many prokaryotic and eukaryotic transcription factors. A similar motif, the EF hand, is found in calcium-binding proteins such as calmodulin, troponin C and parvalbumin (figure 13 b), [40]. The two α-helices of the EF hand are approximately perpendicular with a connecting loop containing 12 residues. Four or five loop residues have oxygen containing side chains, preferably aspartate or glutamate that coordinate with a calcium ion.

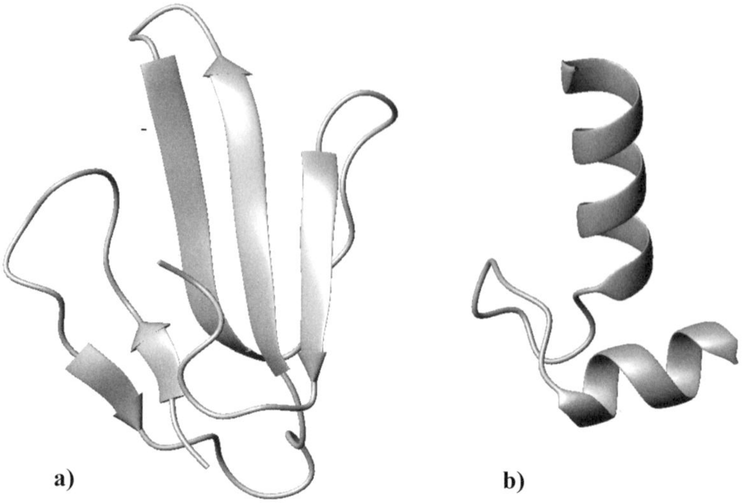


**Figure 13.** a) Erabutoxin , 1era, [41] A small protein formed from two β-hairpins (Greek key) plus one β strand. b) Parvalbumin , 1b8c, [42] Example of an EF hand. The sphere is a $Ca^{2+}$ ion.

## 4. Folds

Simple motifs combine to form folds or domains in which similarity is commonly defined by the chain topology, allowing for insertions and deletions of secondary structure between less closely related structures. Folds can also be defined according to their architecture, which describes how secondary structures pack together irrespective of their connectivity. For example, two β-sheets commonly pack one against the other so they are roughly aligned in a configuration referred to as an aligned β-sandwich; the jellyroll and immunoglobulin folds (figures 23 and 24) have β-sandwich architecture but different topologies.

The prevalence of the two main secondary structures (α-helix and β-sheets), means that domains can be conveniently divided into four classes; mainly-α, mainly-β, α/β in which α and β structures are interspersed and α+β in which α and β structures are segregated. The physical and chemical constraints on secondary packing gives rise to a recurrence of the motifs illustrated in figure 12, which in turn generate groups of similar folds within each class. Moreover, in many cases, an increase in fold size tends to be accomplished by repeating or extending existing motifs suggesting that during evolution the genes encoding the motifs are duplicated [43].

Although it is known that proteins with 30% sequence identity will very likely have the same fold [44], convergent evolution can produce proteins with the same shape but little sequence homology or conversely the sequences of closely related proteins may have diverged over time so that only the structural similarity has remained [45]. Therefore it is necessary to augment sequence alignment methods with automatic structure comparison algorithms such as SSAP, VAST and DALI [46-48] to determine relationships and categorise proteins into a structural hierarchy.

In general, 3D shape comparison software requires: i) A representation of the molecules, usually the xyz coordinates of all the $C_\alpha$ atoms. ii) An objective function, for example, rotate and translate one molecule relative to the other and measure inter-molecular distances between equivalent points on the two chains using iii) a comparison algorithm that requires decision rules derived from statistical analysis of multiple samples. When classifying 3D shape, redundancy is removed by making all sequences that have >25% identity equal and choosing the domain as a fold unit [49]. The advent of this technology has led to the compilation of a number of structural databases such as FSSP [50], SCOP [51] and CATH [52], which are accessible on the World Wide Web.

### *4.1 FSSP*

Fold classification based on Structure-Structure alignment of Proteins (FSSP) uses the fully automated structure comparison algorithm, DALI (Distance ALIgnment algorithm) to calculate a pair-wise structural similarity value between protein chains (S-score). The S scores for all pairs of proteins are evaluated and given statistically meaningful Z scores. Protein pairs with comparable scores are considered to have similar folds and a hierarchical structure, the Dali Domain Dictionary [53] has been created which allows direct comparison with SCOP and CATH. FSSP is accessible on the World Wide Web at http://www.bioinfo.biocenter.helsinki.fi:8080/dali/index.html

### *4.2 The SCOP Database*

The Structural Classification of Proteins (SCOP) database provides a detailed description of the structural and evolutionary relationships of proteins of known structure and is accessible on the World Wide Web at http:// scop.mrc-lmb.cam.ac.uk/scop/. There are two search facilities. One allows the user to enter a sequence to obtain a list of structures with significant sequence homology the other allows the user to enter a keyword to match text in the SCOP database and headers in the Protein Databank (PDB).

SCOP protein classification is a mainly manual process using visual inspection to compare structures but it also employs sequence homology and a variety of automated procedures. The unit of classification is the domain, each being treated separately in multi domain proteins. The hierarchy is described below and the number of entries at each level as of November 2003 is shown in table 2.

- **Family:** Proteins with 30% sequence identity or greater or those with less sequence homology but very similar structures and functions are clustered into families
- **Superfamily:** Superfamilies contain families that have low sequence homology but in which an evolutionary origin is suggested by structural and functional similarities.
- **Fold:** Proteins which have their α-helices and β-sheets in the same topological order and architectural arrangement are defined as having a common fold.
- **Class:** there are seven classes, the four mentioned above (α, β, α/β and α+β), small proteins, multidomain for folds consisting of two or more domains belonging to different classes, and membrane proteins.

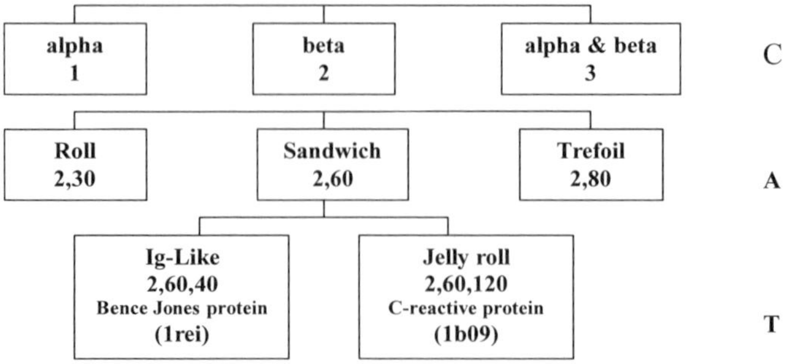


**Figure 14.** Diagram depicting the hierarchical nature of CATH for the three ma classes.

### *4.3 The CATH Database*

CATH is an acronym for the four main levels in the database hierarchy: Class (C), Architecture (A), Topology (T) and Homologous superfamily (H). There is also a fifth level, Sequence family (S). Classification is carried out using sequence alignment methods, the structure comparison algorithm SSAP [46] and human intervention where the automatic processes fail. An entry is assigned a number that correlates to its classification at each level. CATH is accessible on line at http://www.biochem.ucl.ac.uk/bsm/cath/. Users may search by the PDB code, a CATH number or text. The hierarchy is described below and illustrated in figure 14 and the number of entries at each level in September 2003 is shown in table 3.

- **Sequence family:** Proteins with 35% sequence identity or greater are clustered at this level.
- **Homologous superfamily;** Equivalent to the SCOP superfamily where structures are grouped by their functional and structural similarity.
- **Topology:** Similar to the SCOP common fold. Proteins with the same CAT number have the same class, architecture and topology but do not necessarily belong to the same homologous superfamily.
- **Architecture:** This level clusters proteins within the same class by their general shape irrespective of connectivity.
- **Class:** CATH has four classes, mainly-α, mainly-β, α-β and irregular, the latter containing proteins with low secondary structure content. The α/β and α+β classes are distinguishedat the topology level rather than the class level.

### *4.4 Comparison of SCOP, CATH and Dali*

In 2003 an analysis of the three databases [54] found more agreement between the domain definitions of SCOP and CATH than between Dali and either SCOP or CATH. Domain mismatches can occur when part of a protein is excluded from the definition in one database but not in the other. For example in CATH both the N and C terminal domains of MHC class II chains (1iea (A-D)) are classified as one domain whereas SCOP only includes the N-terminus which means that any structure matching the C-terminus will be included in CATH but will not have an equivalent match in SCOP

**Table 2.** SCOP: Structural Classification of Proteins. 1.65 release (1-09-2003) 20619 PDB Entries, 54745 Domains (excluding nucleic acids and theoretical models), http://scop.mrc-lmb.cam.ac.uk/scop/count.html

| **Class** | **Folds** | **Superfamilies** | **Families** |
|---|---|---|---|
| All alpha proteins | 179 | 299 | 480 |
| All Beta proteins | 126 | 234 | 462 |
| a/b proteins | 121 | 192 | 542 |
| a+b proteins | 234 | 330 | 567 |
| Multi domain | 38 | 39 | 53 |
| Membrane and Cell surface proteins | 36 | 64 | 73 |
| Small proteins | 66 | 87 | 150 |
| Total | 800 | 1232 | 2327 |

**Table 3.** CATH, Version 2.5 (11-08-2003) 43229 domains. http://www.biochem.ucl.ac.uk/bsm/cath/releases.html

| **C** | **A** | **T** | **H** | **S** |
|---|---|---|---|---|
| Mainly alpha | 5 | 228 | 433 | 957 |
| Mainly beta | 19 | 139 | 286 | 961 |
| Alpha beta | 12 | 361 | 659 | 2008 |
| Few Secondary structures | 1 | 85 | 89 | 110 |
| Total | 37 | 813 | 1503 | 4036 |

At the level of the fold there is more agreement between Dali and SCOP. CATH is the outsider because of the broader range of structures encompassed by the CATH fold definitions [55]. For example all the structures placed in the two most highly populated folds in CATH, the Rossman fold and the immunoglobulin-like (Ig-like) fold, are also found in the corresponding folds in Dali and SCOP although Dali and SCOP divide these folds into a number of sub families. However the CATH Ig-like and Rossman folds also contain structures that are not found in SCOP and Dali. Nevertheless there is a large amount of agreement between the databases and it is possible to assign CATH and SCOP classifications down to the topology and fold levels respectively from FSSP scores with a high degree of success [56]. This may prove useful since the FSSP database is updated more frequently than the other two.

## 5. Super-folds

Analysis of the CATH database has suggested that there is a limited number of protein folds in nature estimated at one to several thousand [57] and, although there are topologies yet to be sampled, it is evident that fold groups are not uniformly populated. In fact it has been found that there is a bias towards 10 groups at the topology-level called super-folds, which account for approximately one third of all the homologous superfamilies. The super-folds are roughly paralleled by Frequently Occurring Domains (FODs) in the SCOP database and by highly populated regions of fold space in FSSP.

*Predominantly Alpha-Domains α-α Packing* On any surface of an α-helix that runs parallel to the helix axis, the residue side-chains form ridges separated by grooves, and helices pack together so that the ridges of one helix fit into the grooves of the other. The average interaxial distance of packed helices is 9.4Å, which means that inter-helix contact is made by the side chain ends. The relative angle between the helical axes depends upon which ridges and grooves are intercalated. The most common arrangement has the ridges formed by every fourth residue on one helix fitting into the grooves formed from every fourth residue on the other helix in which case the angle is ~50°. Other characteristic angles are observed, the second most common being ~20° which arises when the ridges of one helix formed from each third residue fit into the grooves formed by every forth residue on the other helix [[9].

*Four-helix Bundle* The most frequent α-helical domain in globular proteins is the four-helix bundle, which is made up of two α-hairpins. It occurs in proteins as disparate as cytochrome $b_{562}$ [58] and the tobacco mosaic virus protein coat [59]. Sequential helices are packed together at an angle of ~20° and can be either anti-parallel as in hemerythrin (figure 15) or parallel as in human growth hormone.

*Globin Fold* The globin fold occurs in the mammalian oxygen binding proteins, haemoglobin and myoglobin, and other related proteins such as the phycocyanins [12, 60-61]. It is a bundle of eight helices usually labelled A to H, in which sequential helices are not adjacent (except for G and H) and are arranged to form a pocket for a heme group. The helices are packed at angles of around 50°.

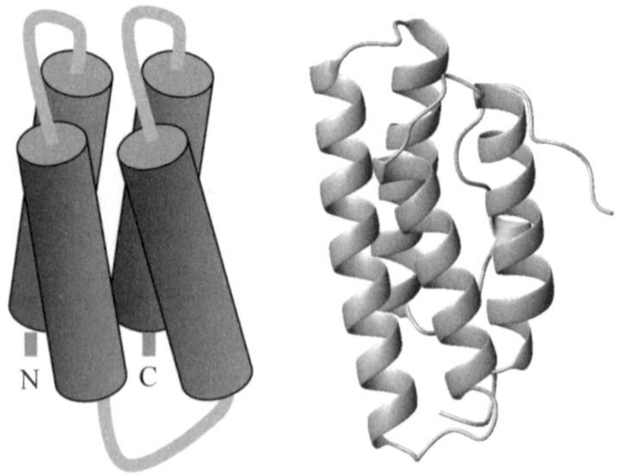


**Figure 15. a)** Schematic diagram of a four-helix-bundle. **b)** Raswin cartoon of Hemerythrin, chain C, 2hmz [62]. *CATH*: 1,20,120,50. (alpha, up-down bundle, 4-helix bundle) *SCOP*: alpha, four-helical up-and-down bundle ( CATH description to topology level, SCOP description to fold level)

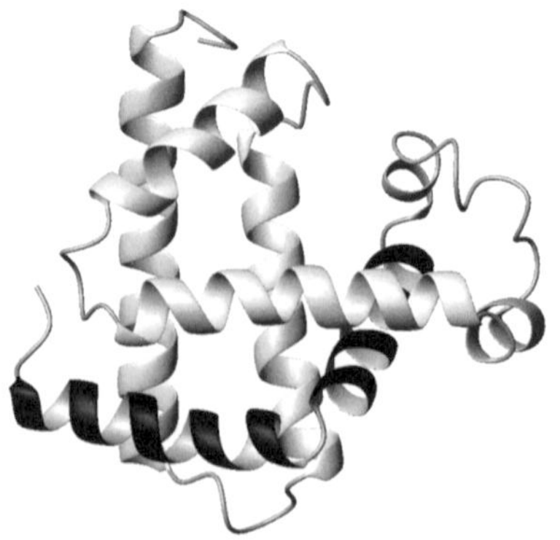

**Figure 16** Raswin cartoon of horse heart myoglobin, 1ymb [12]. For clarity pairs of consecutive helices are shaded differently. *CATH*: **1,10,490,**10 (alpha, orthogonal bundle, globin-like) *SCOP*: Alpha, globin- like.

## 5.2 Alpha / Beta Domains

*αβ-Packing* The super-folds in this class are made up of repeated βαβ motifs. The geometry and energetics of αβ-packing has been extensively studied [63-64]. The α-helix has 3.6 residues per turn, therefore the α-helix face has a right-handed twist which complements the right handed twist of the β-sheet when the structures are parallel. This is the most favoured configuration although another common arrangement has the helix diagonal to the sheet with interactions between the centre of the helix and centre of the sheet or the ends of the helix and corners of the sheet depending on whether the helix is above or below the β-sheet. The helix can also be perpendicular to the sheet in which case contacts can form along the length of the helix. [64-36]. The β-sheet contact surfaces usually comprise small hydrophobic residues such as valine, leucine and isoleucine, which allow for good close packing [63].

*TIM Barrel* The TIM barrel (named after triose phosphate isomerase (figure 17)) has a core of eight twisted, parallel β-strands that form the 'staves' of a barrel surrounded by the connecting α-helices. The parallel β-strands comprise alternate branched and bulky hydrophobic residues, the former pack against the helices, the latter create a tightly packed hydrophobic core. A large proportion of proteins with this structure are enzymes for example, aldolase and tryptophan synthase [65-66].

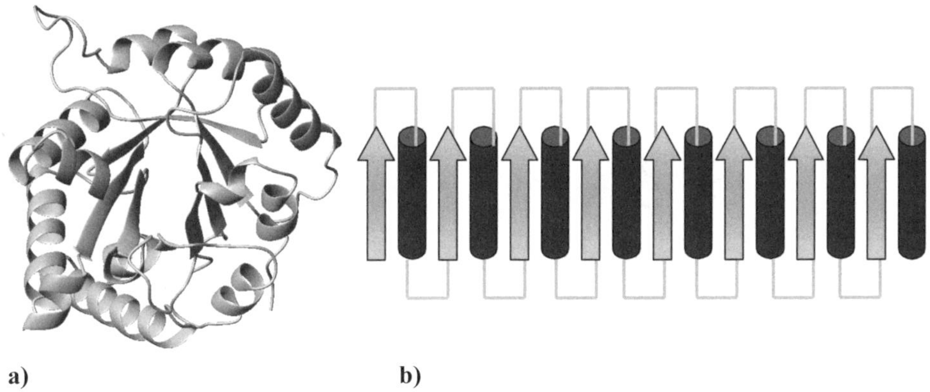

**a)** **b)**

**Figure 17.** a) Triosephosphate isomerase, chain B, 7tim [70]. *CATH*: 3,20,20,90 (alpha beta, barrel, TIM barrel) *SCOP*: a/b, TIM beta/alpha-barrel. b) Topology (arrow = β-strand, cylinder = α-helix)

*Doubly-wound Alpha/Beta Fold (Rossman Fold)* Unlike barrels these structures are open with a central β-sheet flanked by helices to form a 3-layered sandwich. The chain starts in the middle of the β-sheet and travels to the edge, then returns to the centre via a loop or helix and travels outwards to the opposite edge. Proteins with this structure include flavodoxin and adenylate kinase [67-68]. The βαβαβ configuration is named the Rossman fold after Michael Rossman who first described this configuration in nucleotide-binding proteins [69].

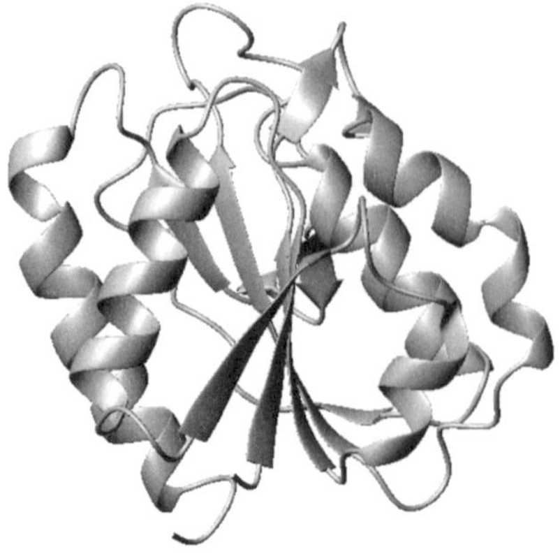

a )

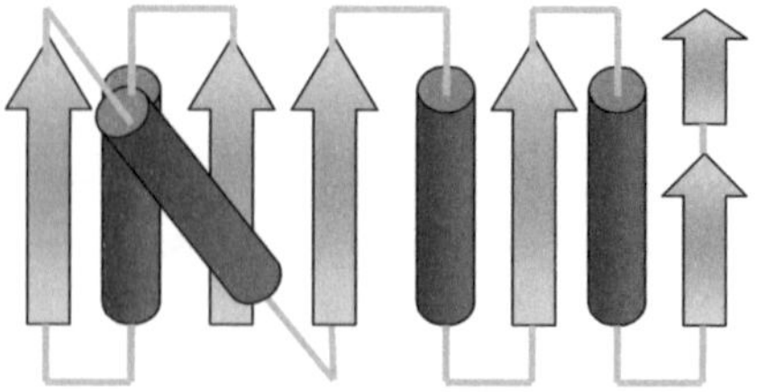

b )

**Figure 18. a)** flavodoxin, 1flv [71]. *CATH*: **3,40,50**,360 (Alpha beta, 3-layer (aba) Sandwich, Rossman fold) *SCOP*: a/b favodoxin-like. **b)** Topology

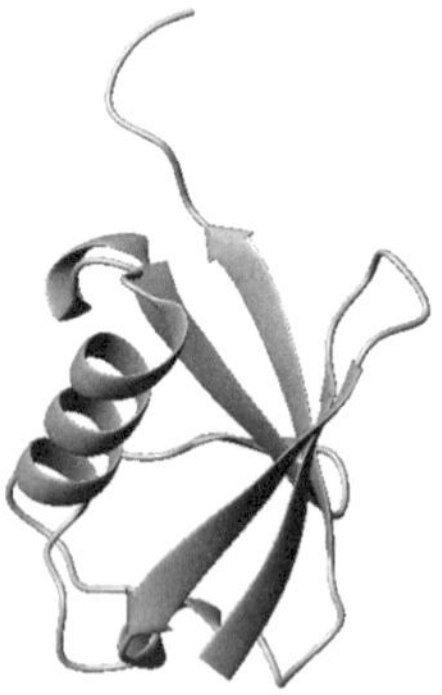

**a)**

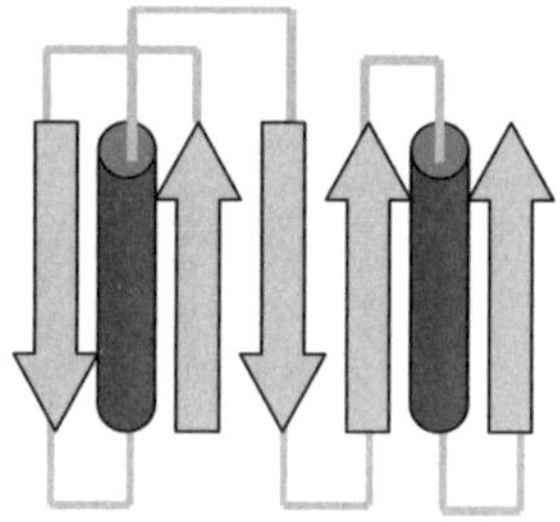

**b)**

**Figure 19.** a) Ubiquitin, 1ubi [76]. Example of a UB roll. *CATH*: 3,10,20,90. (alpha beta, roll, UB roll). *SCOP*: a+b, beta-grasp (ubiquitin-like). b) Topology: ββα motif

*Alpha + Beta Domains* Alpha + beta folds are more diverse than either α or α/β folds and contain many complex folds which cannot be easily clustered into groups. However there are two super-folds in this class made up of repeated ββα units, which form an open anti-parallel or mixed β-sheet with helices on each side.

*UB (Ubiquitin) roll* The UB roll is an open twisted β-sheet with αβ Greek-key topology (ββαβ) packed against α-helices. Examples include ferredoxin and protein G [72-73].

*Plaitfold* The plaitfold is similar to the UB roll comprising a β-sheet packed against α-helices to form a two-layered sandwich containing the αβ Greek-key motif (βαββ). Examples include phosphotransferase and chorismate mutase [74-75].

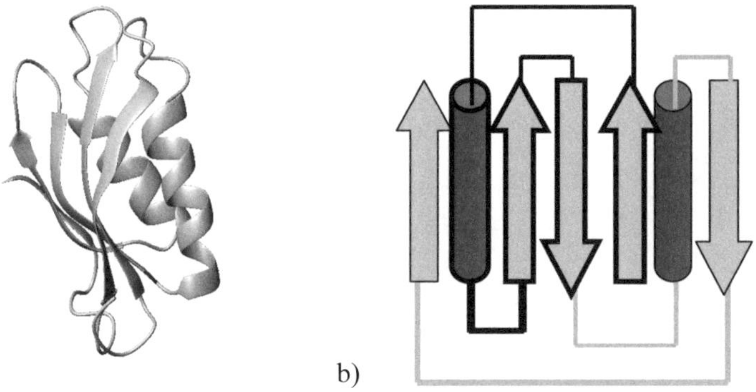

a) b)

**Figure 20.** a) Acylphosphatase, 1aps [77]. Example of a plaitfold domain. *CATH*: 3,30,70,100 (alpha beta, 2-layered sandwich, alpha-beta plaits). *SCOP*: a+b, ferridoxin-like. b) Topology diagram. A βα Greek-Key is outlined in black.

## *5.4 Predominantly Beta-Domains*

*Beta-Beta Packing* There are two common ways of packing β-sheets, aligned and orthogonal. In aligned packing, two β-sheets joined by a non β-segment lie face to face to form a sandwich with the backbone direction of the upper sheet rotated in the clockwise direction with respect to the other by an angle that varies from 20° to 50° depending on the right-handed twist of the β-sheets, but is usually ~ 30°. An aligned β-sheet is illustrated by the immunoglobulin fold in figure 24. Orthogonally packed β-sheets (figure 21) also lie face to face but one is at ~90° to the other and the strands at one corner or two diagonally opposite corners have a bend due to a β-bulge or β-coil and pass uninterrupted from one layer to the next [[9].

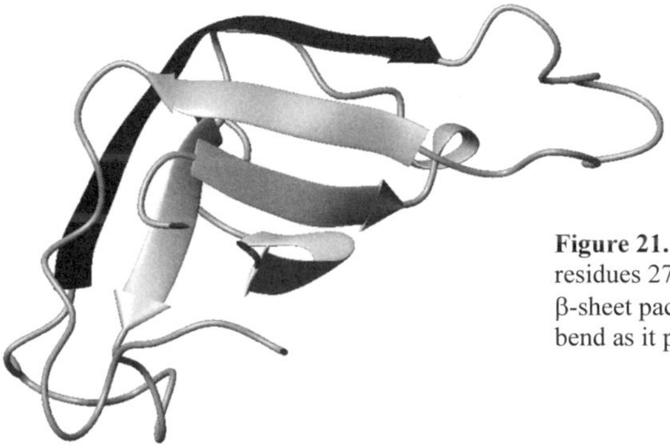

**Figure 21.** α-Chymotrypsin, 5cha, Cain A, residues 27-112 [78]. An example of orthogonal β-sheet packing, the dark shaded strand has a 90° bend as it passes from one layer to the other.

*OB (Oligonucleotid/Oligosaccharide Binding) Roll* The OB roll consists of a five-stranded β-sheet coiled to form a barrel structure, which may be capped by a α-helix. Examples include heat labile enterotoxin and antifungal protein [79-80].
*Jelly Roll.*This fold, which forms a 2-layered sandwich, is made up of Greek-key motifs in which the connection from strand 2 to strand 3 is made between layers and the connection

from strand 3 to strand 4 crosses the other way (figure 23 b). Proteins with this fold include the satellite panicum mosaic virus coat and PAH monooxygenase [81-82].

*Immunoglobulin fold.*The immunoglobulin constant domain consists of a four-stranded β-sheet packed against a three-stranded β-sheet to form an aligned 2-layered β-sandwich with the topography shown in figure 24b. Like the jellyroll, a Greek-key motif is divided between the layers. Examples of proteins with this fold include Bence-Jones protein (figure25) and superoxide dismutase [83-84].

*Trefoil.* The trefoil fold is formed from six two-stranded hairpins, three of which form a three-sided barrel while the others form a triangular array that caps the barrel, giving the fold pseudo three-fold symmetry [85]. The fold is found in several protein families including Kunitz soybean trypsin inhibitors (STIs), ricin-like toxins, plant agglutinins and hisactophilin-like actin-bundling proteins [85-87].

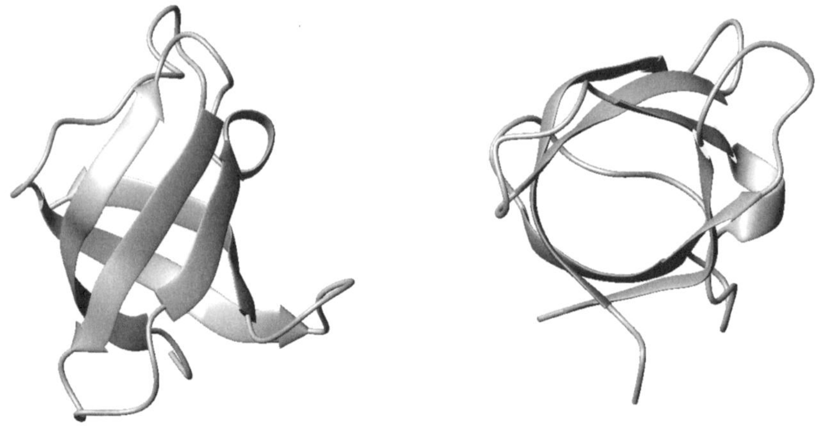

**Figure 22.** Major cold shock protein 7.4, 1mjc [88]. Example of an OB Roll. (Side and end view) *CATH*: 2,40,50,240 (beta, barrel, OB-fold). *SCOP*: beta, OB-fold.

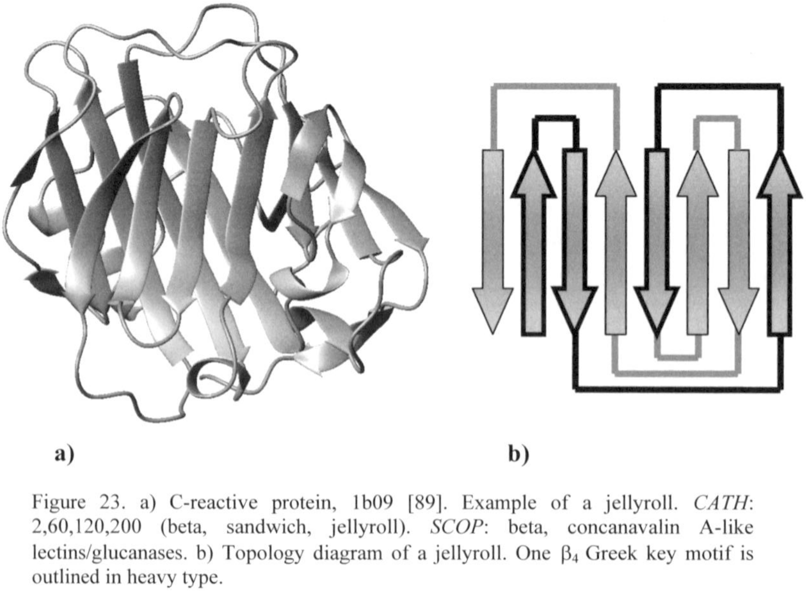

**a)** **b)**

Figure 23. a) C-reactive protein, 1b09 [89]. Example of a jellyroll. *CATH*: 2,60,120,200 (beta, sandwich, jellyroll). *SCOP*: beta, concanavalin A-like lectins/glucanases. b) Topology diagram of a jellyroll. One $\beta_4$ Greek key motif is outlined in heavy type.

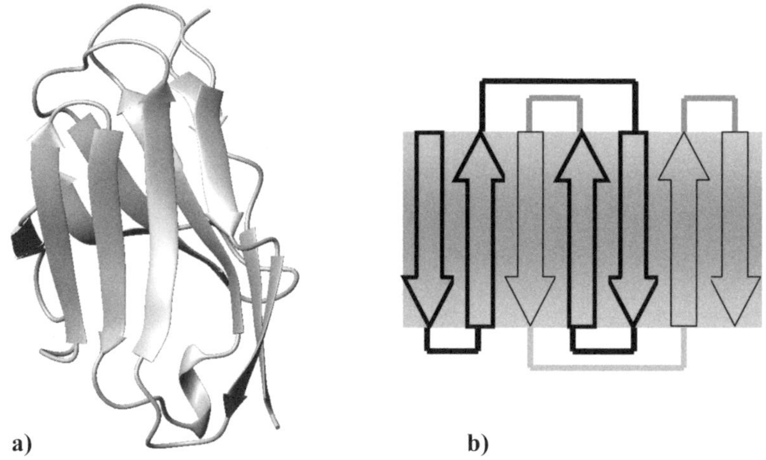


**Figure 24.** a) Bence Jones protein, 1rei [83]. Example of the immunoglobulin fold.*CATH*: 2,60,40,10 (beta, sandwich, immunoglobulin-like) *SCOP*: beta, immunoglobulin-like beta-sandwich. b) Topology diagram. One $\beta_4$ Greek key motif is outlined in heavy type.

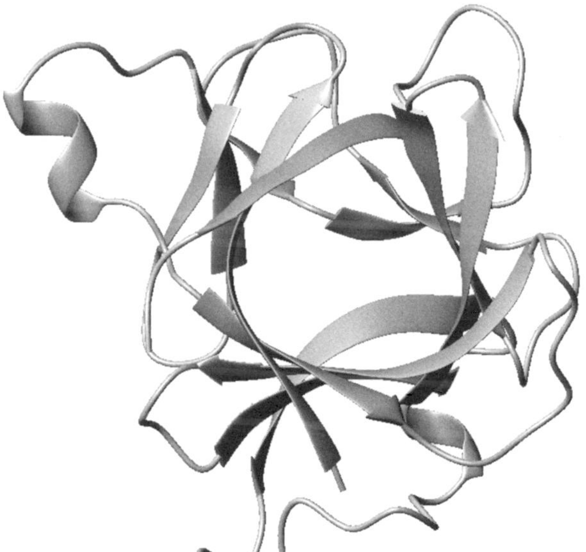

**Figure 25.** Interleukin-1 Beta, 1i1b [90]. Example of a trefoil fold.. *CATH*: 2,80,10, 50 (beta, trefoil, trefoil) *SCOP*: beta, beta-trefoil

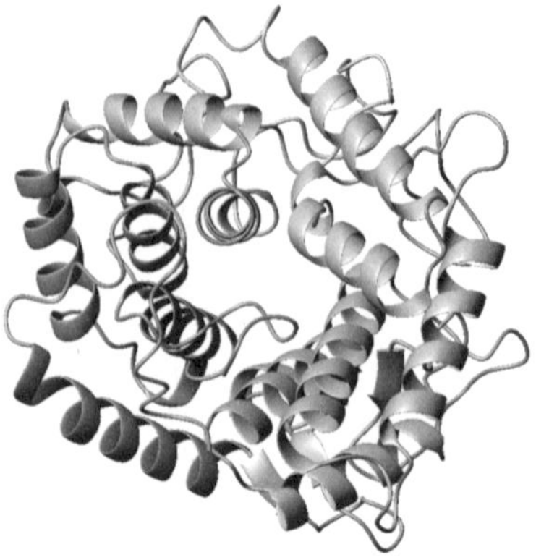

**Figure 26.** Glycosyltransferase, 1cem [92]. Example of an α-Barrel. *CATH:* 1, 50,10,10 (alpha, alpha/alpha barrel, glycosyltransferase). *SCOP*: alpha, alpha/alpha toroid.

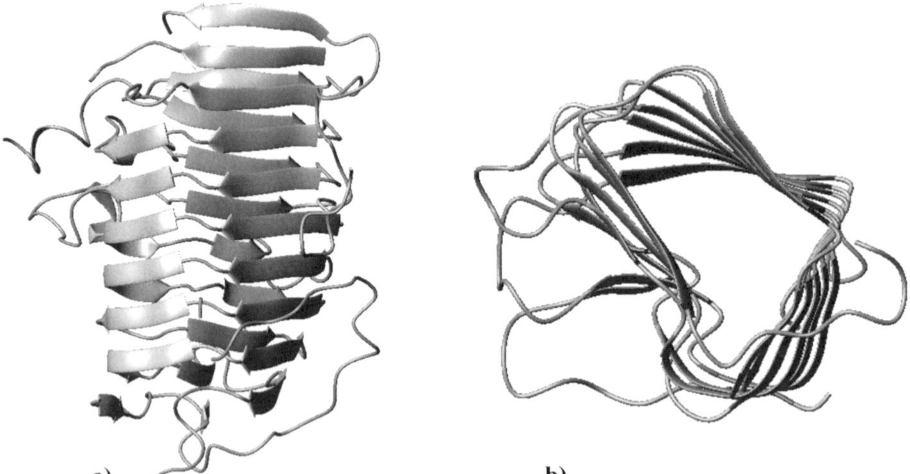

**Figure 27.** a) Pectate lyase C, 2pec [91], Residues 80-280.*CATH:* 2,160,20,10 (beta, 3-solenoid, pectate lyase C-like) *SCOP*: beta, single-stranded right-handed beta-helix. b) Looking down the barrel. Two β-sheets are aligned while the other is perpendicular

## 6. Other Protein Folds

The next section will briefly describe other common folds found in soluble, globular proteins such as those discussed already and folds adopted by structural and membrane proteins.

*6.1 Other Soluble Domains*

Other alpha-helical domains include the alpha-alpha barrel, here illustrated by the catalytic core of glycosyltransferase from *Clostridium thermocellum* shown in figure 26. The barrel is formed by six inner and six outer alpha helices.

Other common β-structures include the β-propellors, β prisms and β-helices. β-propellers comprise 4 to 8 small anti-parallel sheets with identical up-down topology arranged like the blades of a propeller. The example below (figure 28) is the head domain of influenza neuraminidase, which is a six-bladed propeller. As discussed previously, parallel β-sheets are usually formed from repeated βαβ motifs. However pectate lyase and the tailspike protein from P22 phage are β-structures comprising three parallel β-sheets that form a β-helix, classified as a 3-solenoid in the CATH database. The repeat unit that forms one turn of the helix contains three strands and three loops [91]. From figure 27 it can be seen that the parallel β-sheets are almost planar with two packed adjacent to each other while the third sheet is almost perpendicular to the other two. There are related structures with only two sheets packed together found in bacterial extracellular proteases.

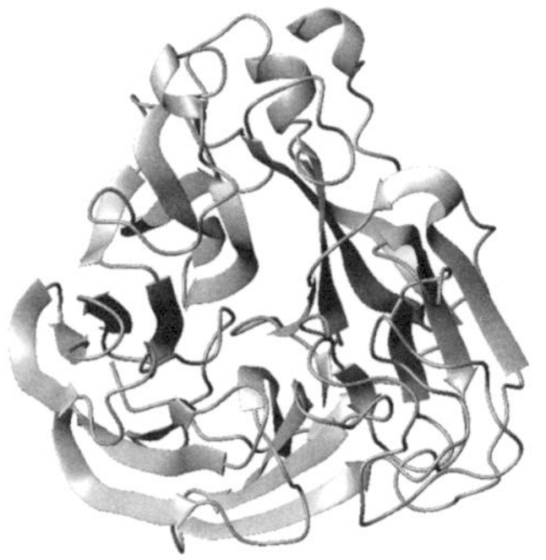

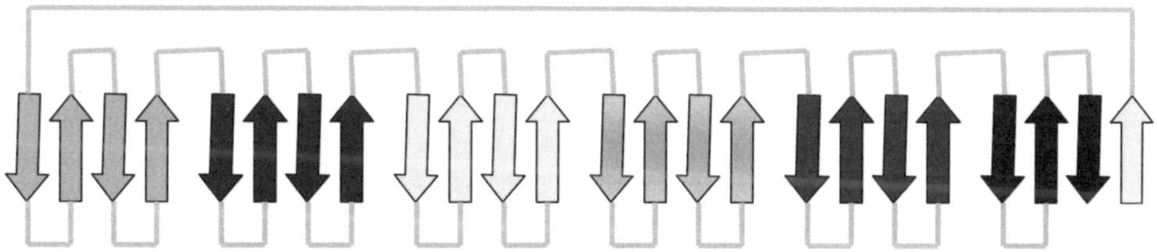

**Figure 28.** a) Neuraminidase from influenza A virus 1f8d, chain A, residues 82-486 [93]. There are 6 anti-parallel β-sheets each shaded differently, forming the 6 blades of a propeller structure. *CATH:* 2.120.10.10 (beta, 6-propeller, neuraminidase) *SCOP*: beta, 6-bladed beta-propeller. b) Topology of neuraminidase (loops not to scale)

### *6.2 Fibrous proteins*

Fibrous proteins, unlike globular proteins, contain repetitive amino acid sequences, giving rise to very regular secondary structures. They can be divided into three main structural groups two of which, the triple helix of collagen and the coiled coils of keratin, myosin and cytoskeleton components, are made up of multiple helices wrapped around each other, Proteins in the third group, which includes spider's silk, are made up of α-helices and β-sheets [94]. The structural units of a fibrous protein are micro-fibrils that aggregate so that the gross structure has a specific strength and elasticity.

The fibrils of collagen consist of three left-handed polyproline II helices running in parallel to form a right handed triple super-helix (figure 29). Each chain, which contains about 1000 residues (~ 3000Å) is made up of repeat sequences Gly-X-Y where X is often proline and Y is often hydroxyproline formed by post-translational modification of proline. The chains are held together by hydrogen bonding between the proline C′O groups of one chain and the glycine NH groups of another. The glycine sidechains point towards the interior of the superhelix where there is not enough space for larger sidechains [95].

Coiled coils are left handed super-helices formed from two α-helices in which the number of residues per turn in each α-helix is reduced from 3.6 to 3.5 and the sequences tend to be repeated every seven residues. The first and fifth residues of the repeat in each helix are hydrophobic and oriented towards the helix axis forming the contact face as the helices coil around each other [96].

**a)** **b)**

**Figure 29.** Backbone trace of a collagen like polypeptide with repeated pro pro gly sequence. 1a3j, [97]. Proline is colored dark grey, glycine is colored light grey. **a)** Side view **b)** Looking down the right-handed triple helix from the N terminal end

### *6.3 Membrane Proteins*

To enable the transfer of a polypeptide into a membrane bilayer its surface residues must be predominantly non-polar and the backbone C′O and NH groups must be internally hydrogen bonded. Consequently many transmembrane protein domains are formed from hydrophobic α-helical bundles connected by hydrophilic loops that project either side of the bilayer. An example is bacteriorhodopsin (figure 30), which comprises seven α-helices that span the membrane to form a channel. The channel contains covalently bound retinal, which undergoes isomerisation upon absorbing a photon and thereby changes the conformation of the protein so that a proton is transferred from the cytosol to the extracellular side of the membrane.

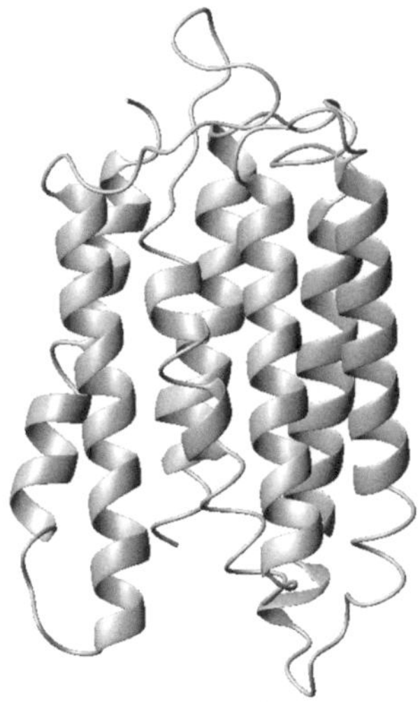

**Figure 30.** Bacteriorhodopsin, 1ap9, [98]. *CATH*: **1.20.1070**.10. (Alpha, up-down bundle, rhopdopsin 7-helix transmembrane proteins) *SCOP*: membrane and cell surface proteins and peptides, family A G protein-coupled receptor-like protein.

Beta-structures can also span the membrane if all the main chain polar groups are engaged in inter-strand hydrogen bonding, a criterion satisfied by the closed topology of a β-barrel. Porins are trimeric proteins in which each sub-unit is made up of between 12 and 18 β-strands that form an up-down β-barrel spanning the outer membrane of gram negative bacteria (figure 31). The radius of each barrel is large enough for the interior to form a pore lined with hydrophilic residues. Within the porin trimer there is a hydrophobic core around the symmetry axis, which usually lends the structure a degree of stability, however dissociation of the porin timer is accompanied by denaturation of the subunits.

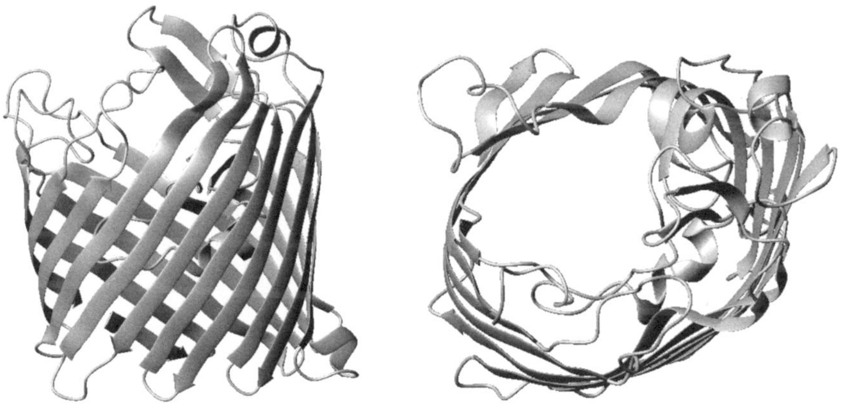

**a)** **b)**

**Figure 31.** Matrix OmpF porin, 1bt9, [99]. *CATH*: 2.40.160.10 (Beta, barrel, porin) *SCOP*: membrane and cell surface proteins and peptides, transmembrane beta-barrels. a) Side view; b) Looking down the barrel.

## 7. Conclusion

The thermodynamics and kinetics of hydrogen bond formation collude with the steric constraints of the peptide bond to favour the formation of α-helices, β-strands and reverse turns along with a few less common secondary structures. Secondary structures are characterized by the dihedral angles of the chain and the backbone hydrogen-bonding pattern. The assignment of structure to a section of the chain by a pattern recognition algorithm depends upon the definition of the hydrogen bonds and the boundaries of $\phi/\varphi$ space assigned for each structure.

The classification of folds in the PDB is achieved by segregating mainly- α, mainly-β and mixed (α/β and α+β) structures. Structures in each class are then clustered by overall shape and topology. Classification depends upon an overlap between structures in the same group and a clear distinction between groups. However, while the mainly α-structures tend to cluster separately, the highly populated architectures of the β-sheet- containing classes tend to adopt similar two or three-layered sandwich-like structures, or barrels. Furthermore, at the topology level the recurrence of certain structural motifs causes a significant overlap between some folds. For example the repeated βαβ motif is found in both TIM barrels and the doubly wound fold while the Greek-key is embedded in the jellyroll and the immunoglobulin fold. Thus in some parts of fold space there is a continuum between structures rather than distinct steps and in these regions, the criteria used for clustering depends upon the purpose of the analysis.

CATH, SCOP and FSSP represent three unique ways of classifying protein structure, FSSP uses a completely automated process, SCOP is principally derived from manual inspection and CATH uses automated and manual procedures. Moreover whereas FSSP and SCOP were created with an eye to evolutionary and functional relationships, CATH was based solely on structural comparisons.

Another system, developed by Harrison *et al.*, [100], classifies folds by their 'gregariousness' which is a measure of how many other folds have significant structural overlap with a particular fold but have a different overall topology. In the analysis, folds in the highly populated architectures, including the 10 so called super-folds, are highly gregarious whereas folds such as β-helices comprising common motifs that are packed in unusual ways, or folds with uncommon motifs, have low gregariousness. This method is implemented by a graph-theoretic program, GRATH [100] that rapidly and accurately, matches a novel structure against a library of domain structures to find the most similar ones. It can be accessed via a server at http://www.biochem.ucl.ac.uk/cgi-bin/cath/Grath.pl. GRATH is relatively fast and provides a reliable front-end filter for the more accurate, but computationally expensive, residue based structure comparison algorithm SSAP, currently used to classify domain structures in the CATH database.

## Acknowledgements

AM is the recipient of an MRC Studentship. This work was supported, in part, by grant B02959 from the BBSRC to BAW.

## References

[1] Aurora, R., Creamer, T.P., Srinivasan, R. & Rose, G.D., (1997) Local Interactions in Protein folding: Lessons From The α-Helix. *J. Biol. Chem.* **272**, 1412-1416

[2] Kendrew, J.C., *et al.*, (1958) A Three-Dimensional Structure of the Myoglobin Molecule Obtained by X-ray Analysis. *Nature* **181**, 662-666

[3] Pauling, L., Cory, R.B. & Branson, H.R., (1951) The Structure of Proteins: Two Hydrogen Bonded Helical configurations of the Polypeptide Chain. *Proc. Natl. Acad. Sci. USA* **37**, 205-211

[4] IUPAC-IUB Commission on Biochemical Nomenclature, 1969., (1970) Abreviations and Symbols for the Description of the Conformation of Polypeptide Chains. *Biochemistry* **9**, 3471-3479

[5] Ramachandran, G.N., Ramakrishnan, C. & Sasisekharan, V., (1963) Stereochemistry of Polypeptide Chain Configuration. *J.Mol.Biol.* **7**, 95-99

[6] Ho, B.K., Thomas, A. & Brasseur, R., (2003) Revisiting the Ramachandran Plot: Hard-Sphere Repulsion, Electrostatics and H-Bonding in the α-Helix. *Protein Sci.* **12**, 2508-2522

[7] Schulz, G.E. & Schirmer, R.H., (1979) *Principals of Protein Structure*, Springer, pp 66-79

[8] Perutz, M.F., (1951) New X-Ray Evidence on the Configuration of Polypeptide Chains. *Nature* **167**, 1053-1054

[9] Chothia, C., (1984) Principles That Determine the Structure of Proteins. *Ann. Rev. Biochem.* 53, 537-572

[10] Hopfinger, A.J., (1973) *Conformational Properties of Macromolecules.* Academic Press, New York.

[11] Adzhubei, A.A. & Sternberg, M.J.E., (1992) Left-handed Polyproline II Helices Commonly Occur in Globular Proteins. *J. Mol. Biol.* **229**, 472-493

[12] Evans, S.V. & Brayer, G.D., (1990) High-Resolution Study of the Three-Dimensional Structure of Horse Heart Metmyoglobin. *Biochemistry* **213**, 885-897

[13] Pauling, L. & Cory, R.B., (1951) Configurations of Polypeptide Chains with Favoured Orientations Around Single Bonds: Two New Pleated Sheets. *Proc. Natl. Acad. Sci. USA* **37**, 729-740

[14] Prztycka, T., Rajeev, A. & Rose, D.G., (1999) A Protein Taxonomy Based on Secondary Structure. *Nat. Struct. Biol.* **6**, 672-682

[15] Chou, K.C., Pottle, M., Nemethy, G., Ueda, Y. & Scheraga, H.A., (1982) Origin of the Right-Handed Twist and of the Increased Stability of β-Sheets. *J. Mol. Biol.* **162**, 89-112

[16] Chou, K.C., Nemethy, G. & Scheraga, H.A., (1983) Role of Interactions in the Stabilisation of the Right-Handed Twist of β-Sheets. *J. Mol. Biol.* **168**, 389-407

[17] Richardson, J.S., Getzoff, E.D. & Richardson, D.C., (1978) The Bulge: A Common Small Unit of Non-Repetitive Protein Structure. *Proc. Nat. Acad. Sci. USA* **75**, 2574-2578

[18] Becker, K., Savvides, S.N., Keese, M., Schirmer, R.H. & Karplus, P.A., (1998) Enzyme Inactivation Through Sulfhydryl Oxidation by Physiologic NO-Carriers. *Nat. Struct. Biol.* **5**, 267-271

[19] Eichhorn, E., Davey, D.A., Sargent, D.F., Leisinger, T. & Richmond, T.J., (2002) Crystal Structure of *Escherichia coli* Alkanesulfonate Monooxygenase. *J. Mol. Biol.* **324**, 457-468

[20] Chou, K., (2000) Prediction of Tight Turns and Their Types in Proteins. *Anal. Biochem.* **268**, 1-16

[21] Venkatachalam, C.M., (1968) Stereochemical Criteria for Polypeptides and Proteins. V. Conformation of a System of Three Linked Peptide Units. *Biopolymers* **6**, 1425-1436

[22] Lewis, P.N., Momany, F.A. & Scheraga, H.A., (1973) Chain Reversals in Proteins. *Biochem. Biophys. Acta.* **303**, 211-229

[23] Richardson, J.S., (1981) The Anatomy and Taxonomy of Protein Structure. *Advan. Prot. Chem.* **34**, 167-339

[24] Wilmot, C.M. & Thornton, J.M., (1988) Analysis and Prediction of the Different Types of β-Turn in Proteins. *J. Mol. Biol.* **203**, 221-232

[25] Hutchinson, E.G. & Thornton, J.M., (1994) A Revised Set of Potentials for Beta-Turn Formation in Proteins, *Protein Sci.* **3**, 2207-2216

[26] Hutchinson, E.G. & Thornton, J.M., (1994) PROMOTIVE - A Program to Identify and Analyse Structural Motifs in Proteins. *Protein Sci.* **5**, 212-220

[27] Guruprasad, K. & Rajkumar, S., (2000) β and γ -Turns in Proteins Revisited: A New Set of Amino Acid Turn-Type Dependent Positional Preferences and Potentials. *J. Biosci.* **25**, 143-156

[28] Pavone, V., Gaeta, G., Lombardi, A., Nastri, F., Maglio, O., Isernia, C. & Saviano, M., (1996) Discovering Protein Secondary Structure: Classification and Description of Isolated α- Turns. *Biopolymers* **38**, 705-721

[29] Rajashankar, K.R. & Ramakumar, S., (1996) π-Turns in Proteins and Peptides: Classification, Conformation, Occurrence, Hydration and Sequence. *Protein Sci.* **5**, 932-946

[30] Richards, F.M. & Kundrot C.E., (1988) Identification of Structural Motifs From Protein Coordinate Data: Secondary Structure and First-Level Supersecondary Structure. *Prot. Struc. Func. Gen.* **3**, 71-84

[31] Kabach, W. & Sander, C., (1983) Dictionary of Protein Secondary Structure: Pattern Recognition of Hydrogen-Bonded and Geometrical Features. *Biopolymers,* **22**, 2577-2637

[32] Frishman, D. & Argos P., (1995) Knowledge-Based Protein Secondary Structure Assignment. *Proteins* **23**, 566-79

[33] King, S.M. & Johnson, W.C. Jr., (1999) Assigning Secondary Structure From Protein Coordinate Data. *Proteins Struct. Func. Gen.* **35**, 313-320

[34] Sun, Z. & Jiang, B., (1996) Patterns and Conformations of Commonly Occurring Supersecondary Structures (Basic Motifs) in the Protein Data Bank. *J. Prot. Chem.* **15**, 675-690

[35] Sibanda, B.L. & Thornton, J.M., (1985) β-Hairpin Families in Globular Proteins. *Nature* **316**, 170-174

[36] Boutonnet, N.S., Kajava, A.V. & Rooman, M.J., (1998) Structural Classification of ααβ and ββα Supersecondary Structure Units in Proteins. *Prot. Struct. Func. Gen.* **30**, 193-212

[37] Zhang, C. & Kim, S-H., (2000) A Comprehensive Analysis of the Greek-Key Motifs in Protein Barrels and Sandwiches. *Prot. Struct. Func. Gen.* **40**, 409-414

[38] Epimov, A., (1995) Structural Similarity Between 2-Layer Alpha/Beta-Proteins and Beta-Proteins. *J. Mol. Biol.* **245**, 402-415

[39] Branden, C. & Tooze, J., (1999) *Introduction to Protein Structure*. 2nd Ed. Garland Publishing, New York.

[40] Kretsinger, R.H., (1980) Structure and Evolution of Calcium-Modulated-Proteins. *Crit. Rev. Biochem.* **8**, 119-174

[41] Hatanaka, H., Oka, M., Kohda, D., Tate, S., Suda, A., Tamiys, N. & Inagaki, F., (1994) Tertiary Structure of Erabutoxin b in Aqueous Solution Elucidated by Nuclear Magnetic Resonance. *J. Mol. Biol.* **240**, 155-166

[42] Cates, M.S., Berry, M.B., Ho, E., Li, J.D., Potter, J.D. & Phillips, G.N. Jr., (1999) Metal Ion Affinity and Specificity in EF- Hand Proteins: Coordination Geometry and Domain Placticity in Parvalbumin. *Structure.* **7**, 1269-1278

[43] Harrison, A., Pearl, F., Mott, R., Thornton, J. & Orengo, C.A., (2002) Quantifying the Similarities Within Fold Space. *J. Mol. Biol.* **323**, 909-926

[44] Flores, T.P., Orengo, C.A. & Thornton, J.M., (1993) Conformational Characteristics of Structurally Similar Proteins. *Protein Sci.* **7**, 31-37

[45] Orengo, C.A., (1994) Classification of Protein Folds. *Curr. Op. Struct. Biol.* **4**, 429-440

[46] Taylor, W.W., & Orengo, C.A., (1989) Protein Structure Alignment. *J. Mol. Biol.* **208**, 1-22

[47] Madej, T., Gibrat, J-F. & Bryant, S.H., (1995) Threading a Database of Protein Cores. *Proteins Struct. Func. Genet.* **23**, 356-359

[48] Holm, L. & Sander, C., (1993) Protein Structure Comparison by Alignment of Distance Matrices. *J. Mol. Biol.* **223**, 123-138

[49] Holm, L. & Sander, C., (1996) Mapping the Protein Universe. *Science* **273**, 595-602

[50] Holm, L. & Sander, C., (1997) Dali/FSSP Classification of Three-Dimensional Protein Folds. *Nucleic Acids Res.* **25**, 231-234

[51] Murzin, A. G., Lesk, A. M. & Chothia, C., (1992). β-Trefoil Fold. Patterns of Structure and Sequence in the Kunitz Inhibitors, Interleukins-1b and 1a and Fibroblast Growth Factors. *J. Mol. Biol.* **223**, 531-543

[52] Orengo, C.A., Michie, A.D., Jones, S., Jones, D.T., Swindells, M.B. & Thornton, J.M., (1997) CATH- A Hierarchic Classification of Protein Domain Structures. *Structure.* **5**, 1093-1108

[53] Dietmann, S., Park, J., Notredame, C., Hegar, A., Lappe, M. & Holm, L., (2001) A Fully Automatic Evolutionary Classification of Protein Folds: Dali Domain Dictionary Version 3. *Nucleic Acids Res.* **29**, 55-57

[54] Day, R., Beck, D.A.C., Roger, S. & Daggett, V., (2003) A Consensus View of Fold Space: Combining SCOP, CATH and the Dali Domain Dictionary. *Protein Sci.* **12**, 2150-2160

[55] Hadley, C. & Jones, D.T., (1999) A Systematic Comparison of Protein Structure Classifications: SCOP, CATH and FSSP. *Structure* **7**, 1099-1112

[56] Getz, G., Vendruscolo, M., Sachs, D. & Domany, E., (2002) Automated Assignment of SCOP and CATH protein structure classifications from FSSP scores. *Proteins: Stuct. Func. Gen.* **46**, 405-415

[57] Orengo, C.A., Jones, D. & Thornton, J.M., (1994) Protein Superfamilies and Domain Super-folds. *Nature* **372**, 631-634

[58] Hamada K., Bethge P.H., Mathews F.S., (1995) Refined Structure of Cytochrome b562 from *Escherichia coli* at 1.4-A Resolution. *J. Mol. Biol.* **247**, 947-962

[59] Bhyravbhatla, B., Watowich, S.J. & Caspar D.L., (1998) Refined Atomic Model of the Four-Layer Aggregate of the Tobacco Mosaic Virus Coat Protein at 2.4 Å Resolution. *Biophys. J.* 7**4**, 604- 615

[60] Safo M.K. & Abraham D.J., (2001) The X-ray Structure Determination of Bovine Carbonmonoxy Hemoglobin at 2.1 Å Resolution and its Relationship to the Quaternary Structures of Other Hemoglobin Crystal Forms. *Protein Sci.* **10**, 1091-1099

[61] Duerring, M., Schmidt, G.B. & Huber R., (1991) Isolation, Crystallization, Crystal Structure Analysis and Refinement of Constitutive C-Phycocyanin from the Chromatically Adapting Cyanobacterium *Fremyella diplosiphon* at 1.66-A Resolution. *J. Mol. Biol.* **217**, 577-592

[62] Holmes, M.A. & Stenkamp, R.E., (1991) The Structures of Met and Azidomet Hemerythrin at 1.66 Angstroms Resolution. *J. Mol. Biol.* **220**, 723- 737

[63] Janin, J. & Chothia, C., (1980) Packing of α-Helices onto β-Pleated Sheets and the Anatomy of α/β Proteins. *J. Mol. Biol.* **143**, 95-128

[64] Chou, K.C., Nemethy, G., Rumsey, S., Tuttle, R.W. & Scheraga, H.A., (1985) Interactions Between an α-Helix and a β-Sheet: Energetics of α/β Packing in Proteins. *J. Mol. Biol.* **186**, 591-609

[65] Blom, N. & Sygusch, J., (1997) Product Binding and Role of the C-Terminal Region in Class I D-Fructose 1,6-Bisphosphate Aldolase. *Nat. Struct. Biol.* **4**, 36-39

[66] Hyde, C.C., Ahmed, S.A., Padlan, E.A., Miles, E.W. & Davies, D.R., (1988) Three-Dimensional Structure of the Tryptophan Synthase Alpha 2 Beta 2 Multienzyme Complex from *Salmonella typhimurium*. *J. Biol. Chem.* **263**, 17857-17871

[67] Drennan, C.L., Pattridge, K.A., Weber, C.H., Metzger, A.L., Hoover, D.M. & Ludwig, M.L., (1999) Refined structures of Oxidized Flavodoxin from *Anacystis nidulans*. *J. Mol. Biol.* **294**, 711-724

[68] Schlauderer, G.J. & Schulz, G.E., (1996) The Structure of Bovine Mitochondrial Adenylate Kinase: Comparison With Isoenzymes in Other Compartments. *Protein Sci.* **5**, 434-441

[69] Rao, S.T., & Rossman, M.G., (1973) Comparison of Super-Secondary Structures in Proteins *J. Mol. Biol.* **76**, 241-256

[70] Davenport, R.C., Bash, P.A., Seaton, B.A., Karplus, M., Petsko, G.A. & Ringe, D. (1991) Structure of the Triose Phosphate Isomerase-Phosphoglycolohydroxamate Complex: An Analogue of the Intermediate on the Reaction Pathway. *Biochemistry* **30**, 5821-5826

[71] Rao, S.T., Shaffie, F., Yu, C., Satyshur, K.A. & Stockman, B.J., (1993) Structure of the Oxidized Long Chain Flavodoxin From Anabaena at 2 Angstroms Resolution. *Protein Sci.* **1**, 1413- 1427

[72] Derrick, J. P. & Wigley, D. B., (1994) The Third IgG-Binding Domain From Streptococcal Protein G. An Analysis By X-Ray Crystallography of the Structure Alone and in a Complex With Fab. *J. Mol. Biol.* **243**, 906-918

[73] Fukuyama, K., Ueki, N., Nakamura, H., Tsukihara, T. & Matsubara, H., (1995) Tertiary Structure of [2Fe-2S] Ferredoxin From *Spirulina platensis* Refined at 2.5 A Resolution: Structural Comparisons of Plant-Type. *J. Biochem.* **117**, 1017-1023.

[74] Jia, Z., Vandonselaar, M., Hengstenberg, W., Quail, J.W. & Delbaere LT., (1994) The 1.6 A Structure of Histidine-Containing Phosphotransfer Protein HPr From *Streptococcus faecalis*. *J. Mol. Biol.* **236**, 1341-1355

[75] Chook, Y.M., Ke, H. & Lipscomb, W.N., (1993) Crystal Structures of the Monofunctional Chorismate Mutase from *Bacillus subtilis* and its Complex with a Transition State Analog. *Proc. Nat. Acad. Sci. USA* **90**, 8600-8603

[76] Ramage, R., Green. J., Muir. T.W., Ogunjobi, O.M., Love, S., & Shaw. K., (1994) Synthetic, Structural and Biological Studies of the Ubiquitin System: The Total Chemical Synthesis of Ubiquitin. *Biochem. J.* **299**, 151-158

[77] Pastore, A., Saudek. V., Ramponi, G. & Williams, R.J.P., (1992) Three-dimensional structure of Acylphosphatase. Refinement and Structure Analysis *J. Mol. Biol.* **224**, 427-440

[78] Blevins, R.A. & Tulinsky, A., (1985) The Refinement and the Structure of the Dimer of Alpha-Chymotrypsin at 1.67 Angstroms Resolution. *J. Biol. Chem.* **260**, 4264-4275

[79] van den Akker, F., Sarfaty, S., Twiddy, E..M., Connell, T.D., Holmes, R.K., & Hol, W.G.J. (1996) Crystal Structure of a New Heat-Labile Enterotoxin, LT-Iib *Structure* **4**, 665-678

[80] Campos-Olivas, R., Bruix, M., Santoro, J., Lacadena, J., Martinez del Pozo, A., Gavilanes J.G. & Rico M., (1995) NMR Solution Structure of the Antifungal Protein from *Aspergillus giganteus*: Evidence for Cysteine Pairing Isomerism. *Biochemistry* **34**, 3009-3021

[81] Ban, N. & McPherson, A., (1995) The Structure of Satellite Panicum Mosaic Virus at 1.9Å Resolution. *Nat. Struct. Biol.* **10**, 882-890

[82] Prigge, S.T., Kolhekar, A.S., Eipper, B.A., Mains, R.E. & Amzel, L.M., (1997) Amidation of Bioactive Peptides: The Structure of Peptidylglycine Alpha-Hydroxylating Monooxygenase. *Science* **278**, 1300-1305

[83] Epp, O., Lattman, E.E., Schiffer, M., Huber, R. & Palm, W., (1975) The Molecular Structure of a Dimer Composed of the Variable Portions of the Bence-Jones Protein REI Refined at 2.0-A Resolution. *Biochemistry* **14**, 4943-4952

[84] Rypniewski, W.R., Mangani, S., Bruni, B., Orioli P.L., Casati, M. & Wilson, K.S., (1995) Crystal Structure of Reduced Bovine Erythrocyte Superoxide Dismutase at 1.9-A Resolution. *J. Mol. Biol.* **251**, 282-296

[85] Murzin, A.G., Lesk, A.M. & Chothia, C., (1992) Patterns of Structure and Sequence in the Kunitz Inhibitors, Interlukins and Fibroblast Growth Factors. *J. Mol. Biol.* **223**, 531-543

[86] Van Deutekom, J. C., Lemmers, R. J., Grewal, P. K., van Habazetti, J., Gondol, D., Wiltscheck, R., Otlewski, J., Schleicher, M. & Holak, T. A. (1992) Structure of Hisactophilin is Similar to Interleukin-1b and Fibroblast Growth Factor. *Nature* **359**, 855-858

[87] Swindells, M. B. & Thornton, J. M., (1993) A Study of Structural Determinants in the Interleukin-1 Fold. *Protein Eng.* **6**, 711-715

[88] Schindelin, H., Jiang, W. & Heinemann, U., (1994) Crystal Structure of CspA, the Major Cold Shock Protein of *Escherichia coli, Proc. Nat. Acad. Sci. USA* **91**, 5119-5123

[89] Thompson, D., Pepys, M.B. & Wood, S.P., (1999) The Physiological Structure of Human C-Reactive Protein and its Complex with Phosphocoline *Structure. Fold. Des.* **7**, 169-177

[90] Finzel, B.C., Clancy L.L., Holland, D.R., Muchmore, S.W.,Watenpaugh, K.D. & Einspahr, H.M., (1989) Crystal Structure of Recombinant Human Interleukin-1Beta at 2.0 Angstroms Resolution. *J. Mol. Biol.* **209**, 779-791

[91] Yoder, M.D. & Jurnak, F., (1994) Protein Motifs. 3. The Parallel Beta Helix and Other Coiled Folds. *FASEB J.* **5**, 335-42

[92] Alzari, P.M., Souchon, H. & Dominguez, R., (1996) The Crystal Structure of Endoglucanase CelA, a Family 8 Glycosyl Hydrolase from *Clostridium thermocellum. Structure* **15**, 265-75

[93] Smith, B.J., Colman, P.M.,Von Itzstein, M., Danylec, B. & Varghese, J.N., (2001) Analysis of Inhibitor Binding in Influenza Virus Neuraminidase *Protein Sci.* **10**, 689-696

[94] Hinman, M.B., Jones. J.A. & Lewis. R.V., (2000) Synthetic Spider Silk: A Modular Fiber. *Trends. Biotec.* **18**, 374-379

[95] Millar, A., (1982) Molecular Packing in Collagen Fibrils. *TIBS.* **7**, 13-18

[96] Talbot, J.A., & Hodges, R.S., (1982) A Model Protein for Studying coiled-Coil and α-Helix Stabilization. *Acc. Chem. Res.* **15**, 224-230

[97] Kramer, R.Z., Vitagliano, L., Bella, J., Berisio, R., Mazzarella, L., Brodsky, B., Zagari, A. & Berman, H.M., (1998) X-ray Crystallographic Determination of a Collagen-Like Peptide with the Repeating Sequence (Pro-Pro-Gly). *J. Mol. Biol.* **280**, 623-638

[98] Pebay-Peyroula, E., Rummel, G., Rosenbusch, J.P. & Landau, E.M., (1997) X-ray Structure of Bacteriorhodopsin at 2.5 Angstroms From Microcrystals Grown in Lipidic Cubic Phases *Science* **277**, 1676-1681

[99] Cowan, S.W., Schirmer, T., Rummel, G., Steiert, M., Ghosh, R., Pauptit, R.A., Jansonius, J.N. & Rosenbusch, J.P., (1992) Crystal Structures Explain Functional Properties of Two *E. coli* Porins. *Nature* **358**, 727-733

[100] Harrison, A., Pearl, F., Sillitoe, I., Slidel, T., Mott, R., Thornton, J. & Orengo, C., (2003) Recognizing The Fold of a Protein Structure *Bioinformatics* 19, 1748-1759

*Essays in Bioinformatics*
*D.S. Moss et al. (Eds.)*
*IOS Press, 2005*




# Macromolecular Structure Databases

Eric W. SAYERS and Stephen H. BRYANT
*National Center for Biotechnology Information, National Library of Medicine*
*National Institutes of Health, 8600 Rockville Pike, Bethesda, MD 20894 USA*

**Abstract** The resources provided by NCBI for studying the three-dimensional (3D) structures of proteins center around two databases: the Molecular Modeling Database (MMDB), which provides structural information about individual proteins; and the Conserved Domain Database (CDD), which provides a directory of sequence and structure alignments representing conserved functional domains within proteins (CDs). Together, these two databases allow scientists to retrieve and view structures, find structurally similar proteins to a protein of interest, and identify conserved functional sites. To enable scientists to accomplish these tasks, NCBI has integrated MMDB and CDD into the Entrez retrieval system. In addition, structures can be found by BLAST, because sequences derived from MMDB structures have been included in the BLAST databases. Once a protein structure has been identified, the domains within the protein, as well as domain "neighbors" (i.e., those with similar structure) can be found. For novel data not yet included in Entrez, there are separate search services available. Protein structures can be visualized using Cn3D, an interactive 3D graphic modeling tool. Details of the structure, such as ligand-binding sites, can be scrutinized and highlighted. Cn3D can also display multiple sequence alignments based on sequence and/or structural similarity among related sequences, 3D domains, or members of a CDD family. Cn3D images and alignments can be manipulated easily and exported to other applications for presentation or further analysis.

## 1. Overview

The Structure homepage[1] (Figure 1) contains links to the more specialized pages for each of the main tools and databases, introduced below, as well as search facilities for the Molecular Modeling Database (MMDB) [1]. MMDB[2] is based on the structures within the Protein Data Bank (PDB) and can be queried using the Entrez search engine, as well as via the more direct but less flexible Structure Summary search (see Figure 1). Once found, any structure of interest can be viewed using Cn3D[3] [2], a piece of software that can be freely downloaded for Mac, PC, and UNIX platforms.

Often used in conjunction with Cn3D is the Vector Alignment Search Tool (VAST) [3, 4]. VAST[4] is used to precompute "structure neighbors" or structures similar

[1] [http://www.ncbi.nlm.nih.gov/Structure]
[2] [http://www.ncbi.nlm.nih.gov/Structure/MMDB/mmdb.shtml]
[3] [http://www.ncbi.nlm.nih.gov/Structure/CN3D/cn3d.shtml]
[4] [http://www.ncbi.nlm.nih.gov/Structure/VAST/vast.shtml]

to each MMDB entry. For those who have a set of 3D coordinates for a protein not yet in MMDB, there is also a VAST search service[5].

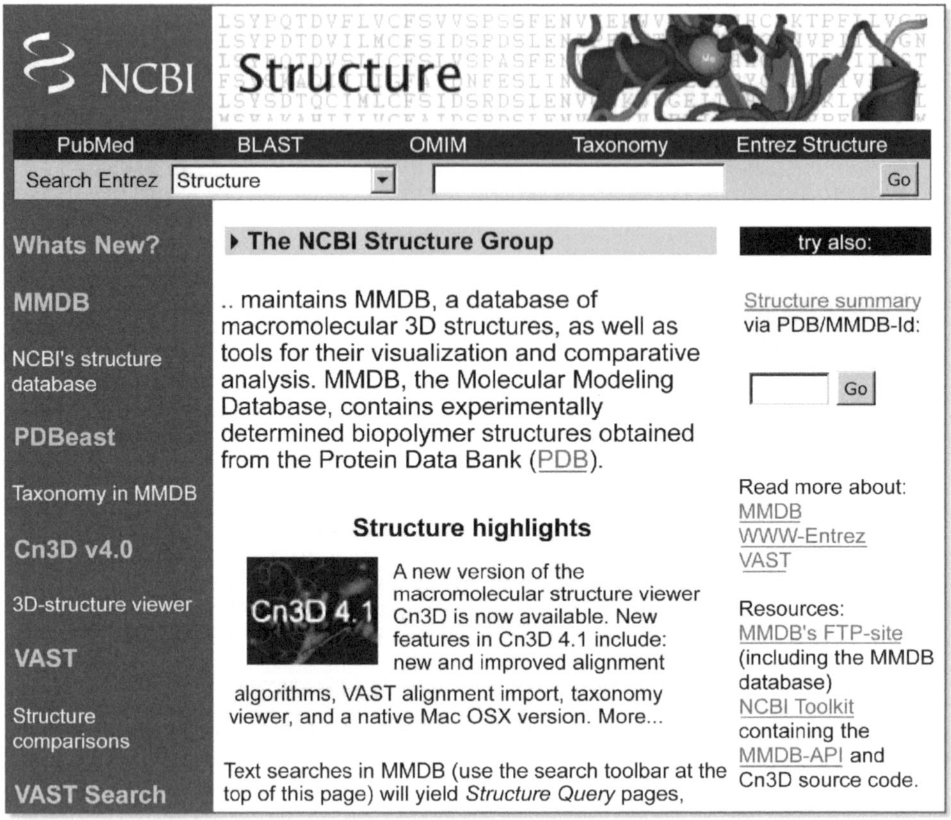


**Figure 1: The Structure homepage.** This page can be found by selecting the *Structure* link on the tool bar atop many NCBI Web pages. Two searches can be performed from this page, an *Entrez Structure* search or a *Structure Summary* search. Both query the MMDB database. The difference is that Entrez Structure can take any text as a query (such as a PDB code, protein name, text word, author, or journal) and will result initially in a list of one or more document summaries, displayed within the Entrez environment, whereas only a PDB code or MMDB ID number can be used for the Structure Summary search, resulting in direct display of the Structure Summary page for that record (Figure 2). Announcements about new features or updates can also be found on this page, as well as links to more specialized pages on the various Structure databases and tools.

The output of the precomputed VAST searches is a list of structure records, each representing one of the Non-Redundant PDB chain sets (nr-PDB)[6], which can also be downloaded. There are four clustered subsets of MMDB that compose nr-PDB, each consisting of clusters having a preset level of sequence similarity.

[5][http://www.ncbi.nlm.nih.gov/Structure/VAST/vast.shtml]

[6]http://www.ncbi.nlm.nih.gov/Structure/VAST/nrpdb.html

The structures within MMDB are also linked to the NCBI Taxonomy database. Known as the PDBeast project[7], this effort makes it possible to find the following: (1) all MMDB structures from a particular organism; and (2) all structures within a node of the taxonomy tree (such as lizards or *Bacillus)*, by launching the Taxonomy Browser showing the number of MMDB records in each node.

The second database within the structure resources is the Conserved Domain Database (CDD) [5], originally based largely on Pfam and SMART, collections of alignments that represent functional domains conserved across evolution. CDD now also contains the alignments of the NCBI COG database along with new curated alignments assembled at NCBI. CDD can be searched from the CDD page …….. in several ways, including by a domain keyword search[8]. Three tools have been developed to assist in analysis of CDD: (1) the CD-Search[9], which uses a BLAST-based algorithm to search the position-specific scoring matrices (PSSM) of CDD alignments; (2) the CD-Browser, which provides a graphic display of domains of interest, along with the sequence alignment; and (3) the Conserved Domain Architecture Retrieval Tool CDART which searches for proteins with similar domain architectures.

All the above databases and tools are discussed in more detail in other parts of this document, including tips on how to make the best use of them.

## 2. Content of the Molecular Modeling Database (MMDB)

### *2.1 Sources of Primary Data*

To build MMDB [1], 3D structure data are retrieved from the PDB database [6] administered by the Research Collaboratory for Structural Bioinformatics (RCSB). In all cases, the structures in MMDB have been determined by experimental methods, primarily X-ray crystallography and Nuclear Magnetic Resonance (NMR) spectroscopy. Theoretical structure models are omitted. The data in each record are then checked for agreement between the atomic coordinates and the primary sequence, and the sequence data are then extracted from the coordinate set. The resulting association between sequence and structure allows the record to be linked efficiently into searches and alignment displays involving other NCBI databases.

The data are converted into ASN.1 [7], which can be parsed easily and can also accept numerous annotations to the structure data. In contrast to a PDB record, a MMDB record in ASN.1 contains all necessary bonding information in addition to sequence information, allowing consistent display of the 3D structure using Cn3D. The annotations provided in the PDB record by the submitting authors are added, along with uniformly defined secondary structure and domain features. These features support structure-based similarity searches using VAST. Finally, two coordinate subsets are added to the record: one containing only backbone atoms, and one representing a single-conformer model in cases where multiple conformations or structures were present in the PDB record. Both of these additions further simplify viewing both an individual structure and its alignments

[7] [http://www.ncbi.nlm.nih.gov/Structure/PDBEAST/pdbeast.shtml]

[8] [http://www.ncbi.nlm.nih.gov/Structure/cdd/cdd.shtml]

[9] [http://www.ncbi.nlm.nih.gov/Structure/cdd/wrpsb.cgi]

with structure neighbors in Cn3D. When this process is complete, the record is assigned a unique Accession number, the MMDB-ID (Appendix 1), while also retaining the original four-character PDB code.

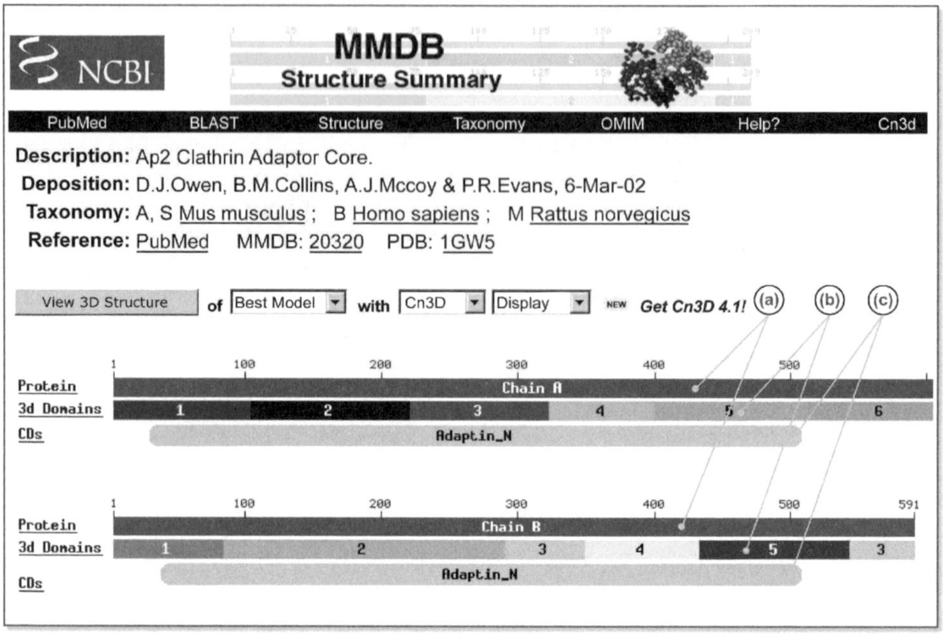


**Figure 2: The Structure Summary page.** The page consists of three parts: the header, the view bar, and the graphic display. The header contains basic identifying information about the record: a description of the protein (*Description:*), the author list (*Deposition:*), the species of origin (*Taxonomy:*), literature references (*Reference:*), the MMDB-ID (*MMDB:*), and the PDB code (*PDB:*). Several of these data serve as links to additional information. For example, the species name links to the Taxonomy browser, the literature references link to PubMed, and the PDB code links to the PDB Web site. The view bar allows the user to view the structure record either as a graphic with Cn3D or as a text record in either ASN.1, PDB (RasMol), or Mage formats. The latter can also be downloaded directly from this page. The graphic display contains a variety of information and links to related databases: (*a*) The Chain bar. Each chain of the molecule is displayed as a *dark bar* labeled with residue numbers. To the *left* of this bar is a *Protein* hyperlink that takes the user to a view of the protein record in Entrez Protein. The bar itself is also a hyperlink and displays the VAST neighbors of the chain. If a structure contains nucleotide sequences, they are displayed in the order contained in the PDB record. A *Nucleotide* hyperlink to their *left* takes the user to the appropriate record in Entrez Nucleotide. (*b*) The VAST (3D) Domain bar. The *colored bars* immediately below the Chain bar indicate the locations of structural domains found by the original MMDB processing of the protein. In many cases, such a domain contains unconnected sections of the protein sequence, and in such cases, discontinuous pieces making up the domain will have bars of the same color. To the *left* of the Domain bar is a 3D Domains hyperlink (*3d Domains*) that launches the 3D Domains browser in Entrez, where the user can find information about each constituent domain. Selecting a colored segment displays the VAST Structure Neighbors page for that domain. (*c*) The CD bar. Below the VAST Domain bar are *rounded, rectangular bars* representing conserved domains found by a CD-Search. The bars identify the best scoring hits; overlapping hits are shown only if the mutual overlap with hits having better scores is less than 50%. The *CDs* hyperlink to the *left* of the bar displays the CD records in Entrez Domains. Each of the colored bars is also a hyperlink that displays the corresponding CD Summary page configured to show the multiple alignment of the protein sequence with members of the selected CD.

### 2.2 Annotation of 3D Domains

After initial processing, 3D domains are automatically identified within each MMDB record. 3D domains are annotations on individual MMDB structures that define the boundaries of compact substructures contained within them. In this way, they are similar to secondary structure annotations that define the boundaries of helical or β-strand substructures. Because proteins are often similar at the level of domains, VAST compares each 3D domain to every other one and to complete polypeptide chains. The results are stored in Entrez as a 3D Domain Neighbors link.

To identify 3D domains within a polypeptide chain, MMDB's domain parser searches for one or more breakpoints in the structure. These breakpoints fall between major secondary structure elements such that the ratio of intra- to interdomain contacts remains above a set threshold. The 3D domains identified in this way provide a means to both increase the sensitivity of structure neighbor calculations and also present 3D superpositions based on compact domains as well as on complete polypeptide chains. They are not intended to represent domains identified by comparative sequence and structure analysis, nor do they represent modules that recur in related proteins, although there is often good agreement between domain boundaries identified by these methods.

### 2.3 Links to Other NCBI Resources

After initially processing the PDB record, structure staff add a number of links and other information that further integrate the MMDB record with other NCBI resources.

To begin, the sequence information extracted from the PDB record is entered into the Entrez Protein and/or Nucleotide databases as appropriate, providing a means to retrieve the structure information from sequence searches. As with all sequences in Entrez, precomputed BLAST searches are then performed on these sequences, linking them to other molecules of similar sequence. For proteins, these BLAST neighbors may be different than those determined by VAST; whereas VAST uses a conservative significance threshold, the structural similarities it detects often represent remote relationships not detectable by sequence comparison. The literature citations in the PDB record are linked to PubMed so that Entrez searches can allow access to the original descriptions of the structure determinations. Finally, semiautomatic processing of the "source" field of the PDB record provides links to the NCBI Taxonomy database. Although these links normally follow the genus and species information given, in some cases this information is either absent in the PDB record or refers only to how a sample was obtained. In these cases, the staff manually enters the appropriate taxonomy links.

### 2.4 The MMDB Record

The Structure Summary page for each MMDB record summarizes the database content for that record and serves as a starting point for analyzing the record using the NCBI structure tools (Figure 2).

### *2.5 VAST Structure Neighbors*

Although VAST itself is not a database, the VAST results computed for each MMDB record are stored with this record and are summarized on a separate page for the whole polypeptide chain as well as for each 3D domain found in the protein (Figure 3). These pages can be accessed most easily by clicking on either the chain bar or the 3D Domain bar in the graphic display of the Structure Summary page (Figure 2).

### *2.6 nr-PDB*

The non-redundant PDB database (nr-PDB) is a collection of four sets of sequence-dissimilar cluster PDB polypeptide chains assembled by NCBI Structure staff. The four sets differ only in their respective levels of non-redundancy. The staff assembles each set by comparing all the chains available from PDB with each other using the BLAST algorithm. The chains are then clustered into groups of similar sequence using a single-linkage clustering procedure. Chains within a sequence-similar group are automatically ranked according to the quality of their structural data (nr-PDB[10]).

## 3. Content of the Conserved Domain Database (CDD)

### *3.1 What Is a Conserved Domain (CD)?*

CDs are recurring units in polypeptide chains (sequence and structure motifs), the extents of which can be determined by comparative analysis. Molecular evolution uses such domains as building blocks and these may be recombined in different arrangements to make different proteins with different functions. The CDD contains sequence alignments that define the features that are conserved within each domain family. Therefore, the CDD serves as a classification resource that groups proteins based on the presence of these predefined domains. CDD entries often name the domain family and describe the role of conserved residues in binding or catalysis. Conserved domains are displayed in MMDB Structure summaries and link to a sequence alignment showing other proteins in which the domain is conserved, which may provide clues about protein function.

### *3.2 Sources of Primary Data*

The collections of domain alignments in the CDD are imported either from two databases outside of the NCBI, named Pfam [8] and SMART [9]; from the NCBI COG database; or from a database curated by the CDD staff. The first task is to identify the underlying sequences in each collection and then link these sequences to the corresponding ones in

[10] [http://www.ncbi.nlm.nih.gov/Structure/VAST/nrpdb.html]

Entrez. If the CDD staff cannot find the Accession numbers for the sequences in the records from the source databases, they locate appropriate sequences using BLAST. Particular attention is paid to any resulting match that is linked to a structure record in MMDB, and the staff substitute alignment rows with such sequences whenever possible.

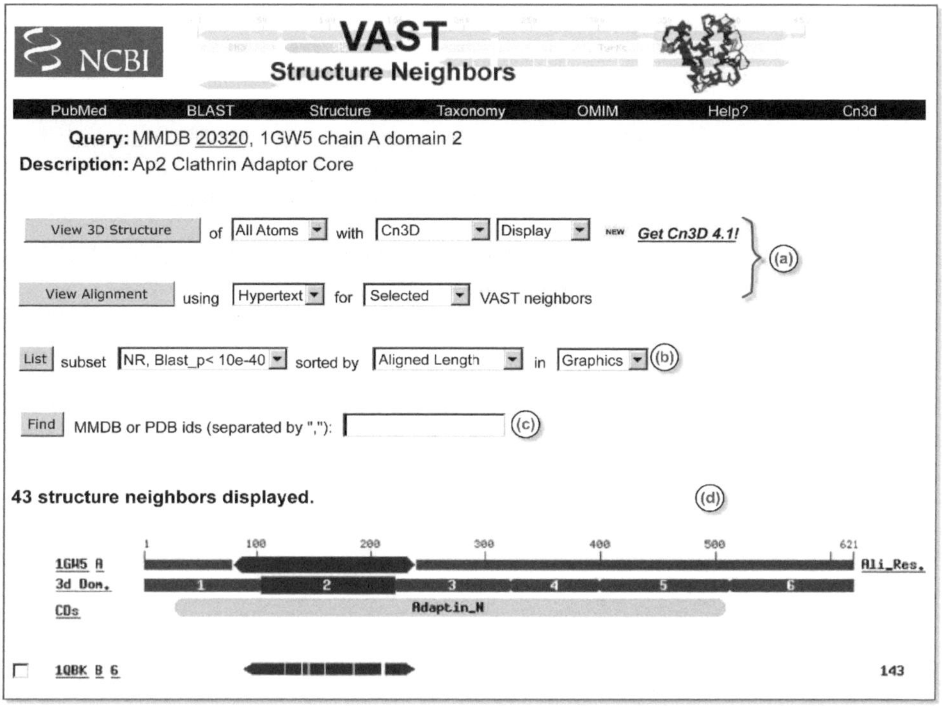


**Figure 3: VAST Structure Neighbors page.** The *top portion* of the page contains identifying information about the 3D Domain, along with three functional bars. (*a*) The View bar. This bar allows a user to view a selected alignment either as a graphic using Cn3D or as a sequence alignment in HTML, text, or mFASTA format. The user may select which chains to display in the alignment by checking the boxes that appear to the *left* of each neighbor in the *lower portion* of the page. (*b*) The nr-PDB bar. This bar allows a user to either display all matching records in MMDB or to limit the displayed domains to only representatives of the selected nr-PDB set. The user may also select how the matching domains are sorted in the display and whether the results are shown as graphics or as tabulated data. (*c*) The Find bar. This bar allows the user to find specific structural neighbors by entering their PDB or MMDB identifiers. (*d*) The *lower portion* of the page displays a graphical alignment of the various matching domains. The *upper three bars* show summary information about the query sequence: the *top bar* shows the maximum extent of alignment found on all the sequences displayed on the current page (users should note that the appearance of this bar, therefore, depends on which hits are displayed); the *middle bar* represents the query sequence itself that served as input for the VAST search; and the *lower bar* shows any matching CDs and is identical to the CD bar on the Structure Summary page. Listed below these three summary bars are the hits from the VAST search, sorted according to the selection in the nr-PDB bar. The bars represent aligned regions, with gaps indicating unaligned regions. To the *left* of each domain accession is a check box that can be used to select any combination of domains to be displayed either on this page or using Cn3D. Moreover, each of the bars in the display is itself a link, and placing the mouse pointer over any bar reveals both the extent of the alignment by residue number and the data linked to the bar.

After the staff imports a collection, they then choose a sequence that best represents the family. Whenever possible, the staff chooses a representative that has a structure record in MMDB.

### *3.3 The Position-specific Score Matrix (PSSM)*

Once imported and constructed, each domain alignment in CDD is used to calculate a model sequence, called a consensus sequence, for each CD. The consensus sequence lists the most frequently found residue in each position in the alignment; however, for a sequence position to be included in the consensus sequence, it must be present in at least 50% of the aligned sequences. Aligned columns covered by the consensus sequence are then used to calculate a PSSM, which memorizes the degree to which particular residues are conserved at each position in the sequence. Once calculated, the PSSM is stored with the alignment and becomes part of the CDD. The RPS-BLAST tool locates CDs within a query sequence by searching against this database of PSSMs.

### *3.4 Reverse Position-specific BLAST (RPS-BLAST*

RPS-BLAST is a variant of the popular Position-specific Iterated BLAST (PSI-BLAST) program. PSI-BLAST finds sequences similar to the query and uses the resulting alignments to build a PSSM for the query. With this PSSM the database is scanned again to draw in more hits and further refine the scoring model. RPS-BLAST uses a query sequence to search a database of precalculated PSSMs and report significant hits in a single pass. The role of the PSSM has changed from "query" to "subject"; hence, the term "reverse" in RPS-BLAST. RPS-BLAST is the search tool used in the CD-Search service.

### *3.5 The CD Summary*

Analogous to the Structure Summary page, the CD Summary page displays the available information about a given CD and offers various links for either viewing the CD alignment or initiating further searches (Figure 4). The CD Summary page can be retrieved by selecting the CD name on any page.

### *3.6 CD Records Curated at NCBI*

In 2002, NCBI released the first group of curated CD records, a new and expanding set of annotated protein multiple sequence alignments and corresponding structure alignments. These new records have Accession numbers beginning with "cd" and have been added to the default CD-Search database. Most curated CD records are based on existing family descriptions from SMART and Pfam, but the alignments may have been revised extensively by quantitatively using three-dimensional structures and by re-examining the

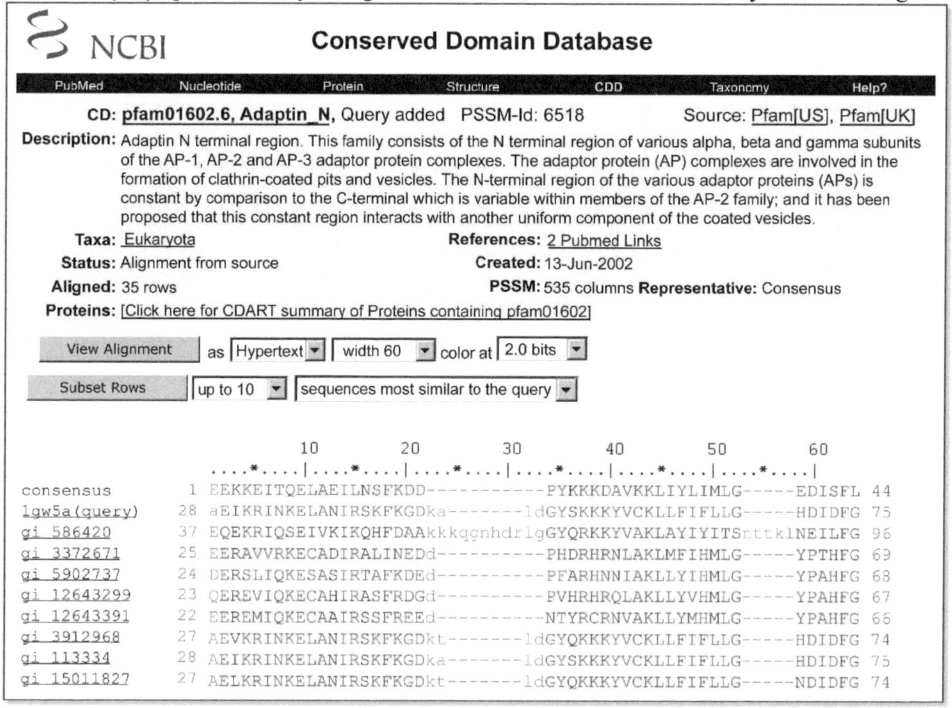


**Figure 4: CD summary page.** The *top* of the page serves as a header and reports a variety of identifying information, including the name and description of the CD, other related CDs with links to their summary pages, as well as the source database, status, and creation date of the CD. A taxonomic node link (*Taxa:*) launches the Taxonomy Browser, whereas a Proteins link (*Proteins:*) uses CDART to show other proteins that contain the CD. *Below* the header is the interface for viewing the CD alignment, which can be done either graphically with Cn3D (if the CD contains a sequence with structural data) or in HTML, text, or mFASTA format. It is also possible to view a selected number of the top-listed sequences, sequences from the most diverse members, or sequences most similar to the query. In addition, users may now select sequences with the NCBI Taxonomy Common Tree tool. The *lower portion* of the page contains the alignment itself. Members with a structural record in MMDB are listed first, and the identifier of each sequence links to the corresponding record.domain extent. In addition, CDD curators annotate conserved functional residues,

ligands, and co-factors contained within the structures. They also record evidence for these sites as pointers to relevant literature or to three-dimensional structures exemplifying their properties. These annotations may be viewed using Cn3D and thus provide a direct way of visualizing functional properties of a protein domain in the context of its three-dimensional structure. (See Appendix 3 and Figure 7.)

### *3.7 The Distinction between 3D Domains and CDs*

The term "domain" refers in general to a distinct functional and/or structural unit of a protein. Each polypeptide chain in MMDB is analyzed for the presence of two classes of domains, and it is important for users to understand the difference between them. One class, called 3D Domains, is based solely on similar, compact substructures, whereas the second class, called Conserved Domains (CDs), is based solely on conserved sequence motifs. These two classifications often agree, because the compact substructures within a protein often correspond to domains joined by recombination in the evolutionary history of a protein. Note that CD links can be identified even when no 3D structures within a family are known. Moreover, 3D Domain links may also indicate relationships either to structures not included in CDD entries or to structures so distantly related that no significant similarity can be found by sequence comparisons.

## 4. Finding and Viewing Structures

For an example query on finding and viewing structures, see Appendix 2.

### *4.1 Why Would I Want to Do This?*

- To determine the overall shape and size of a protein
- To locate a residue of interest in the overall structure
- To locate residues in close proximity to a residue of interest
- To develop or test chemical hypotheses regarding an enzyme mechanism
- To locate or predict possible binding sites of a ligand
- To interpret mutation studies
- To find areas of positive or negative charge on the protein surface
- To locate particularly hydrophobic or hydrophilic regions of a protein
- To infer the 3D structure and related properties of a protein with unknown structure from the structure of a homologous protein
- To study evolutionary processes at the level of molecular structure
- To study the function of a protein
- To study the molecular basis of disease and design novel treatments

### *4.2 How to Begin*

The first step to any structural analysis at NCBI is to find the structure records for the protein of interest or for proteins similar to it. One may search MMDB directly by entering search terms such as PDB code, protein name, author, or journal in the Entrez Structure *Search* box on the Structure homepage[11]. Alternative points of entry are shown below.

---

[11] [http://www.ncbi.nlm.nih.gov/Structure]

By using the full array of Entrez search tools, the resulting list of MMDB records can be honed, ideally, to a workable list from which a record can be selected. Users should note that multiple records may exist for a given protein, reflecting different experimental techniques, conditions, and the presence or absence of various ligands or metal ions. Records may also contain different fragments of the full-length molecule. In addition, many structures of mutant proteins are also available. The PDB record for a given structure generally contains some description of the experimental conditions under which the structure was determined, and this file can be accessed by selecting the PDB code link at the top of the Structure Summary page.

### *4.3 Alternative Points of Entry*

Structure Summary pages can also be found from the following NCBI databases and tools:

- Select the Structure *links* to the right of any Entrez record found; records with Structure links can also be located by choosing *Structure links* from the *Display* pull-down menu.
- Select the *Related Sequences* link to the right of an Entrez record to find proteins related by sequence similarity, and then select *Structure links* in the *Display* pull-down menu.
- Choose the PDB database from a blastp (protein-protein BLAST) search; only sequences with structure records will be retrieved by BLAST. The *Related Structures* link provides 3D views in Cn3D.
- Select the *3D Structures* button on any BLink report to show those BLAST hits for which structural data are available.
- From the results of any protein BLAST search, click on a red 'S' linkout to view the sequence alignment with a structure record.

### *4.4 Viewing 3D Structures*

*3D Domains.* The 3D domains of a protein are displayed on the Structure Summary page. It is useful to know how many 3D domains a protein contains and whether they are continuous in sequence when viewing the full 3D structure of the molecule.

*Secondary Structure.* Knowing the secondary structure of a protein can also be a useful prelude to viewing the 3D structure of the molecule. The secondary structure can be viewed easily by first selecting the *Protein* link to the left of the desired chain in the graphic display. Finding oneself in Entrez Protein, selecting *Graphics* in the Display pull-down menu presents secondary structure diagrams for the molecule.

*Full Protein Structures.* Cn3D is a software package for displaying 3D structures of proteins. Once it has been installed[12] and the Internet browser has been configured correctly, simply selecting the *View 3D Structure* button on a Structure Summary page launches the application. Once the structure is loaded, a user can manipulate and

---

[12] [http://www.ncbi.nlm.nih.gov/Structure/CN3D/cn3dinstall.shtml]

annotate it using an array of options as described in the Cn3D Tutorial[13]. By default, Cn3D colors the structure according to the secondary structure elements. However, another useful view is to color the protein by domain (see *Style* menu options), using the same color scheme as is shown in the graphic display on the Structure Summary page. These color changes also affect the residues displayed in the Sequence/Alignment Viewer, allowing the identification of domain or secondary structure elements in the primary sequence. In addition to Cn3D, users can also display 3D structures with RasMol or Mage. Structures can also be saved locally as an ASN.1, PDB, or Mage file (depending on the choice of structure viewer) for later display.

## 5. Finding and Viewing Structure Neighbors

For an example query on finding and viewing structure neighbors, see Appendix 2

### *5.1 Why Would I Want to Do This?*

- To determine structurally conserved regions in a protein family
- To locate the structural equivalent of a residue of interest in another related protein
- To gain insights into the allowable structural variability in a particular protein family
- To develop or test chemical hypotheses regarding an enzyme mechanism
- To predict possible binding sites of a ligand from the location of a binding site in a related protein
- To identify sites where conformational changes are concentrated
- To interpret mutation studies
- To find areas of conserved positive or negative charge on the protein surface
- To locate conserved hydrophobic or hydrophilic regions of a protein
- To identify evolutionary relationships across protein families
- To identify functionally equivalent proteins with little or no sequence conservation.

### *5.2 How to Begin*

The Vector Alignment Search Tool (VAST) is used to calculate similar structures on each protein contained in the MMDB. The graphic display on each Structure Summary page (Figure 2) links directly to the relevant VAST results for both whole proteins and 3D domains:

- The 3D Domains link transfers the user to Entrez 3D Domains, showing a list of the VAST neighbors.

---

[13] [http://www.ncbi.nlm.nih.gov/Structure/CN3D/cn3dtut.shtml]

- Selecting the chain bar displays the VAST Structure Neighbors page for the entire chain.
- Selecting a 3D Domain bar displays the VAST Structure Neighbors page for the selected domain.

### *5.3 Alternative Points of Entry*

- From any Entrez search, select **Related 3D Domains** to the right of any record found to view the Vast Structure Neighbors page.
- *5.4 Viewing a 2D Alignment of Structure Neighbors*
- A graphic 2D HTML alignment of VAST neighbors can be viewed as follows:
- On the lower portion of the VAST Structure Neighbors page (Figure 3), select the desired neighbors to view by checking the boxes to their left.
- On the *View/Save* bar, configure the pull-down menus to the right of the *View Alignment* button.
- Select *View Alignment.*

### *5.5 Viewing a 3D Alignment of Structure Neighbors*

Alignments of VAST structure neighbors can be viewed as a 3D image using Cn3D.

- On the lower portion of the VAST Structure Neighbors page (Figure 3), select the desired neighbors to view by checking the boxes to their left.
- On the *View/Save* bar, configure the pull-down menus to the right of the *View 3D Structure* button.
- Select *View 3D Structure.*

Cn3D automatically launches and displays the aligned structures. Each displayed chain has a unique color; however, the portions of the structures involved in the alignment are shown in red. These same colors are also reflected in the Sequence/Alignment Viewer. Among the many viewing options provided by Cn3D, of particular use is the *Show/Hide* menu that allows only the aligned residues to be viewed, only the aligned domains, or all residues of each chain.

## 6. Finding and Viewing Conserved Domains

For an example query on finding and viewing conserved domains, see Appendix 3.

### *6.1 Why Would I Want to Do This?*

- To locate functional domains within a protein
- To predict the function of a protein whose function is unknown
- To establish evolutionary relationships across protein families
- To interpret mutation studies

- To predict the structure of a protein of unknown structure

### *6.2 How to Begin*

Following the Domains link for any protein in Entrez, one can find the conserved domains within that protein. The CD-Search[14] (or Protein BLAST, with CD-Search option selected) can be used to find conserved domains (CDs) within a protein. Either the Accession number, gi number, or the FASTA sequence can be used as a query.

### *6.3 Alternative Points of Entry*

Information on the CDs contained within a protein can also be found from these databases and tools:

- From any Entrez search: select the *Domains* link to the right of a displayed record.
- From the Structure Summary page of a MMDB record: this page displays the CDs within each protein chain immediately below the 3D Domain bar in the graphic display. Selecting the *CDs* link shows the CD-Search results page.
- From an Entrez Domains search: choose *Domains* from the Entrez *Search* pull-down menu and enter a search term to retrieve a list of CDs. Clicking on any resulting CD displays the CD Summary page. The location of this CD in each aligned protein is indicated by green numerals in the alignment in the bottom portion of this page.
- From the CDD page: locate CDs by entering text terms into the search box and proceed as for an Entrez CD search.
- From a BLink report: select the *CDD-Search* button to display the CD-Search results page.
- From the BLAST main page: follow the RPS-BLAST link to load the CD-Search page.

### *6.4 Viewing Conserved Domains*

Results from a CD search are displayed as colored bars underneath a sequence ruler. Moving the mouse over these bars reveals the identity of each domain; domains are also listed in a format similar to BLAST summary output. Pairwise alignments between the matched region of the target protein and the representative sequence of each domain are shown below the bar. Red letters indicate columns with sequence conservation scores higher than the bits setting in the View Alignment controls, whereas blue letters indicate residues with conservation scores less than the bit setting.

---

[14] [http://www.ncbi.nlm.nih.gov/Structure/cdd/wrpsb.cgi]

### *6.5 Viewing Multiple Alignments of a Query Protein with Members of a Conserved Domain*

These can be displayed by clicking a CD bar within a MMDB Structure Summary page or from a hyperlinked CD name on a CD-Search results page.

### *6.6 Viewing CD Alignments in the Context of 3D Structure*

If members of a CD have MMDB records, one of these records can be viewed as a 3D image along with the sequence alignment using Cn3D (launched by selecting the pink dot on a CD-Search results page). As in other alignment views, colored capital letters indicate aligned residues, allowing the sequence of the protein sequence of interest to be mapped onto the available 3D structure.

## 7. Finding and Viewing Proteins with Similar Domain Architectures

For an example query on finding and viewing proteins with similar domain architectures, see Appendix 3.

### *7.1 Why Would I Want to Do This?*

- To locate related functional domains in other protein families
- To gain insights into how a given CD is situated within a protein relative to other CDs
- To explore functional links between different CDs
- To predict the function of a protein whose function is unknown
- To establish evolutionary relationships across protein families

### *7.2 How to Begin*

Following the *Domain Relatives* link for any protein in Entrez, one can find other proteins with similar domain architecture. The Conserved Domain Architecture Retrieval Tool (CDART[15]) can take an Accession number or the FASTA sequence as a query to find out the domain architecture of a protein sequence and list other proteins with related domain architectures.

### *7.3 Alternative Points of Entry*

- From a CD-Search results page, click *Show Domain Relatives*

---

[15] [http://www.ncbi.nlm.nih.gov/Structure/lexington/lexington.cgi?cmd=rps]

- From a CD-Summary page, click the *Proteins* link
- From an Entrez Domains search, click the *Proteins* link in the Links menu

### *7.4 Results of a CDART Search*

These are described in Figure 5. The protein "hits", which have similar domain architectures to the query sequence, can be further refined by taxonomic group, in which the results can be limited to selected nodes of the taxonomic tree. Furthermore, search results may be limited to those that contain only particular conserved domains.

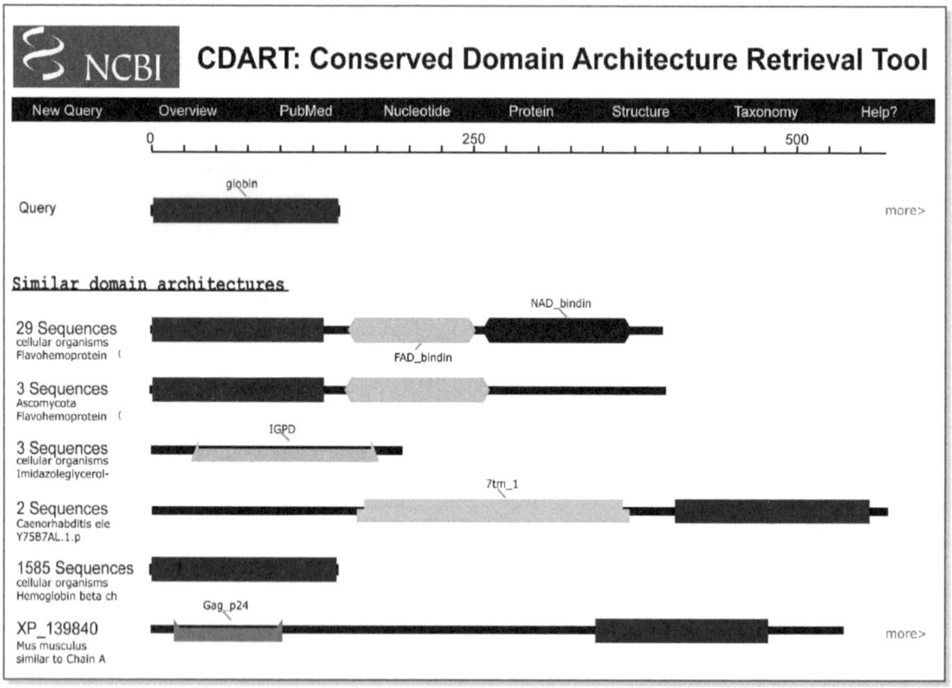


**Figure 5: A CDART results page.** At the *top* of the CDART results page, the query sequence CDs are represented as "beads on a string". Each CD had a unique color and shape and is labeled both in the display itself and in a legend located at the *bottom* of the page. The shapes representing CDs are hyperlinked to the corresponding CD summary page. The matching proteins to the query are listed *below* the query, ranked according to the number of non-redundant hits to the domains in the query sequence. Each match is either a single protein, in which case its Accession number is shown, or is a cluster of very similar proteins, in which case the number of members in the cluster is shown. Cluster members can be displayed by selecting the logo to the *left* of its diagram. Selecting any protein Accession number displays the flat file for that protein. To the *right* of any drawing for a single protein (either on the main results page or after expanding a protein cluster) is a *more>* link, which displays the CD-Search results page for the selected protein so that the sequence alignment, e.g., of a CDART hit with a CD contained in the original protein of interest, can be examined.

## 8. Links Between Structure and Other Resources

### *8.1 Integration with Other NCBI Resources*

As illustrated in the sections above, there are numerous connections between the Structure resources and other databases and tools available at the NCBI. What follows is a listing of major tools that support connections.

*Entrez.* Because Entrez is an integrated database system, the links attached to each structure give immediate access to PubMed, Protein, Nucleotide, 3D Domain, CDD, or Taxonomy records.

*BLAST.* Although the BLAST service is designed to find matches based solely on sequence, the sequences of Structure records are included in the BLAST databases, and by selecting the PDB search database, BLAST searches only the protein sequences provided by MMDB records. A new *Related Structure* link provides 3D views for sequences with structure data identified in a BLAST search.

*BLink.* The BLink report represents a precomputed list of similar proteins for all sequences in Entrez Protein. The *3D Structures* option on any BLink report shows the BLAST hits that have 3D structure data in MMDB, whereas the *CDD-Search* button displays the CD-Search results page for the query protein.

*Microbial Genomes.* A particularly useful interface with the structural databases is provided on the Microbial Genomes page[16] [10]. To the left of the list of genomes are several hyperlinks, two of which offer users direct access to structural information. The red *[D]* link displays a listing of every protein in the genome, each with a link to a BLink page showing the results of a BLAST pdb search for that protein. The *[S]* link displays a similar protein list for the selected genome, but now with a listing of the conserved domains found in each protein by a CD-Search.

### *8.2 Links to Non-NCBI Resources*

*The Protein Data Bank (PDB).* As stated elsewhere, all records in the MMDB are obtained originally from the Protein Data Bank (PDB) [6]. Links to the original PDB records are located on the Structure Summary page of each MMDB record. Updates of the MMDB with new PDB records occur once a month.

*Pfam and SMART.* The CDD staff imports CD collections from both the Pfam and SMART databases. Links to the original records in these databases are located on the appropriate CD Summary page. Both Pfam and SMART are updated several times per year in roughly bimonthly intervals, and the CDD staff update CDD accordingly.

---

[16] [http://www.ncbi.nlm.nih.gov/PMGifs/Genomes/micr.html]

## 9. Saving Output from Database Searches

### *9.1 Exporting Graphics Files from Cn3D*

Structures displayed in Cn3D can be exported as a Portable Network Graphics (PNG) file from within Cn3D (the Export PNG command in the *File* menu). The structure file itself, including all annotations and aligned sequences and structures then present in Cn3D, can also be saved for later launching in Cn3D.

### *9.2 Saving Individual MMDB Records*

Individual MMDB records can be downloaded to a local computer directly from the Structure Summary page for that record. *Save File* in the *View* bar downloads the file in a choice of three formats: ASN.1 (select *Cn3D*); PDB (select *RasMol*); or Mage (select *Mage*).

### *9.3 Saving VAST Alignments*

Alignments of VAST neighbors can be downloaded from the VAST Structure Neighbors page of any MMDB record. By selecting options in the View Alignment row, the alignment data can be formatted as HTML, text, or mFASTA, and then saved. By selecting "save file" from the View 3D Structure row, the full ASN.1 alignment file can be downloaded, including rotational matrices for producing the VAST alignment from the original PDB files.

## 10. FTP

### *10.1 MMDB*

Users can download the NCBI Structure databases from the NCBI FTP site: ftp://ftp.ncbi.nih.gov/mmdb. A Readme file contains descriptions of the contents and information about recent updates. Within the mmdb directory are four subdirectories that contain the following data:

- mmdbdata: the current MMDB database (NOTE: these files can not be read directly by Cn3D)
- vastdata: the current set of VAST neighbor annotations to MMDB records
- nrtable: the current non-redundant PDB database
- pdbeast: table listing the taxonomic classification of MMDB records

### *10.2 CDD*

CDD data can be downloaded from ftp://ftp.ncbi.nih.gov/pub/mmdb/cdd. A Readme file contains descriptions of the data archives. Users can download the PSSMs for each CD record, the sequence alignments in mFASTA format, or a text file containing the accessions and descriptions of all CD records.

## 11. Frequently Asked Questions

- Cn3D [http://www.ncbi.nih.gov/Structure/CN3D/cn3dfaq.shtml]
- VAST searches [http://www.ncbi.nih.gov/Structure/VAST/vastsearch_faq.html]
- CDD [http://www.ncbi.nlm.nih.gov/Structure/cdd/cdd_help.shtml]

## References

[1] Wang Y, Anderson JB, Chen J, Geer LY, He S, Hurwitz DI, Liebert CA, Madej T, Marchler GH, Marchler- Bauer A, et al. MMDB: Entrez's 3D-structure database. Nucleic Acids Res 30:249–252; 2002. (PubMed)

[2] Wang Y, Geer LY, Chappey C, Kans JA, Bryant SH. Cn3D: sequence and structure views for Entrez.

[3] Madej T, Gibrat J-F, Bryant SH. Threading a database of protein cores. Proteins 23:356–369; 1995. (PubMed)

[4] Gibrat J-F, Madej T, Bryant SH. Surprising similarities in structure comparison. Curr Opin Struct Biol 6:377–385; 1996. (PubMed)

[5] Marchler-Bauer A, Panchenko AR, Shoemaker BA, Thiessen PA, Geer LY, Bryant SH. CDD: a databaseof conserved domain alignments with links to domain three-dimensional structure. Nucleic Acids Res 30:281–283; 2002. (PubMed)

[6] Westbrook J, Feng Z, Jain S, Bhat TN, Thanki N, Ravichandran V, Gilliland GL, Bluhm W, Weissig H,Greer DS, et al. The Protein Data Bank: unifying the archive. Nucleic Acids Res 30:245–248; 2002. (PubMed) [7. Ohkawa H, Ostell J, Bryant S. MMDB: an ASN.1 specification for macromolecular structure. Proc Int Conf Intell Syst Mol Biol 3:259–267; 1995. (PubMed)

[8] Bateman A, Birney E, Cerruti L, Durbin R, Etwiller L, Eddy SR, Griffiths-Jones S, Howe KL, Marshall M, Sonnhammer ELL. The Pfam proteins family database. Nucleic Acids Res 30:276–280; 2002. (PubMed)

[9] Letunic I, Goodstadt L, Dickens NJ, Doerks T, Schultz J, Mott R, Ciccarelli F, Copley RR, Ponting CP,Bork P. SMART: a Web-based tool for the study of genetically mobile domains. Recent improvements to the SMART domain-based sequence annotation resource. Nucleic Acids Res 30:242–244; 2002. (PubMed)

[10] Wang Y, Bryant S, Tatusov R, Tatusova T. Links from genome proteins to known 3D structures. Genome Res 10:1643–1647; 2000. (PubMed)

## Appendix 1: Accession numbers

MMDB records have several types of Accession numbers associated with them, representing the following data types:

- Each MMDB record has at least three Accession numbers: the PDB code of the corresponding PDB record (e.g., 1CYO, 1B8G); a unique MMDB-ID (e.g., 645, 12342); and a gi number for each protein chain. A new MMDB-ID is assigned whenever PDB updates either the sequence or coordinates of a structure record, even if the PDB code is retained.
- If an MMDB record contains more than one polypeptide or nucleotide chain, each chain in the MMDB record is assigned an Accession number in Entrez Protein or

Nucleotide consisting of the PDB code followed by the letter designating that chain (e.g., 1B8GA, 3TATB, 1MUHB).

- Each 3D Domain identified in an MMDB record is assigned a unique integer identifier that is appended to the Accession number of the chain to which it belongs (e.g., 1B8G A 2). This new Accession number becomes its identifier in Entrez 3D Domains. New 3D Domain identifiers are assigned whenever a new MMDB-ID is assigned.
- For conserved domains, the Accession number is based on the source database:
  Pfam: pfam00049
  SMART: smart00078
  CD: cd00101
  COG: COG5641

## Appendix 2: Example query: finding and viewing structural data of a protein

*Finding the Structure of a Protein.* Suppose that we are interested in the biosynthesis of aminocyclopropanes and would like to find structural information on important active site residues in any available aminocyclopropane synthases. To begin, we would go to the *Structure* main page and enter "aminocyclopropane synthase" in the *Search* box. Pressing Enter displays a short list of structures, one of which is 1B8G, 1-aminocyclopropane-1-carboxylate synthase. Perhaps we would like to know the species from which this protein was derived. Selecting the *Taxonomy* link to the right shows that this protein was derived from *Malux x domestica*, or the common apple tree. Going back to the Entrez results page and selecting the PDB code (1B8G) opens the Structure Summary page for this record. The species is again displayed on this page, along with a link to the *Journal of Molecular Biology* article describing how the structure was determined. We immediately see from this page that this protein appears as a dimer in the structure, with each chain having three 3D domains, as identified by VAST. In addition, CD-Search has identified an "aminotran_1_2" CD in each chain. Now we are ready to view the 3D structure.

*Viewing the 3D Structure.* Once we have found the Structure Summary page, viewing the 3D structure is straightforward. To view the structure in Cn3D, we simply select the *View 3D Structure* button. The default view is to show helices in green, strands in brown, and loops in blue. This color scheme is also reflected in the Sequence/Alignment Viewer.

*Locating an Active Site.* Upon inspecting the structure, we immediately notice that a small molecule is bound to the protein, likely at the active site of the enzyme. How do we find out what that molecule is? One easy way is to return to the Structure Summary page and select the link to the PDB code, which takes us to the PDB Structure Explorer page for 1B8G. Quickly, we see that pyridoxal-5′-phosphate (PLP) is a HET group, or heterogen, in the structure. Our interest piqued, we would now like to know more about the structural domain containing the active site. Returning to Cn3D, we manipulate the structure so that PLP is easily visible and then use the mouse to double-click on any PLP atom. The molecule becomes selected and turns yellow. Now from the *Show/Hide* menu, we choose *Select by distance* and *Residues only* and enter 5 Angstroms for a search radius. Scanning the Sequence/Alignment Viewer, we see that seven residues are now highlighted: 117-119, 230, 268, 270, and 279. Glancing at the 3D Domain display in the Structure Summary page, we note that all of these residues lie in domain 3. We now focus our attention on this domain.

*Viewing Structure Neighbors of a 3D Domain.* Given that this enzyme is a dimer, we arbitrarily choose domain 3 from chain A, the accession of which is thus 1B8GA3. By clicking on the 3D Domain bar at a point within domain 3, we are taken to the VAST Structure Neighbors page for this domain, where we find nearly 200 structure neighbors.

*Restricting the Search by Taxonomy.* Perhaps we would now like to identify some of the most evolutionarily distant structure neighbors of domain 1B8GA3 as a means of finding conserved residues that may be associated with its binding and/or catalytic function. One powerful way of doing this is to choose structure neighbors from phylogenetically distant organisms. We therefore need to combine our present search with a Taxonomy search. Given that 1B8G is derived from the superkingdom Eukaryota, we would like to find structure neighbors in other superkingdom taxa, such as Eubacteria and Archaea. Returning to the Structure Summary page, select the 3D Domains link in the graphic display to open the list of 3D Domains in Entrez. Finding 1B8GA3 in the list, selecting the *Related 3D Domains* link shows a list of all the structure neighbors of this domain. From this page, we select *Preview/Index*, which shows our recent queries. Suppose our set of related 3D Domains is #5. We then perform two searches:

1. #5 AND "Archaea"[Organism]
2. #5 AND "Eubacteria"[Organism]

Looking at the Archaea results, we find among them 1DJUA3, a domain from an aromatic aminotransferase from *Pyrococcus horikoshii.* Concerning the Eubacteria results, we find among the several hundred matching domains 3TATA2, a tyrosine aminotransferase from *Escherichia coli.*

*Viewing a 3D Superposition of Active Sites.* Returning to the VAST Structure Neighbors page for 1B8GA3, we want to select 1DJUA3 and 3TATA2 to display in a structural alignment. One way to do this is to enter these two Accession numbers in the *Find* box and press *Find.* We now see only these two neighbors, and we can select *View 3D Structure* to launch Cn3D.

Cn3D again displays the aligned residues in red, and we can highlight these further by selecting *Show aligned residues* from the *Show/Hide* menu. The excellent agreement between both the active site structures and the conformations of the bound ligands is readily apparent. Furthermore, by selecting *Style/Coloring Shortcuts/Sequence Conservation/Variety*, we can easily see that the most highly conserved residues are concentrated near the binding site (Figure 6).

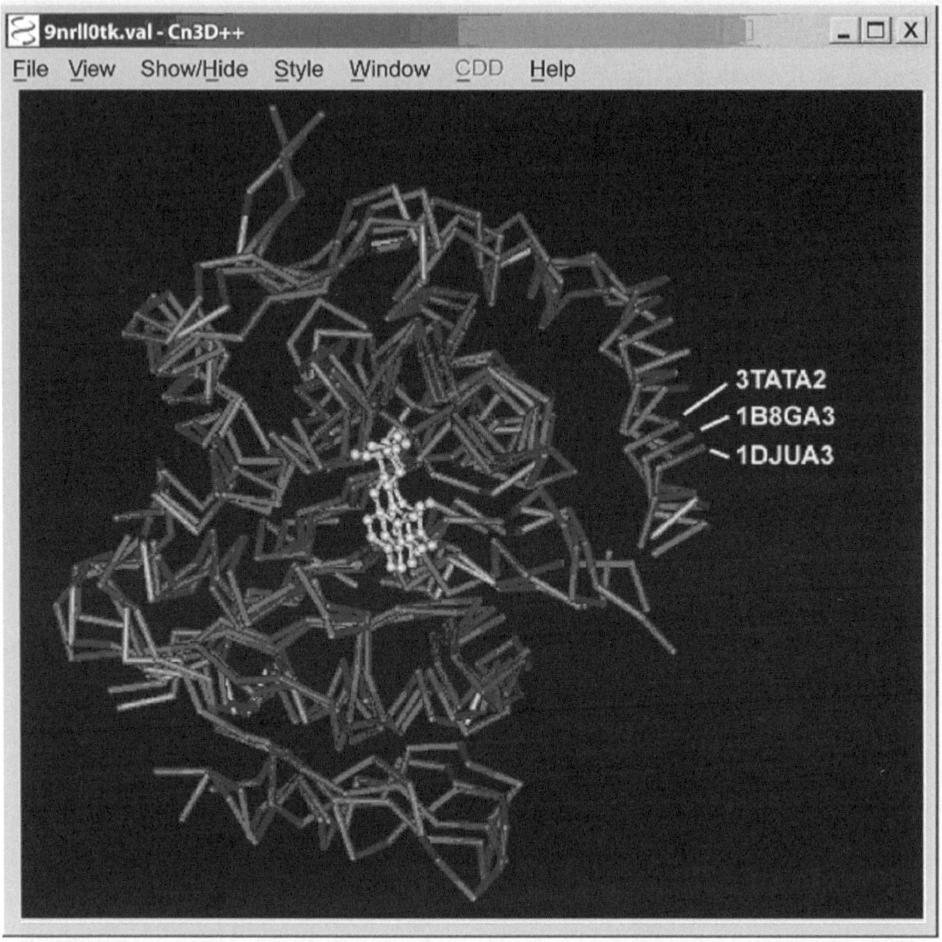


**Figure 6: VAST structural alignment of 1B8GA3, 3TATA2, and 1DJUA3.** The backbone atoms of the aligned residues of the three structures are shown. The bound pyridoxal phosphate ligands (center) are shown in a lighter shade of grey.

## Appendix 3: Example query: finding and viewing CDs in a protein

*Finding CDs in a Protein.* Suppose that we are interested in topoisomerase enzymes and would like to find human topoisomerases that most closely resemble those found in eubacteria and thus may share a common ancestor. Further suppose that through a colleague, we are aware of a recent and particularly interesting crystal structure of a topoisomerase from *Escherichia coli* with PDB code 1I7D. How can we identify the conserved functional domains in this protein and then find human proteins with the same domains? From the Structure main page, we enter the PDB code 1I7D in the Structure *Summary* search box and quickly find the Structure Summary page for this record. We see that in this crystal structure, the protein is complexed with a single-stranded oligonucleotide. We also see that the protein has five 3D Domains. Three CDs align to the sequence as well, and two of these overlap with one another at the N-terminus of the protein in the region corresponding to 3D domains 1-3.

*Analyzing CDs Found in a Protein.* The Structure summary page displays only the CDs that give the best match to the protein sequence. To see all of the matching CDs, we can easily perform a full CD-Search. Select the *Protein* link to the left of the graphic to reveal the flat file for the record. Then follow the *Domains* link in the *Link* menu on the right to view the results of the CD-Search. Select *Show Details* to see all CDs matching the query sequence. We find that nine CDs match this sequence, and that the statistics of each match are shown below the alignment graphic. The CDs are listed by database, with curated CDs at the top, followed by SMART, Pfam, and COG records. We see that the protein contains two domains, an N-terminal TOPRIM domain and a C-terminal Topoisomerase, subtype IA domain. We can learn more about these CDs by studying the pairwise alignments at the bottom of the page and by studying their CD Summary pages, reached by selecting the links to their left.

*Finding Other Proteins with Similar Domain Architecture.* We now would like to find human proteins that have these same CDs. To perform a CDART search, simply select the *Show Domain Relatives* button. To limit these results to human proteins, we select the *Subset by Taxonomy* button. A taxonomic tree is then displayed, and we next check the box for *Mammal*, the lowest taxa including *Homo sapiens*. Selecting *Choose* then displays a Common Tree, and by clicking on the appropriate "scissor" icons, we can cut away all branches except the one leading to *H. sapiens*. We can execute this taxonomic restriction by selecting *Go back*, and we now find a much shorter list of CDART results. In the second group of proteins, we find two members, one of which is NP_004609. Selecting the *more>* link for this record shows the CD-Search results for this human protein. Interestingly, we find that the topoisomerase is very well conserved, whereas only a portion of the TOPRIM domain has been retained.

*Viewing a CD Alignment with a 3D Structure* . We now would like to view the alignment of the topoisomerase in the human protein to other members of this CD. On the CD-Search page, select the colored bar of this CD to see a CD-Browser window displaying the alignment. Because this is a curated CD record, we are able to view functional features of the protein domain on a structural template. The rightmost menu in the View Alignment bar shows the available features for this domain, whereas the topmost row in the alignment itself marks the residues involved in this feature with # symbols. The second row of the alignment is the consensus sequence of the CD record, whereas the third row contains the NP_004609 sequence, labeled "query". At the bottom of the page, buttons allow Cn3D to be launched with various structural features highlighted. For example, if we are interested in nucleotide binding site II, Cn3D will launch with the view depicted in Figure 7, showing the bound nucleotide in orange. Additonal Cn3D windows not shown in Figure 7 allow one to highlight the binding site residues yellow as shown, and these highlights also appear in the sequence window. In this figure, the NP_004609 sequence has been merged into the alignment (bottom row) using tools within Cn3D, and the result shows that this human protein closely conserves these important functional residues.

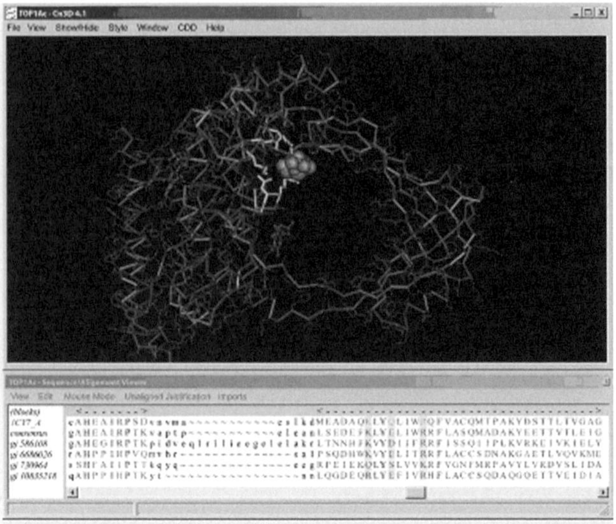

**Figure 7**: Sequence and structure views of the TOP1Ac conserved domain common to type III bacterial and eukaryotic DNA topoisomerases. The *upper window* displays the structure of the domain with the residues involved in the nucleotide binding site colored light grey. The nucleotide bound at site II is shown as a space-filling model. The *lower window* displays the sequence alignment for the domain with aligned residues shown as capital letters. The sequence for NP_004609 (gi 10835218) occupies the bottom row

*Essays in Bioinformatics*
*D.S. Moss et al. (Eds.)*
*IOS Press, 2005*



# Protein Secondary Structure Prediction: Comparison of Ten Common Prediction Algorithms Using a Neural Network

Jorn R. DE HAAN[1] and Jack A.M. LEUNISSEN[2]
*[1]Laboratory of Analytical Chemistry, Radboud University Nijmegen, Toernooiveld, 6525 ED Nijmegen, the Netherland, and [2]Laboratory of Bioinformatics, Wageningen University, Dreijenlaan 3, 6703 HA Wageningen, the Netherlands*

**Abstract.** Protein secondary structure prediction is believed to improve by combining different predictions into a consensus secondary structure prediction. Ten different protein secondary structure prediction programs were compared and given weights by a feed forward neural network. A dataset of approximately 6000 proteins was taken from the DSSP database and was used to train the neural network. The resulting weights indicate that the secondary structure prediction programs PHD and Predator performed better than the other methods. However training of the neural network with a smaller but more stringently selected dataset did not support these results for the Predator program. The performance of the program PHD remained the same when the smaller dataset was used to train the neural network.

## 1. Introduction

### *1.1. Secondary structure prediction*

The "Holy Grail" in bioinformatics for years was (and still is) the *ab initio* prediction of protein 3D structure, i.e. constructing the folding structure of a protein based upon the amino acid sequence alone. One important step to attaining this goal is the prediction of protein secondary structure from the primary structure. Several methods have been developed to make and improve secondary structure predictions for proteins; these are amongst the oldest algorithms used in bioinformatics, the oldest ones dating back to the early seventies (e.g. Chou & Fasman, 1974, Lim 1974, Garnier 1978) [1-5]. Improvement of secondary structure prediction is relevant and interesting because secondary structure predictions allow for a wide variety of conclusions on the fold classification and function of a protein and, in particular, provide important information for 3D-structure prediction [6]. Furthermore the results of secondary structure prediction have been an aid for designing new proteins [7], predicting the effect of point mutations, identifying the protein class, for instance, all-α or all-β proteins, and predicting epitopes [8]. In this report we research the possibilities of combining predictions from secondary structure prediction methods to form a consensus prediction. The goal of a consensus method is to improve the final prediction result in comparison with the individual predictions.

The field of secondary structure prediction for proteins can be divided in two ways of predicting. First there is secondary structure class prediction, in which a protein is characterized as an all α-, all β- or α/β class protein. Second there is 'normal' secondary structure prediction, in which the secondary structure state (α-helix, β-sheet or other) is predicted for each residue of a protein. In this study the predictions of computer programs, or methods, of the 'normal' secondary structure prediction were used. It is therefore noteworthy to mention that experiments were done to look into the difference in prediction of α-helix or β-sheet between secondary structure prediction methods. When prediction of α-helix or prediction of β-sheet are mentioned further down one should remind that this means 'normal' secondary structure prediction per residue and not protein class prediction.

### *1.2. Methods in secondary structure prediction*

Methods in protein secondary structure prediction are designed and work on the basis of different underlying prediction principles. Some of these principles and methods using this particular principle are mentioned here below in no particular order: statistical analysis [1,2]; simple linear statistics, information theory [5,8,9]; neural networks and machine learning [10,11]; k-way nearest neighbour [12,13]; linear discrimination [14]; hydrogen bonding propensities [15]; conservation number weighted prediction [16]; and hybrid methods, a combination of principles [17-19].

In the section below we will briefly describe the main characteristics of these algorithms:

- The Chou-Fasman method uses statistical analysis to predict secondary structure [1]. In the first implementation of this method only 15 proteins of known 3D structure were analysed and residues were assigned according to their ability to initiate or terminate particular secondary structure elements. Residues were classified into strong formers, weak formers, formers, indifferent formers, strong breakers and breakers. In later updates of the algorithm a more elaborate database of protein tertiary structures were used [2].
- The Garnier method [5] uses simple linear statistics and information theory to make secondary structure predictions. Besides information theory the algorithm, like Chou-Fasman, uses statistical data extracted from structural databases. Furthermore Garnier also takes into account the accuracy of the data: the likelihood for each residue and neighbouring residues to be in a certain conformation was obtained by examining data collected from 8 residues on either side of each amino acid residue. This way a protein can be scanned with a 17 residues long window, which predicts the likelihood of each residue to assume a specific secondary structure. The algorithm has seen several revisions, GOR4 being the fourth and more recent version of the Garnier secondary structure prediction method, based on information theory [8]. In this algorithm the prediction of beta turns and random coil structure have been abandoned.
- The program DSC (Discrimination of protein Secondary structure Class) of King & Sternberg [14] combines several secondary structure prediction principles. DSC applies Garnier residue attributes, amino acid hydrophobicity values and amino acid positional information. Also information from a multiple sequence alignment is used to perform the secondary structure prediction. Simple and linear statistical methods are applied to filter the different prediction concepts and to remove false predictions.
- PREDATOR2 [15] is a secondary structure prediction method, which predicts secondary structure on the basis of hydrogen bonding propensities and non-local interaction statistics. These propensities were calculated for each of the possibly 400 amino-acid pairs. Furthermore local pairwise alignments are used to incorporate information from homologous proteins.
- SIMPA96 [20] is a nearest neighbour secondary structure prediction method, which uses a similarity matrix, similarity threshold and information from a database of known secondary structures.

- NNpredict [21] is a program that predicts the secondary structure type for each residue in an amino acid sequence by using a two-layer, feed-forward neural network.

Examples of hybrid methods are the programs PHD and PSIPRED. The program PHD [17-19] uses a combination of multiple alignment and several cascading neural networks. The program may generate its own alignment with the submitted sequence and is composed of several cascading neural networks (previously trained on proteins of known structures). PSIPRED [10] incorporates two feed-forward neural networks, which perform an analysis on output obtained from PSI-BLAST [22]. PHD and PSIPRED are currently considered to be amongst the best performing methods. They are both hybrid methods and this suggests that it could be profitable to combine principles than to use one method [10,14].

### *1.3. Consensus secondary structure predictions*

Different ways of combining prediction principles into a hybrid secondary structure prediction program are known. There is a "standard approach" in which the most appropriate strategy (or principle) is applied to a specific task. The predicting problem has to be broken down into different tasks. For each task the best strategy is used to improve the results. Another approach is "ensemble learning". Here the focus is on a single prediction task and multiple predictors or classifiers are built for that task. The different predictors are combined either by voting or by training a classifier to combine them.

A consensus method is using the last principle of ensemble learning to improve the prediction results. The results of several secondary structure prediction programs can be compared and combined by a classifier. Therefore in case of a secondary prediction consensus method the multiple predictors are already built and predictions can be used to make a consensus predicted sequence.

As mentioned before a consensus method looks at the results of several different prediction programs. In order to choose when to use the results of which program(s) a decision mechanism or classifier has to be implemented in the method. Three of those consensus method classifiers are discussed below, i.e. decision tree, majority wins (winner takes it all), and neural network.

A decision tree is a representation of a decision procedure in order to attain classification for a given example [6]. At each node of the tree, there is a question, and a branch corresponding to each of the possible outcomes of this question. At each leaf node, there is a classification. Decision trees have many uses, particularly for solving problems that can be formulated in terms of producing a single answer in the form of a class name. Decision trees are constructed from examples that are already labelled. Decision trees could be used to apply rules for determination of secondary structure for a specific residue. In fact the next classifier could be viewed as a very short decision tree with few questions.

The consensus program JPRED [23,33,34] uses the majority wins principle. Despite all the efforts and different methods, the $Q_3$ (percentage of correct prediction) of protein secondary structure prediction for all the methods mentioned before is 60 to 80 percent. The makers of a consensus secondary structure server called JPRED aimed to improve this percentage by combining six different secondary prediction programs like the ones mentioned before. The server is available through a web-interface and no neural network is used in making the consensus prediction. JPRED builds a consensus prediction by comparing the results of these programs and JPRED takes the predicted state, which is most abundant. The majority wins and therefore this principle is also called the "winner takes it all method". The correct prediction of protein secondary structure of JPRED is 72.9 percent.

### *1.4. Evaluation of prediction results*

In order to compare the results of different secondary structure programs an objective score of prediction accuracy is required. The most used index is the three-state per-residue accuracy ($Q_3$). The formula below gives the percentage of residues predicted correctly for α-helix ($q_a$), β-strand ($q_b$) and other ($q_c$) of the total number of residues (N).

$$Q_3 = \frac{q_a + q_\beta + q_c}{N} \text{ x } 100\% = \text{percentage correctly predicted residues}$$

A closer look at the $Q_3$ value learns that it is not very convenient when the target class is present in only a relative small part of the data. This is because in that case correct prediction of the non-regular class tends to dominate the three-state accuracy. A more precise method avoiding this is the Matthew Correlation Coefficient (C) [24], which is defined by the formula shown below.

$$C = \frac{t_p t_n - f_p f_n}{\sqrt{(t_p + f_p)(t_p + f_n)(t_n + f_p)(t_n + f_n)}}$$

The value of the Matthews Correlation Coefficient is between 1 and 0 and can be calculated from the number of true positive- ($t_p$), true negative- ($t_n$), false positive- ($f_p$) and false negative predicted residues ($f_n$).

### *1.5. Neural Networks*

To address the function of a consensus method classifier the decision tree principle and "majority wins" principle seem fairly simple and crude principles. A neural network could be a more complex and possibly better classifier in a consensus secondary prediction method. More information from the methods could be used by a neural network to determine when to use what method. But what is a neural network?

A definition in the *DARPA Neural Network Study [25] states* "... a neural network is a system composed of many simple processing elements operating in parallel whose function is determined by network structure, connection strengths, and the processing performed at computing elements or nodes". Another slightly more recent definition reads "artificial neural systems, or neural networks, are physical cellular systems which can acquire, store, and utilise experiential knowledge" [26].

As mentioned in both definitions, a neural network consists of computing units (processing elements, nodes or cells). These units can be grouped in layers, an input layer, a variable number of hidden layers and an output layer. These layers can be interconnected (see figure 1). Each unit receives input, which is transformed by a transfer function to output. Biases can be imported in these transfer functions. The output can be conducted to a next computing unit or several units. Thus the connected units form a network. Each connection between units has a weight attached to it. Building and programming the units in different conformations can make various types of neural networks. These types listen to illustrious names like Bolzmann machine, Hebbian network and Hopfield network. An

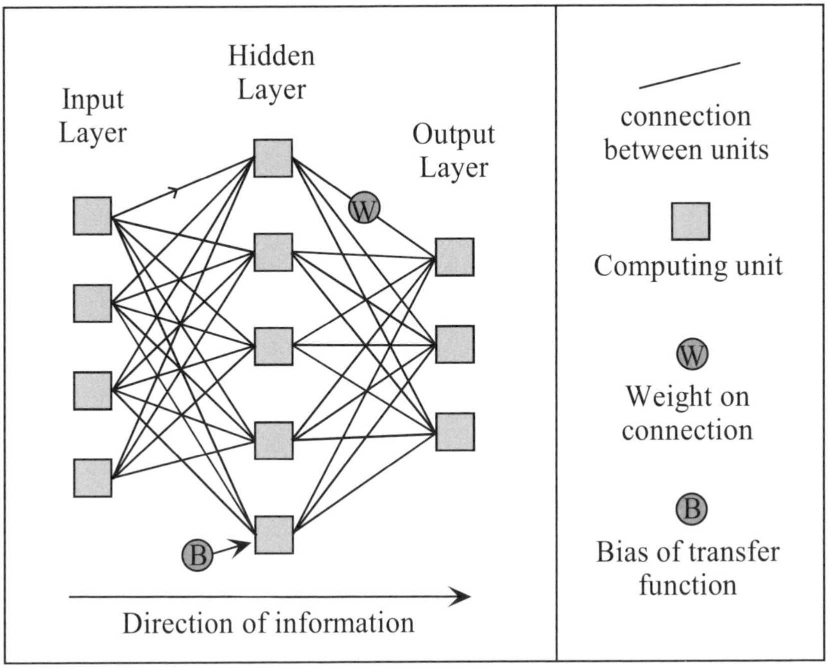


**Figure 1.** Example of a feed-forward neural network, showing the connections between the computing units. The network consists of an input layer, one hidden layer and an output layer. In this example one weight is set on the upper right connection and a bias is put into the transfer function of the lowest computing unit of the hidden layer.

example of a "feed forward" neural network can be seen in figure 1. It is called feed forward because the direction of all the connections is forward.

An important feature of neural networks is their ability to learn from experimental data. Neural networks can be trained on experimental data in order to make predictions about the future. By changing weights, constants and biases the output of the network can be influenced. The weights and biases can be changed by the learning rules of the neural network. These learning rules, which are applied to attain learning are various and are another distinguishing feature of different neural networks.

An example of such an error function, the summed square error, is defined as:

$$E_{sum-squared} = \sum_x [o_x - t_x]^2$$

In this equation the square of the error between the output (O) and target (t) for each of the elements x from the training set is added, resulting in the summed square error

(Esum-squared). The network attempts to minimize this squared error by adjusting weights and biases.

In the experiments described in this report a simple feed forward neural network was used. The network had a different function from a consensus classifier. Instead it was used to assign weights to secondary structure prediction methods.

## 2. Materials and methods

### *2.1. Preparation of the datasets*

This suggests In order to attain the goal of improving the consensus prediction feature of Seccons, a test set of training data for the neural network was built first. This data set was composed of 6000 proteins. For each protein predictions by ten different secondary prediction methods and the verified secondary structure were collected. The true secondary structure of a protein was extracted from the PDB by Kabsch & Sander's DSSP program [27]. This is to allow the neural network to learn from predictions in the data set by comparing them with the true secondary structure. The sequences of the proteins were also taken from the DSSP files and used as input for the prediction programs. Al sequences in the dataset had a length of at least 25 residues. Also sequences with errors were excluded from the dataset.

A second data set was made from a selection of proteins, which complied with the following terms:

1. The protein was added to the database after the programs were released (1997). This was checked using the local SRS (Sequence Retrieval Server) database server.
2. The protein is not similar to other proteins in the database (less than 30 percent sequence homology). To verify this the PDBSelect algorithm was used [28]. The algorithm picked structures from the PDB and used the program WHAT IF [29] to do pairwise alignments. If there was a match higher than 30%, the structure with the lower resolution was removed from the list.
3. The protein is present in the aforementioned data set of 6000 proteins.

These criteria rendered a data set of 301 proteins.

### *2.2. Creating the target output in the data files*

As mentioned before the target files, which contained the verified secondary structure, were taken from DSSP files. The definition of secondary structure itself differs in the number of defined secondary structure states. In DSSP for instance, the states coil (C) (or turn (T)), bend (S), 3-10 helix (G), short beta bridge (B) and pi helix (I) are also known, besides the structure elements α-helix (H), β-sheet (E).

Furthermore, some of the secondary structure prediction programs used for predicting secondary structure also predict the secondary structure elements coil or turn, while others only predict the elements α-helix and β-sheet. Because the other programs do not have this feature it was left out of the predictions. The states viewed in this report are reduced to a-helix, b-sheet and "other". Therefore the states 3-10 helix (G) and pi helix (I) in the DSSP file were converted to α-helix in the target sequence of the data files. Also the short beta bridge (B) element from DSSP was translated to b-sheet in the target sequences. This conversion was performed automatically by the SecCons program (see below), which also converted the DSSP elements bend (S) and (s) to Turn (T) and Coil (C) respectively.

### *2.3. Secondary structure prediction programs*

The format of the input files for the secondary structure prediction methods was changed to the different acquired file types. Predictions of eight out of the ten secondary structure prediction programs for proteins could be done locally on a Silicon Graphics Origin2000 computer at the CMBI in Nijmegen.

The program Pepplot [30] was used to make Chou-Fasman [1,2] predictions. Predictions this method made by Pepplot will be referred to as Chou-Fasman predictions. Slightly adapted Chou-Fasman predictions are produced by the program PeptideStructure [31,32]. These Chou-Fasman predictions will be referred to as CFpred from this point. PeptideStructure uses a modified version of the previously mentioned method of Chou and Fasman: for $\alpha$-helix predictions not all conditions are used, and for $\beta$-sheet predictions a minimum length of five residues is obligatory.

PeptideStructure also predicts secondary structure according to a modified version of the Garnier prediction method [5]. Predictions from this method will be referred to as Garnier predictions. The alterations to the Garnier method by PeptideStructure consist of the following rules: the minimum length of a helix is six and of a beta-sheet is four, and regions without adequate predictions are replaced by the conformational state of the next best probability.

Secondary structure predictions by more recent version of the Garnier secondary structure prediction method were performed using the program GOR4 [8].

The program DSC (Discrimination of protein Secondary structure Class) combines several secondary structure prediction principles [14]. From the output file of DSC the program SecCons (see below) extracts another secondary structure prediction, which uses slightly different rules. This prediction is called DSC-l to distinguish it from the normal DSC prediction.

PREDATOR2 [15] is a secondary structure prediction method, which predicts secondary structure on the basis of hydrogen bonding propensities and non-local interaction statistics. These propensities were calculated for each of the possibly 400 amino-acid pairs. Furthermore local pairwise alignments are used to incorporate information from homologous proteins.

SIMPA96 [20] is a nearest neighbour secondary structure prediction method, which uses a similarity matrix, similarity threshold and information from a database of known secondary structures.

The predictions of the two remaining secondary prediction programs were obtained by making use of e-mail or HTML servers. NNpredict [21] is available via the NNpredict web server. Sequences were submitted to the server and the retrieved HTML files were later processed.

From the PredictProtein web and e-mail server predictions of the aforementioned PHD program [17-19] were obtained. An e-mail message containing the protein sequence and name was sent to this server, which returned an e-mail with the secondary structure prediction.

### *2.4. Converting secondary structure predictions to Neural Network input*

Every program returned its predictions in a distinct file format. In order to use the predictions and the verified secondary structure as input for a neural network all the prediction files for a certain protein were gathered by the program SecCons (JAML, unpublished data). SecCons can compare outputs of different secondary structure prediction programs in one (text or graphical) view. The output files of SecCons were converted to format of the neural network software successively. Finally this resulted in a data set with one text file for each of the 6000 proteins, containing both ten predictions and the true secondary structure.

For each protein the text file was converted to Matlab scripting language to make it suitable for input in the neural network. In the script the predictions were declared first in a matrix of normalised numbers ranging between 0 and 1. These indicated the likelihood for a residue to be in a particular secondary structure state. The neural network would compare these figures with the target matrix (secondary structure taken from DSSP), which was declared consecutively.

### *2.5. The neural network*

Next a neural network was programmed in the neural network toolbox of Matlab 6.0. The network was composed of an input layer (10 units), one layer of hidden units (10 units) and an output layer (10 units). It uses the standard 'errorsqr' error function from Matlab. The number of learning iterations for one protein was optimized to 300 iterations to save time without losing the learning performance of the network. The transfer function of the hidden layer is the Matlab standard function 'tanh' and for the output layer 'softmax'. The network was used in an implementation in Matlab 6.0, which was written by Tom Heskes (dept. of Medical Physics & Biophysics, Nijmegen University).

In a training session one by one the proteins were put through the neural network. After training of the network on the dataset a weight matrix containing the weights between hidden and output layer was extracted with the implementation mentioned above. These are the weights for the concerning secondary structure prediction methods. The higher the weight, the better the performance of the prediction method.

## 3. Results

After all data were collected and transformed to Matlab scripts, weights were assigned to all methods for predicting α-helix, β-sheet and combined prediction of both α-helix and β-sheet (three different training sessions) on a test set of 1000 randomly assigned proteins.

Table 1 shows the results of this experiment. It is clear that PREDATOR 2 and PHD have been assigned the highest weights in comparison to other methods. Careful observation of the data reveals another remarkable feature: though PHD has a weight of 6.1 for predicting β-sheet and a weight of 10.0 for predicting α-helix it has a weight of 9.4 for predicting both. One would expect the weight for the prediction of both α-helix and β-sheet to be lower.

This can be accounted for by the percentages of α-helix and β-sheet residues in the DSSP database and in the 6000 proteins used in the experiments. The percentage β-sheet (20.6% in DSSP, 20.8% in our set of 6000 proteins) is much less than the percentage α-helix in these data sets (38.0% in DSSP, 36.2% in the training set; this explains why the lower weight for predicting β-sheet is less reflected in the weight of overall prediction of both α-helix and β-sheet for the method PHD.

**Table 1.** Results of a training session showing the weights attributed to each method indicating the relative performance of this method in comparison to others. The three columns show results in predicting helix,sheet or both. (set consisted of 1000 randomly selected proteins).

| number | name | β-sheet | α-helix | both |
|---|---|---|---|---|
| 1 | ChouFasman | 0.7 | 0.9 | 0.7 |
| 2 | PREDATOR 2 | 10.0 | 9.6 | 10.0 |
| 3 | Garnier | 0.4 | 0.6 | 0.3 |
| 4 | Simpa96 | 1.4 | 1.6 | 1.2 |
| 5 | GOR4 | 0.7 | 0.6 | 0.6 |
| 6 | DSC | 0.7 | 0.6 | 0.7 |
| 7 | DSC-l | 0.5 | 0.6 | 0.5 |
| 8 | CFpred | 0.4 | 0.6 | 0.3 |
| 9 | PHD | 6.1 | 10.0 | 9.4 |
| 10 | NNpredict | 0.6 | 0.6 | 0.5 |

The results in table 1 also show that there is a difference in performance in predicting α-helix or β-sheet between different methods. This effect can only be clearly seen in some of the methods (for instance the method PHD). In the following experiments the difference between prediction of α-helix or β-sheet is no longer taken into account. In these experiments the weights are for the prediction of both α-helix and β-sheet.

Next a neural network training session was done on a training set of 6000 proteins. The weights acquired in this session can be seen in table 2 (column labelled "weight 1"). Again the methods PREDATOR 2 and PHD are getting far better scores compared to other methods. To investigate a possible suppressive effect of the two aforementioned methods on the weights of other methods another experiment was performed. Data of the methods PREDATOR 2 and PHD was excluded from the training. The test set consisted of 2000 proteins and again a neural network was trained to find the weights for the remaining methods. The results are shown in table 3. Although the weights seem to be much higher, they are not. This is because the weights represent the relative weight for a method. The table shows us that indeed PREDATOR 2 and PHD suppress other methods, but by leaving them out there is no real change on our views of the other methods. The method, which is the best now in comparison with the other methods, Simpa96, was also better than these methods in former experiments (except in comparison with PREDATOR 2 and PHD of course). Also the methods with bad scores, like Garnier and CFpred, stay at the bottom.

In order to investigate whether the weights from the last experiment have changed at the same rate for all the remaining methods in comparison with one and another, the ratio of the weights from the experiments with and without the programs PREDATOR 2 and PHD is calculated.

When the ratios of the weights of the last two experiments are compared a striking difference between better valued methods from the last experiment and less valued methods emerges. Table 4 shows that Simpa96, GOR 4, DSC and DSC-l all have a ratio of approximately 9. Chou-Fasman, Garnier, Cfpred and NNpredict all have a ratio of about 4 to 5. Assuming that the neural network favours better methods by assigning higher weights,

**Table 2.** Results of a training session showing the weights attributed to each method indicating the relative performance of this method in comparison to others. (the test set consisted of 6000 proteins, Weight 1). Weight 2 show results of a training session on a selection of proteins, which were added to the DSSP database after 1997. These proteins had less then 30 percent sequence homology with other proteins already present in the database. (test set consisted of 301 proteins).

| Number | Name | Weight 1 | Weight 2 |
|---|---|---|---|
| 1 | ChouFasman | 0.8 | 0.9 |
| 2 | PREDATOR2 | 7.6 | 1.7 |
| 3 | Garnier | 0.3 | 0.3 |
| 4 | Simpa96 | 1.1 | 1.2 |
| 5 | GOR4 | 0.7 | 0.8 |
| 6 | DSC | 0.7 | 0.8 |
| 7 | DSC-1 | 0.6 | 0.6 |
| 8 | CFpred | 0.3 | 0.3 |
| 9 | PHD | 10.0 | 10.0 |
| 10 | NNpredict | 0.5 | 0.3 |

this suggests that the newer methods Simpa96, GOR4 and DSC perform significantly better than the classical prediction methods.

Altogether this indicates that PREDATOR 2 and PHD would be the main source of information if the neural network would have to make a consensus sequence if it was assigned as classifier.

Finally an experiment was done to be certain that the methods performing well really predict secondary structure well, instead of "cheating" by using known structures, used in the training session. The resulting weights can be seen in table 2 (in the column labelled "weight 2").

## 4. Discussion

The objective of this project was to obtain a set of weights to make a better consensus sequence in SecCons. To a certain extent this object is achieved, because a neural network could be trained on the data set and a set of weights was found. The question remains whether these weights are useful. The consensus sequence would improve by using these weights but this improvement would rely on the use of a single method (PHD) to achieve this. That is undesirable when speaking of a consensus method. The object of a consensus method is to combine methods to improve the predictions and not to use the best one.

The results of the experiments give little information about the consensus method for secondary structure prediction. Whether or not the combining of prediction results of different prediction methods improves secondary structure predictions cannot be concluded from this experiment. It still remains a fact that no method is perfect but the best methods are hybrid prediction methods [36].

**Table 3.** Results of a training session from which the methods PREDATOR2 and PHD were excluded. (test set consisted of 2000 proteins)

| number | name | weight |
|---|---|---|
| 1 | ChouFasman | 3.0 |
| 2 | PREDATOR2 | - |
| 3 | Garnier | 1.1 |
| 4 | Simpa96 | 10.0 |
| 5 | GOR4 | 6.8 |
| 6 | DSC | 6.2 |
| 7 | DSC-l | 5.6 |
| 8 | CFpred | 1.5 |
| 9 | PHD | - |
| 10 | NNpredict | 2.3 |

**Table 4.** Comparing the weights assigned without PREDATOR 2 and PHD (column A) or with both methods (column B). The last column shows the ratio of columns A and B.

| number | name | A(no PHD/Pr.) | B(all methods) | ratio(A/B) |
|---|---|---|---|---|
| 1 | ChouFasman | 3.0 | 0.8 | 3.8 |
| 3 | Garnier | 1.1 | 0.3 | 3.7 |
| 4 | Simpa96 | 10.0 | 1.1 | 9.1 |
| 5 | GOR4 | 6.8 | 0.7 | 9.7 |
| 6 | DSC | 6.2 | 0.7 | 8.9 |
| 7 | DSC-l | 5.6 | 0.6 | 9.3 |
| 8 | CFpred | 1.5 | 0.3 | 5 |
| 10 | NNpredict | 2.3 | 0.5 | 4.6 |

The weight assigned to the PREDATOR 2 program drops dramatically in comparison with its former weight when the second smaller dataset is used in the training of the neural network. This is because the program PREDATOR 2 uses a database, which contains sequences with "prototype" secondary structure predictions. When the sequences to be predicted are not "known" by this database it can be seen in the results of the secondary structure predictions accordingly.

Also the question remains if the information needed to improve the overall prediction rate is present in the present secondary prediction methods. If the results of the predicting programs are not complementary it is improbable to find improvement with consensus or hybrid methods. Possible causes of the problems in protein secondary structure prediction could be long range interactions and Cys-Cys disulfide bridges.

Another problem is the definition of secondary structure from 3D co-ordinates. This definition is not exact because different algorithms to determine secondary structure are used. For instance, DSSP and Stride agree in 96% of all residues [35], which leaves 4% of the residue assignments open for interpretation.

The results of this work suggest that better methods should be used to construct a consensus method that outperforms the best algorithm in the selection (PHD). For a consensus method to work well we need more methods competing with each other in performance. And clearly this was not the case in this study. One method turned out to be far better than the rest, causing the weights of the other prediction methods to be marginal: instead of an agreement one method takes all the decisions. To make a useful consensus method, prediction methods should be used which, ideally, have slightly complementary predictions because they are based on different principles.

This could be expanded to assigning different weights to distinct secondary structure elements. The results of the experiments show difference in performance in predicting α-helix or β-sheet between different methods. This could be translated into different sets of weights for prediction of α-helix or β-sheet accordingly, thus increasing the overall performance.

## Acknowledgements

The authors like to thank dr. T. Heskes (Dept. of Medical Physics and Biophysics, University of Nijmegen, NL) for his expert help on neural networks and programming in Matlab. Part of this work was performed as a student thesis project of JRdH under supervision of JAML at the Centre for Molecular and Biomolecular Informatics (CMBI) of the University of Nijmegen. The CMBI is gratefully acknowledged for the use of their computing facilities.

## References

[1] P.Y. Chou and G.D. Fasman, Prediction of protein conformation, Biochemistry **13** (1974), 222-245.
[2] P.Y. Chou and G.D. Fasman, Prediction of the secondary structure of proteins from their amino acid sequence, Advanced Enzymology **47** (1978) 45-148.
[3] V.I. Lim, Structural principles of the globular organisation of protein chains. A stereochemical theory of globular protein secondary structure, Journal of Molecular Biology **88** (1974) 857-872.
[4] V.I. Lim, Algorithms for prediction of alpha-helical and beta-structural regions in globular proteins, Journal of Molecular Biology **88** (1974) 873-894.
[5] J. Garnier, D.J. Osguthorpe and B. Robson, Analysis of the accuracy and implications of simple methods for predicting the secondary structure of globular proteins, Journal of Molecular Biology **120** (1978) 97-120.
[6] J. Selbig, T. Mevissen and T. Lengauer, Decision tree-based formation of consensus protein secondary structure prediction, Bioinformatics **15** (1999) 1039-1046.
[7] S. Pongor and A. Szalay, Prediction of homology and divergence in the secondary structure of polypeptides, Proceedings of the National Academy of Science U.S.A. **82** (1985) 366-370.
[8] J. Garnier, J.-F. Gibrat and B. Robson, GOR method for predicting protein secondary structure from amino acid sequence, Methods in Enzymology **266** (1996) 540-553.
[9] J-F. Gibrat, J. Garnier and B. Robson, Further developments of protein secondary structure prediction using information theory, Journal of Molecular Biology **198** (1987) 425-443.
[10] D.T. Jones, Protein secondary structure prediction based on position-specific scoring matrices, Journal of Molecular Biology **292** (1999) 195-202.
[11] J.-M. Chandonia and M. Karplus, New methods for accurate prediction of protein secondary structure, Proteins (structure, function and genetics) **35** (1999) 293-306.
[12] A.A. Salamov and V.V. Solovyev, Prediction of protein secondary structure by combining nearest-neighbour algorithms and multiply sequence alignments, Journal of Molecular Biology **247** (1995) 11-15.
[13] T.M. Yi and S. Lander, Protein secondary structure prediction using nearest-neighbour methods, Journal of Molecular Biology **232** (1993) 1117-1129.
[14] R.D. King and M.J.E. Sternberg, Identification and application of the concepts important for accurate and reliable protein secondary structure prediction, Protein Science **5** (1996) 2298-2310.
[15] D. Frishman and P. Argos, 75% accuracy in protein secondary structure prediction, Proteins **27** (1997) 329-335.

[16] M. Zvelebil, G. Barton, W. Taylor and M. Sternberg, Prediction of protein secondary structure and active sites using the alignment of homologous sequences, Journal of Molecular Biology **195** (1987) 957-961.
[17] B. Rost and C. Sander, Improved prediction of protein secondary structure by use of sequence profiles and neuronal networks, Proceedings of the National Academy of Science U.S.A. **90** (1993) 7558-7562.
[18] B. Rost and C. Sander, Combining evolutionary information and neural networks to predict protein secondary structure, Proteins **19** (1994) 55-72.
[19] B. Rost, PHD: predicting one-dimensional protein structure by profile based neural networks, Methods in Enzymology **266** (1996) 525-539.
[20] J.M. Levin, Exploring the limits of nearest neighbour secondary structure prediction, Protein Engineering **7** (1997) 771-776.
[21] D.G. Kneller, F.E. Cohen and R. Langridge, Improvements in Protein Secondary Structure Prediction by an Enhanced Neural Network, Journal of Molecular Biology **214** (1990) 171-182.
[22] S.F. Altschul, T.L. Madden, A.A. Schaffer, J. Zhang, Z. Zhang, W. Miller and D.J. Lipman, Gapped BLAST and PSI-BLAST: a new generation of protein database search programs, Nucleic Acids Research **25** (1997) 3389-3402.
[23] J.A. Cuff, M.E. Clamp, A.S. Siddiqui, M. Finlay and J.F. Barton, Jpred: a consensus secondary structure prediction server, Proteins (structure, function and genetics) **14** (1998) 892-893.
[24] B.W. Matthews, Comparison of the predicted and observed secondary structure of T4 phage lysozyme, Biochimica Biophysica Acta **405** (1975) 442-451.
[25] Defense Advanced Research Projects Agency (DARPA), DARPA Neural Network Study. AFCEA International Press, Fairfax, 1988.
[26] J.M. Zurada, Introduction to Artificial Neural Systems. West Publishing Company, St. Paul, Minnesota, 1992.
[27] W. Kabsch and C. Sander, Dictionary of protein secondary structure: pattern recognition of hydrogen-bonded and geometrical features, Biopolymers **22** (1983) 2577-637
[28] R.W.W. Hooft, C. Sander and G. Vriend, The PDBFINDER database: a summary of PDB, DSSP and HSSP information with added value, Computer Applications in the Biosciences **12** (1996) 525-529.
[29] G. Vriend, WHAT IF: a molecular modelling and drug design program, Journal of Molecular Graphics **8** (1990) 52-56.
[30] M. Gribskov, R.R. Burgess and J. Devereux, PEPPLOT, a protein secondary structure analysis program for the UWGCG sequence analysis software package, Nucleic Acids Research **14** (1986) 327-334.
[31] B.A. Jameson and H. Wolf, The antigenic index: a novel algorithm for predicting antigenic determinants, Computer Applications in the Biosciences **4** (1988) 181-186.
[32] H. Wolf, S. Modrow, M. Motz, B.A. Jameson, G. Hermann and B. Fortsch, An integrated family of amino acid sequence analysis programs, Computer Applications in the Biosciences 4 (1988) 187-191
[33] Cuff J.A., Barton G.J. (1999) Evaluation and improvement of multiple sequence methods for protein secondary structure prediction. Proteins (structure, function and genetics), 34, 508-519.
[34] Cuff J.A., Barton G.J. (2000) Application of multiple sequence alignment profiles to improve protein secondary structure prediction. Proteins (structure function and genetics), 40, 502-511.
[35] C.A.F. Andersen and B. Rost, Secondary structure assignment. In Structural Bioinformatics. Philip Bourne and Helge Weissig (editors.), Wiley, 2002.
[36] King R.D., Ouali M., Strong A.T., Aly A., Elmaghraby A., Kantardzic M., Page D. (2000) Is it better to combine predictions? Protein Engineering, 13(1), 15-19.

Essays in Bioinformatics
D.S. Moss et al. (Eds.)
IOS Press, 2005


# Predicting Protein Function and Structure Using Bioinformatics Protocols: A Case Study of the SAND Protein Family

Amanda COTTAGE[1], Lisa J. MULLAN[2], Miriam B.D. PORTELA[1], Elizabeth HELLEN[1], Tim J. CARVER[3], Sunil PATEL[4], Tanya VAVOURI[1], Greg ELGAR[1], Yvonne J.K. EDWARDS[5]
*[1]MRC Rosalind Franklin Centre for Genomic Research, Genome Campus, Hinxton, Cambridge, CB10 1SB, UK. [2]EMBL - European Bioinformatics Institute, Genome Campus, Hinxton, Cambridge, CB10 1SD, UK. [3]Wellcome Trust Sanger Institute, Genome Campus, Hinxton, Cambridge, CB10 1SA, UK. [4]Accelrys Inc., 334 Cambridge Science Park, Milton Road, Cambridge, CB4 OWN, UK. [5]Comparative Genomics & Bioinformatics, School of Biological and Chemical Sciences, Queen Mary, University of London, Mile End Road, London E1 4NS, UK*

**Abstract.** In this chapter, bioinformatics techniques are used to gain some insights into the structure and function of a largely uncharacterised protein family called SAND. From a phylogenomics analysis, we determine SAND as a eukaryotic gene and show that a duplication event gave rise to two SAND genes in vertebrates. SAND was found to be absent from archea and bacteria. From a phylogenetic analysis, we characterise a number of subfamilies. With the use of multiple sequence alignments, we highlight amino acids and sequence motifs conserved in SAND proteins plus those invariant in subfamilies or taxonomical groups. In addition, we predict a secondary structure and solvent accessibility profile and carry out protein fold predictions for the SAND proteins.

## Introduction

Predicting protein structure from sequence often involves tailored sequence similarity searches against specialised databases. For example, carrying out a BLASTP search against NRL3D (a databank of protein sequences of known structures), or a PSI-BLAST search against a non-redundant protein databank, or a HMMER search against PFAM (Tables 1-3). Protein structure prediction could also include performing multiple sequence alignments, secondary structure predictions, solvent accessibility predictions, protein fold recognition, constructing models to atomic resolution and model validation. In many cases, not all protein structure prediction projects involve the use of all these techniques. The key or most central part of a typical protein structure prediction is to identify a structural target from which to extrapolate three-dimensional information for a query sequence. If this central part is in error, the whole prediction will be incorrect. This is the most crucial part of the project.

**Table 1.** Tools for sequence similarity searches and the sequence retrieval system (SRS). The servers permit searches against one or more databases.

| **Software** | **Reference** | **URL** |
|---|---|---|
| BLAST | [1] | http://www.ncbi.nlm.nih.gov/BLAST/ |
| PSI-BLAST | [1] | http://www.ncbi.nlm.nih.gov/BLAST/ |
| HMMER | [2] | UK (http://www.sanger.ac.uk/ Software/Pfam/)<br>USA (http://pfam.wustl.edu/)<br>France (http://pfam.jouy.inra.fr/)<br>Sweden (http://Pfam.cgb.ki.se/) |
| SRS | [3] | http://srs.ebi.ac.uk/ |

**Table 2.** Servers to perform secondary structure, solvent accessibility and fold prediction.

| **Software** | **Reference** | **URL** |
|---|---|---|
| **Secondary structure** | | |
| JPRED | [4] | http://www.compbio.dundee.ac.uk/~www-jpred/ |
| PHD | [5] | http://www.embl-heidelberg.de/predictprotein/predictprotein.html |
| **Protein fold prediction** | | |
| PHD (TOPITS) | [5] | http://www.embl-heidelberg.de/predictprotein/predictprotein.html |
| GenThreader | [6] | http://bioinf.cs.ucl.ac.uk/psipred/psiform.html |
| FUGUE | [7] | http://www-cryst.bioc.cam.ac.uk/servers.html |
| 3D-PSSM | [8] | http://www.sbg.bio.ic.ac.uk/~3dpssm/ |
| PRIDE | [9] | http://www.icgeb.org/pride/ |
| MetaServer | [10] | http://bioinfo.pl/meta/ |
| MetaServer | [11] | http://genesilico.pl/meta/ |

**Table 3.** Tools for comparative modelling of protein structures. The key to the symbols used in the table are as follows; * refers to restraint based molecular modelling methods and refers to rigid body fragment assembly methods.

| **Software** | **Reference** | **URL** |
|---|---|---|
| **Academic versions** | | |
| COMPOSER^ | [12,13] | http://www-cryst.bioc.cam.ac.uk/ |
| Modeller* | [14] | http://salilab.org/modeller/ |
| WhatIF¶ | [15] | http://www.cmbi.kun.nl/whatif/ |
| SwissModel¶ | [16] | http://swissmodel.expasy.org |
| **Commercial** | | |
| Modeller* | - | http://www.accelrys.com/ |
| Homology¶ | - | http://www.accelrys.com/ |
| QUANTA | - | http://www.accelrys.com/ |
| SYBYL | - | http://www.tripos.com/ |
| COMPOSER^ | - | http://www.tripos.com/ |

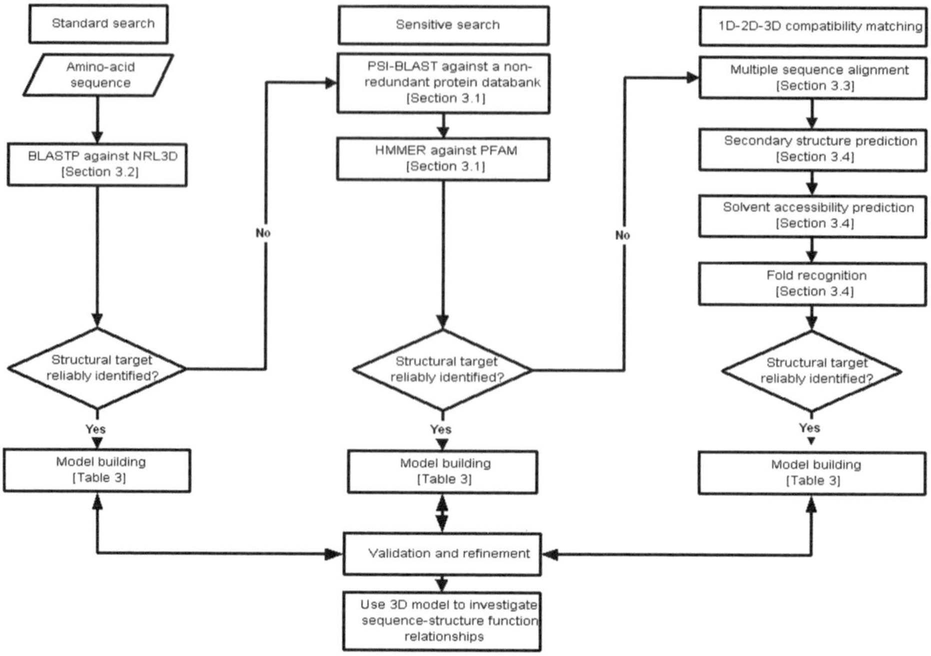


**Figure 1.** A flowchart for predicting 3D structures from protein sequences by using bioinformatics techniques. Predictions using "standard searches" are the most accurate. The sensitive searches and "1D-2D-3D" compatibility matching methods are non-trivial methodologies that can sometimes add value to the sequences where the standard techniques do not identify a structural template for the query sequence.

The first step in a typical protein structure prediction is to establish if a protein sequence or part of a protein sequence has any homologues of known structure in the Protein Data Bank (PDB) [17, 18]. Typically, protein structures are experimentally determined and classified at the level of the domain [19, 20]. Comparative molecular modelling or homology modelling is currently the most successful and accurate method for protein structure prediction [21]. If a protein structure prediction can be based on comparative molecular modelling (Table 3), this should be the method of choice (Figure 1). In the absence of high sequence identity between sequence and structural homologues, deciding what constitutes significant sequence similarity is not straightforward. This type of prediction then becomes "non-trivial". The most promising methods for solving this type of problem involves performing sensitive sequence searches and characterising sequence compatibility with the structural properties of known secondary and tertiary protein structure (also known as "1D-2D-3D" compatibility matching methods). Sensitive searches help identify weak similarities between the sequence of interest and homologues that have had their structures experimentally determined to atomic resolution. The "1D-2D-3D" compatibility matching methods include, secondary structure and solvent accessibility predictions as well as protein fold recognition. Such methods can be useful in predicting common structural folds for proteins that share little or no sequence similarity (Figure 1). However, at low levels of sequence similarity the structures of proteins sharing a common fold diverge to such an extent that the accuracy of models built by comparative techniques are significantly reduced [21, 22].

**Table 4a.** Twenty-three SAND sequences identified in protoctista, fungi, plants and invertebrates.

| Organism | SAND identifier | Accession Number |
|---|---|---|
| **Fungi** | | |
| *Gibberella zeae* | GZ_SAND | EMBL:AACM01000298 |
| *Neurospora crassa* | NC_SAND | SPTR:Q870Q4 |
| *Schizosaccharomyces pombe* | SP_SAND | SPTR:Q10150 |
| *Eremothecium gossypii* | EG_SAND | SPTR:Q75EA2 |
| *Aspergillus nidulans* | AN_SAND | EMBL:EAA64925 |
| *Saccharomyces cerevisiae* | SC_SAND | SPTR:P53129 |
| *Yarrowia lipolytica* | YL_SAND | EMBL:CAG81815 |
| *Candida albicans* | CA_SAND | * |
| **Plasmodia** | | |
| *Plasmodium falciparum* | PF_SAND | SPTR:Q8IDH2 |
| *Plasmodium yoelli trophy* | PY_SAND | SPTR:Q7RL16 |
| **Slime mould** | | |
| *Dictyostelium discoideum* | DD_SAND | EMBL:BJ377438 EMBL:C24407 EMBL:BJ330011 |
| **Nematodes** | | |
| *Caenorhabditis elegans* | CE_SAND | SPTR:Q9B189 |
| *Caenorhabditis briggsae* | CB_SAND | EMBL:AC084558 |
| **Insects** | | |
| *Drosophila melanogaster* | DM_SAND | SPTR:Q9VR38 |
| *Apis mellifera* | AM_SAND | REFSEQ:XP_396160 |
| *Anopheles gambiae* | AG_SAND | SPTR:Q7Q176 |
| ***Sea squirt*** | | |
| *Ciona intestinalis* | CI_SAND | EMBL:BW166332 EMBL:BW295692 |
| **Plants** | | |
| *Arabidopsis thaliana* | AT_SAND | SPTR:Q9SKN1 |
| *Lycopersicon esculentum* | LE_SAND | EMBL:BI927128 EMBL:BI930515 EMBL:AW222182 |
| *Glycine max* | GM_SAND | EMBL:BE474111 EMBL:BM522384 EMBL:CA851897 |
| *Oryza sativa* | OS_SAND | SPTR:Q94CS8 |
| *Triticum aestivum* | TA_SAND | EMBL:AL826200 EMBL:CD896369 |
| *Saccharum officinarum* | SO_SAND | EMBL:CA79484 EMBL:CA097914 |

Unfinished sequences for *C. albicans* was obtained from the NCBI (http://www.ncbi.nlm.nih.gov/BLAST). The *S. cerevisiae* comprises 644 residues; our analysis would indicate that an intron is present in this prediction.

In this chapter, we use these protocols to study a gene, first reported as open reading frame G2889, on chromosome VII of *Saccharomyces cerevisiae*. At the time, the translated ORF G2889 showed no significant sequence similarity to other proteins in the databank [23]. Three years later, a homologue was identified from an analysis of the plasminogen related growth factor receptor 3 (PRGFR) gene locus in *Fugu rubripes* (FR_SAND1; Table 4). The homologue was named SAND because it is neighbouring a PRGFR gene that is an orthologue of SEA [24].

Additional SAND homologues were found as eukaryotic genome sequences, such as *Caenorhabditis elegans*, *Drosophila melanogaster* and *Arabidopsis thaliana*, became available [25]. Whilst one SAND homologue was found in each of these genomes, two copies were identified from searches of the then unassembled human genome [25]. In addition, the protein SAND (known as Mon1p in yeast) was shown to function at the tethering/docking stage of vesicle/vacuole fusion as a critical component of the vacuole SNARE complex [26, 27]. In this chapter, we describe a multidisciplinary bioinformatics approach by using comparative genomics, structure prediction [28] and phylogenomics [29], to shed light on the possible structure and function of various members of the SAND protein family. Resources for various protein structure prediction techniques are described

(Tables 1-3). The expected accuracy and the strengths and weaknesses of the methods are highlighted. These methods outlined can be of value in protein structure predictions.

**Table 4b.** Seventeen SAND sequences identified in nine vertebrate species.

| Organism | SAND identifier | Accession Number |
|---|---|---|
| **Mammals** | | |
| *Homo sapiens* | HS_SAND1 | SPTR:Q9BRF3 |
| *Homo sapiens* | HS_SAND2 | SPTR:O94949 |
| *Mus musculus* | MM_SAND1 | SPTR:Q9CYS2 |
| *Mus musculus* | MM_SAND2 | SPTR:Q8BMQ8 |
| *Rattus norvegicus* | RN_SAND1 | REFSEQ:XP_236627 |
| *Rattus norvegicus* | RN_SAND2 | REFSEQ:XP_226493 |
| *Macaca fascicularis* | MF_SAND1 | SPTR:Q95KG9 |
| **Birds** | | |
| *Gallus gallus* | GG_SAND1 | EMBL:GGA395913 |
| *Gallus gallus* | GG_SAND2 | EMBL:BU209213 EMBL:BU258474 |
| **Amphibians** | | |
| *Xenopus tropicalis* | XT_SAND1 | EMBL:AL849442 |
| *Xenopus tropicalis* | XT_SAND2 | EMBL:BQ388616 EMBL:AL779783 EMBL:BJ072986 |
| **Fish** | | |
| *Danio rerio* | DR_SAND1 | EMBL:BX293991 |
| *Danio rerio* | DR_SAND2 | EMBL:BX927379 |
| *Fugu rubripes* | FR_SAND1 | SPTR:Q9YGN1 |
| *Fugu rubripes* | FR_SAND2 | EMBL:CAAB01003001 |
| *Tetraodon nigroviridis* | TN_SAND1 | EMBL:CAF96888 |
| *Tetraodon nigroviridis* | TN_SAND2 | EMBL:CAG07009 |

## 2. Material and Methods

### *2.1. Identification of SAND Homologues*

Previously reported SAND protein sequences [25] were used to query public databases using version 2.2.6 of the NCBI BLAST algorithm [1]. BLASTP was used to search protein databases with the SAND protein sequences and BLASTX was used to search these databases with translated SAND nucleotide sequences. Protein databases searched included SWISSPROT release 42 and SWISSPROT TrEMBL release 25 [30]. Nucleotide databases searched included EMBL release 77 [31], ENSEMBL release 19 [32] and unfinished genomic sequences (http://www.ncbi.nlm.nih.gov/BLAST). Translations of these database sequences were also searched with translations of known SAND sequences using TBLASTN. Putative SAND gene sequences were verified by comparisons with EST data using BLASTN.

### *2.2. BLASTP against NRL3D and Other Protein Sequence Characterisation*

An investigation of the SAND sequences was carried out using the web-based analysis tool PIX (http://www.hgmp.mrc.ac.uk/Registered/Webapp/pix/). PIX helps to identify regions of interest in a protein sequence. PIX runs several protein analysis programs on a query sequence and notifies the user via e-mail when the results are ready to be inspected. The user can visualise the results of the analysis programs. PIX includes BLASTP searches against NRL3D [33]. The transmembrane prediction programs PHDhtm [5], TopPred2 [34], MEMSAT2 [35] TMPred [36] and DAS [37] were used.

### *2.3. Generation of Multiple Sequence Alignments and Phylogenetic Analysis*

The C-termini of forty SAND sequences (Tables 4a and 4b) were aligned in preparation for phylogenetic analyses using the alignment program Clustalw (version 1.83) [38]. The N-termini sequences were not included as they were too heterogeneous across the species. Phylogenetic analysis was performed using PHYLO_WIN (version 1.2) [39]. SEAVIEW was used to convert the alignment from MSF format to MASE format. PHYLO_WIN was used to obtain a phylogenetic tree in ASCII format using the neighbour joining method, with observed divergence, pairwise gap removal and 500 bootstrap replicates. The character-based tree from PHYLO_WIN was rendered using the phylip drawtree program (Figure 2). Based on this tree a subset of eleven of these sequences were chosen, as being representative of distant taxa (Figure 3), and these were used for further protein sequence analysis and structure prediction. The JEMBOSS Alignment Editor was used to view and annotate sequence alignments (Figure 3) and to generate a percentage pairwise sequence identity matrix (Table 5). JEMBOSS [40, 41] is the graphical interface to EMBOSS [42]. This suite of programmes is freely available at the following site http://emboss.sourceforge.net/.

### *2.4. Secondary Structure, Solvent Accessibility and Fold Prediction*

The secondary structure and solvent accessibility predictions were carried out using the Jpred server [4,43]. The ClustalW alignment of the eleven representative SAND members (Figure 3) was used as input to the Jpred server. The three fragments defined at the end of Section 3.4 were analysed using the protein structure prediction MetaServer at http://BioInfo.PL/meta [10]. This server submits the query-sequence to several servers that perform structural fold predictions, the results are collated, summarised and consensus fold predictions provided. SeqFold [44-45] and profiles-3D [45-46] were used to predict the protein fold of the C-terminal section of SAND.

**Table 5.** A matrix showing the pairwise percentage sequence identity of the SAND proteins in Figure 3. The percentages are calculated using the JEMBOSS alignment editor.

| | HS_1 | FR_1 | HS_2 | FR_2 | CI | DM | AT | OS | CE | PY | SP |
|---|---|---|---|---|---|---|---|---|---|---|---|
| HS_1 | 100.0 | 77.5 | 55.0 | 59.4 | 58.4 | 53.8 | 41.4 | 39.3 | 39.7 | 35.1 | 41.6 |
| FR_1 | | 100.0 | 53.6 | 56.7 | 58.4 | 51.9 | 42.0 | 40.3 | 41.4 | 37.2 | 42.2 |
| HS_2 | | | 100.0 | 63.6 | 47.1 | 42.8 | 36.5 | 34.0 | 32.4 | 25.8 | 33.8 |
| FR_2 | | | | 100.0 | 49.6 | 43.9 | 35.5 | 33.4 | 33.6 | 29.4 | 33.6 |
| CI | | | | | 100.0 | 51.5 | 43.1 | 42.2 | 39.1 | 36.6 | 42.0 |
| DM | | | | | | 100.0 | 38.4 | 37.8 | 40.5 | 35.1 | 37.2 |
| AT | | | | | | | 100.0 | 74.2 | 31.9 | 33.2 | 34.2 |
| OS | | | | | | | | 100.0 | 31.3 | 34.0 | 34.5 |
| CE | | | | | | | | | 100.0 | 30.2 | 33.6 |
| PY | | | | | | | | | | 100.0 | 33.8 |
| SP | | | | | | | | | | | 100.0 |

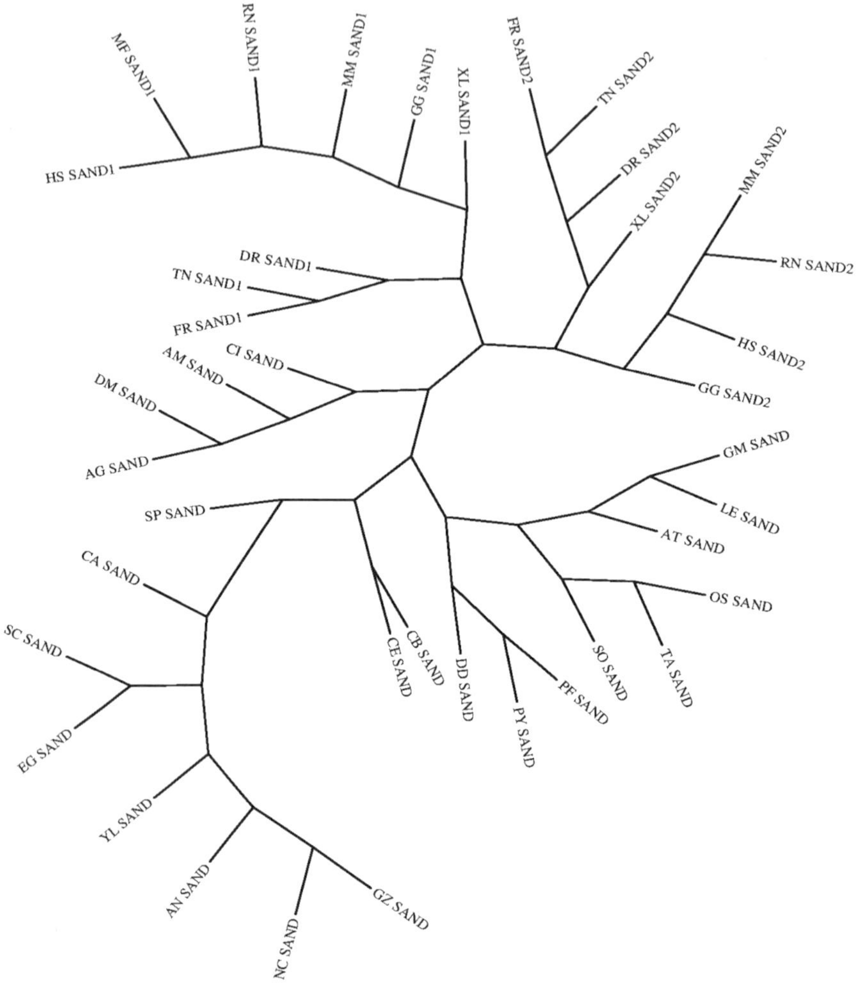


**Figure 2.** A phylogenetic tree generated from 40 C-termini SAND amino-acid sequences. Table 4 provides a key relating the sequence name to species.

## 3. Results

### *3.1. Identification of SAND Homologues and Phylogenetic Analysis*

Our sequence database searches identified 40 SAND sequences in 32 species of eukaryote (Tables 4a and 4b). A single copy of the SAND gene exists in plants, invertebrates, protoctista (single celled eukaryotes) and fungi (Table 4a). In vertebrates, where the full genome sequence was available, two SAND sequences were always identified. We designated these SAND1 and SAND2 (Table 4b). Two full length SAND sequences were found in the following mammals; human, mouse, rat (Table 4b) and chimpanzee (data not shown). Partial SAND sequences were found in pig, cow, sheep and dog from EST searches (data not shown). Two full-length sequences were identified in each of the teleost fishes *Fugu rubripes*, *Danio rerio* and *Tetraodon nigroviridis* (Table 4b). Two partial SAND sequences were found in frog and chicken from EST searches (Table 4b). Subfamily divisions of the SAND family can be seen from the phylogenetic tree, with SAND1, SAND2 and the plant SANDs forming distinct clades (Figure 2). This may be indicative of divergence and specialisation in the function within these SAND groups compared to other SAND groups. As mentioned previously, plants, invertebrates, protoctista and fungi have a single copy of SAND and the yeast sand protein is known to function in mediating vesicle/vacuole fusion [26,27]. Vacuoles are organelles characteristic of eukaryotes such as plants, invertebrates, protoctista and fungi; whilst lysosomes are specialised "vacuole-like" organelles found in vertebrates. The SAND gene duplication event is likely to be associated with the evolution of mediating fusion events into the more specialised lysosome in vertebrates and the duplication event leading to SAND1 and SAND2 in vertebrates occurred somewhere between Chordata (chordates) and Gnathostomata (jawed vertebrates). As we are aware from yeast functional studies, SAND functions in mediating vacuole fusion events and in view of the above, we make the hypothesis that the duplication event occurred concurrently with the evolution of lysosomes from vacuoles in early vertebrates.

### *3.2. BLASTP versus NRL3D and Protein Sequence Characterisation*

No homologues with experimentally determined structures were identified by BLASTP searches of NRL3D with the eleven full-length SAND sequences (Table 4). The iterative BLAST algorithm PSI-BLAST can be used to identify homologous protein sequences with known 3D structures even if the subject and query sequences have less than 20% sequence identity. However in this example, using both full and partial SAND sequences, after four successive PSI-BLAST 2 iterations there was a failure to return any similar sequence of known structure.

Profile Hidden Markov Models (HMMs) built from Pfam alignments can be used to determine if a query protein sequence contains an existing characterised protein domain. Pfam HMMs [2] were searched with all SAND sequences and each returned a match of their C-terminus to the domain DUF254. The DUF254 seed alignment contains 26 SAND sequences from 13 species. These are sequences with an SPTR accession number (Table 4). Our analysis reveals 40 members from 32 species. SANDs from an additional 19 species are uncovered through our analysis of the available databases; these are entries with an EMBL or REFSEQ accession number (Table 4).

From the PIX analysis various features were predicted in individual SAND sequences, for example; coiled coils, signal peptides and peptide cleavage sites. Unfortunately the threshold at which these features were determined was not significant. A putative transmembrane domain was reported by TMPred and DAS comprising residues

32-74 in all the SAND C-termini sequences. The residue numbering in this section is taken from the alignment (Figure 3). This region coincides with the solvent inaccessible α-helix A2 shown in Figure 3. Further analysis using TopPred2 and MEMSAT2 corroborates the prediction. A fifth transmembrane prediction algorithm, PredictProtein, does not report a transmembrane domain for this region. Potential transmembrane regions were noted at positions 444-464 in SAND1 and 20 residues further towards the C-terminus in SAND2 sequences. These regions are highly conserved within the paralogous groups. They may be two highly significant hydrophobic regions or potential membrane-spanning regions, related to the protein interaction with organelles. Domain database searches of SBASE [46] ProDom [47], BLOCKS [48], PRINTS [49] and PROSITE [50] return predicted features in many SAND sequences but these were not consistent nor conserved among the individual members of the SAND sequences submitted with the exception of PRINTS. PRINTS returned 13 signature elements designated YEAST73DUF across 6 species. These elements are all found within our alignments.

### *3.3. Multiple Sequence Alignment*

The alignment generated from eleven full-length C-termini of the SAND protein sequences was used as input for Jpred2 to predict the secondary structure and solvent accessibility. This alignment is deposited in the EMBL-Align database [51] and can be retrieved using the accession number ALIGN_000714. The sequence identity between the C-termini varies from 25.76 and 77.48% (Table 5). There is no significant sequence similarity in the N-termini (approximately first 100 amino acids) across the SAND subgroups (SAND, SAND1 and SAND2) (data not shown) and no consensus α-helices or β-strands predicted by Jpred2 for this region [25]. There is detectable sequence similarity within the subgroups of SAND1 and SAND2 N-termini sequences (data not shown).

The amino acid sequences of the SAND subgroups have a high degree of amino-acid sequence identity which is typically >60% (Table 5). However more sequence divergence is observed between the SAND2 proteins than is seen between the SAND1 proteins (Table 5, Figures 2-3). Several conserved motifs including sixteen invariant amino-acid residues are highlighted in the alignment (Figure 3). Motif GKP is in L3; an alanine and a serine are in solvent inaccessible α-helix A2; a leucine is in α-helix A3; an aspartate and a leucine are in L9; a proline is in L11a; an arginine is in α-helix A5; an aspartate is in α-helix A7; a PXCXP signature spans β-strand B7 and loop L16; phenylalanine is in α-helix A10 and in β-strand B13. The alignment shows loops comprising insertions specific to certain SAND protein groups. Loop L5 contains a plasmodium specific insertion, loop L10 contains an insect specific insertion, loops L13, L21 and L27 are specific to SAND2 proteins and loop L20 is plant specific. Conserved plant cysteines are found at alignment positions 98, 102, 133, 215, 448, 473 and 477 (Figure 3).

SeqFold and profiles-3D were used to predict the protein fold of the C-terminal section of SAND. The structure of the PDB entry 1gw5, a small three layer sandwich with solvent inaccessible strands and amphipathic helices, was predicted to match the first 100 amino acids of the C-terminal section of SAND. This prediction matches the secondary structure consensus for that region (Figure 3). 1gw5 is the experimentally determined protein structure of AP2 (a heterotetrameric clatharin adaptor complex). This protein mediates endocytosis. SeqFold and profiles-3D predicted a match to annexins, 1a8a, an all α-helical protein that matches residues 95-190 of the SAND family. This is complementary to the Jpred2 prediction of 7 consecutive α-helices in this region. This α-bundle structure has several solvent inaccessible α-helices that tie in with our secondary structure prediction and solvent accessibility prediction. No conclusive predictions were made for the remaining region.

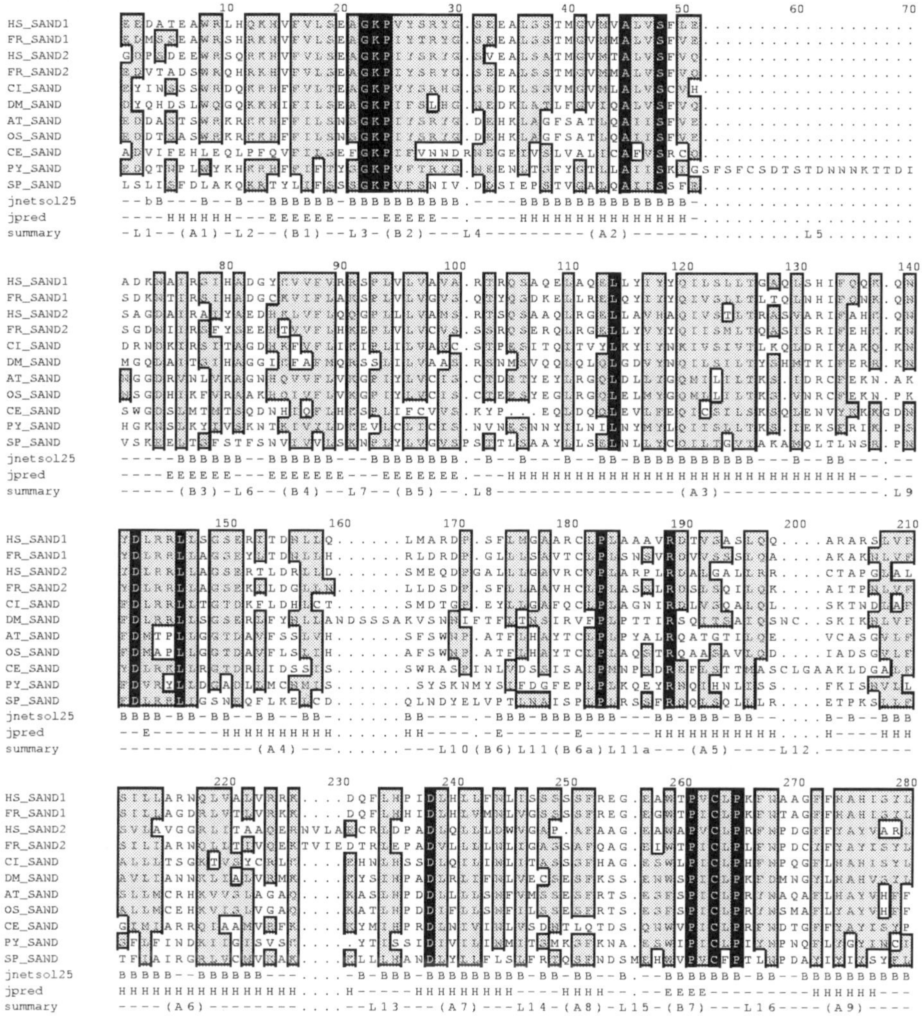

**Figure 3.** (continued on the subsequent page) An alignment of eleven representative C-termini SAND amino acid sequences. Sequence identifiers are defined in Table 4. The Jpred consensus secondary structure prediction (Jpred) is supplied where H is α-helix, E is β-strand and - is loop region. A numbering scheme for the secondary structural elements is provided (Summary) where A is α-helix, B is β-strand and L is loop. The Jnet side-chain solvent accessibility is predicted (Jnetsol25) where "B" denotes that 25% or less of the side-chain is solvent inaccessible and "." or "-" denotes that 25% or more of the side-chain is solvent accessible. Invariant residue positions are formatted with white characters on a black background. The Figure was generated by the JEMBOSS multiple sequence alignment editor.

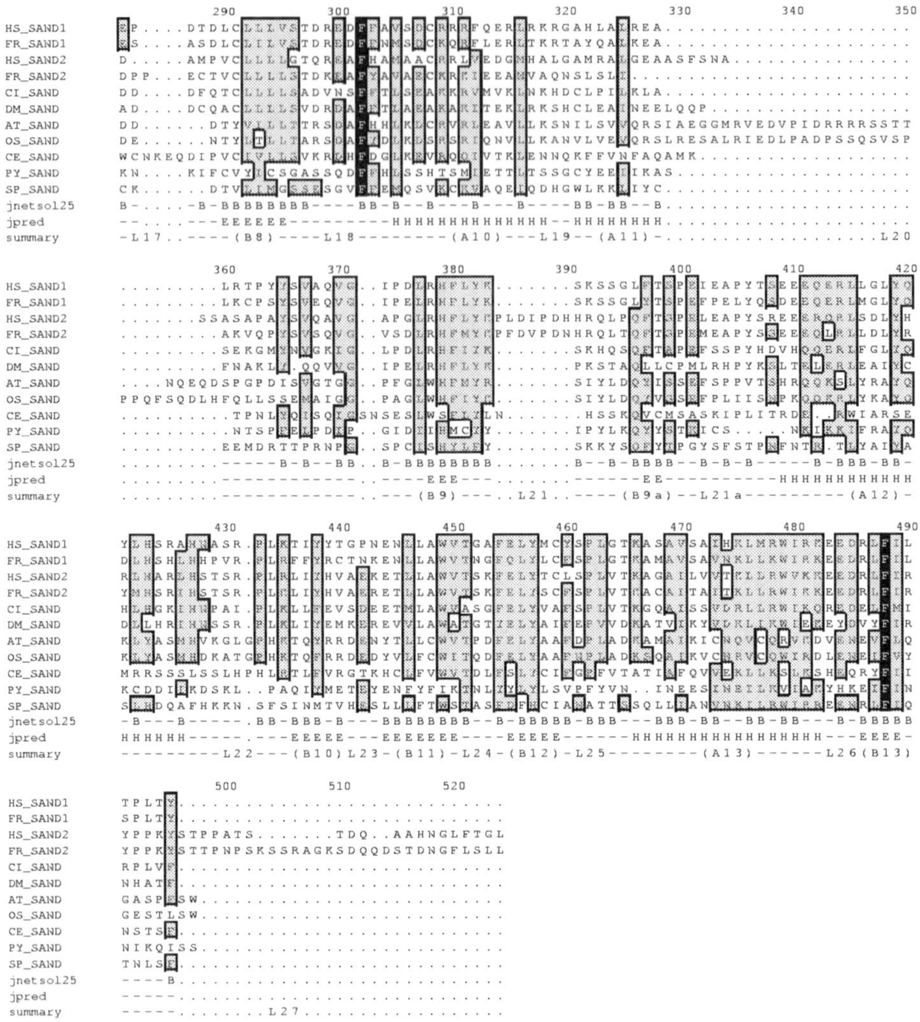

### 3.4. Secondary Structure, Solvent Accessibility and Protein Fold Prediction

The C-termini of the SAND proteins are predicted to contain fifteen β-strands, thirteen α-helices and 29 loops. All β-strands are predicted to be largely solvent inaccessible, as are four α-helices A2, A7, A8 and A9. Eight of the thirteen α-helices display an amphipathic pattern (A3, A4, A5, A6, A10, A11, A12 and A13). These amphipathic α-helices are likely to be located on the outer surface of the protein with one side of the α-helix facing the solvent and the other the hydrophobic interior. The extreme C-termini of the SAND2 sequences are 20-30 residues longer than that of other SANDs (Figure 3) whilst the N-termini of the SAND1 proteins are 40 residues longer (data not shown).

The C-termini of SAND is likely to contain 3 structural domains. These are possibly a layered αβ-sandwich followed by an α-helical bundle structure and a second layered-αβ-sandwich. It is possible that the first and third domains form a non-contiguous TIM barrel with an α-helical domain in the middle of the sequence. To test this hypothesis, the C-termini SAND sequences were split into three protein fragments. The first contains amino-acid residues 1-100, the second spans residues 101-255 and the third comprise residues 256-525 (for numbering see Figure 3). It is generally accepted that many protein fold recognition programs predict more accurately if the domain boundaries are known [21-22, 28]. Each of the three regions was analyzed using the protein structure MetaServer (see section 2.4). The results generated support the predictions obtained from SeqFold and Profiles-3D analyses.

## 4. Discussion and Conclusions

We searched the available genomes, transcriptomes and protein sequence databases and determine that SAND is a eukaryotic gene. We categorised three SAND protein subfamilies. The first subfamily comprises members from protoctista, fungi, plants, invertebrate metazoans. The second and third classes comprise the vertebrate SAND1 and SAND2 proteins respectively. We postulate that the duplication event that gave rise to the SAND1 and SAND2 paralogues is likely to have coincided with the evolution of vacuoles to lysosomes in early vertebrates, therefore providing valuable clues and leads as to the function of SAND1 and SAND2.

We predicted a robust secondary structure for the SAND proteins and have determined amino-acid sequences and motifs that are either invariant or highly conserved across certain subgroups and across the family. The secondary structure prediction on a residue-per-residue level is expected to be 74% accurate [4, 43]. We have made some suggestions as to the type, number and location of structural domains likely to be present in the C-termini of SAND proteins however we did not build these to atomic resolution (Table 3, Figure 1) as these predictions require validation. Bioinformatics techniques are becoming increasingly more effective, more accessible, quicker and simpler to use, whilst the databanks are growing in size and diversity. So these approaches, if used appropriately, should help to close the gap between sequence and structure and complement *in vitro* approaches to investigate molecular structure and function.

## References


[1] Altschul SF, Madden TL, Schaffer AA, Zhang J, Zhang Z, Miller W, Lipman DJ (1997) Gapped BLAST and PSI-BLAST: a new generation of protein database search programs. Nucleic Acids Res 25:3389-3402.

[2] Bateman A, Coin L, Durbin R, Finn RD, Hollich V, Griffiths-Jones S, Khanna A, Marshall M, Moxon S, Sonnhammer ELL, Studholme DJ, Yeats C, Eddy SR (2004). The Pfam protein families database Nucleic Acids Res 32: D138-D141.

[3] Zdobnov EM, Lopez R, Apweiler R, Etzold T. (2002) The EBI SRS server - recent developments. Bioinformatics. 18:368-373.

[4] Cuff JA, Clamp ME, Siddiqui AS, Finlay M, Barton GJ (1998) JPred: a consensus secondary structure prediction server. Bioinformatics 14:892-893

[5] Rost B, Yachdav G, Liu JF (2004). The PredictProteinServer. Nucleic Acids Research 32: W321-W326 Suppl.

[6] Jones DT (1999) GenTHREADER: An efficient and reliable protein fold recognition method for genomic sequences. J Mol Biol 287: 797-815.

[7] Shi JY, Blundell TL, Mizuguchi K (2001) FUGUE: Sequence-structure homology recognition using environment-specific substitution tables and structure-dependent gap penalties. *J. Mol. Biol.* **310**: 243-257.

[8] Kelley LA., MacCallum RM, Sternberg MJE (2000) Enhanced genome annotation using structural profiles in the program 3D-PSSM. *J. Mol. Biol* **299**, 499-520.

[9] Carugo O., Pongor S. (2002) Protein Fold Similarity Estimated by a Probabilistic Approach Based on C(alpha)-C(alpha) Distance Comparison, J Mol Biol., 315:887-898.
[10] Ginalski K, Elofsson A, Fischer D, Rychlewski L (2003) 3D-Jury: a simple approach to improve protein structure predictions Bioinformatics 19:1015-1018
[11] Kurowski MA, Bujnicki JM (2003). GeneSilico protein structure prediction meta-server. Nucleic Acids Res. 31:3305-3307.
[12] Sutcliffe MJ, Haneef I, Carney D, Blundell TL (1987a) Knowledge Based Modeling Of Homologous Proteins.1. 3-Dimensional Frameworks Derived From The Simultaneous Superposition Of Multiple Structures Protein Eng 1: 377-384.
[13] Sutcliffe MJ, Hayes FRF Blundell TL (1987b) Knowledge Based Modeling Of Homologous Proteins .2. Rules For The Conformations Of Substituted Side-Chains Protein Eng 1: 385-392.
[14] Eswar N, John B, Mirkovic N, Fiser A, Ilyin VA, Pieper U, Stuart AC, Marti-Renom MA, Madhusudhan MS, Yerkovich B, Sali A (2003). Tools for comparative protein structure modeling and analysis. Nucleic Acids Res 31: 3375-3380.
[15] Vriend G. (1990) WhatIf: A molecular modeling and drug design program. *J. Mol. Graph* **8**, 52-56.
[16] Schwede T, Kopp J, Guex N, Peitsch MC (2003) SWISS-MODEL: an automated protein homology-modeling server. Nucleic Acids Research 31 (13): 3381-3385.
[17] Golovin A, Oldfield TJ, Tate JG, Velankar S, Barton GJ, Boutselakis H, Dimitropoulos D, Fillon J, Hussain A, Ionides JM, John M, Keller PA, Krissinel E, McNeil P, Naim A, Newman R, Pajon A, Pineda J, Rachedi A, Copeland J, Sitnov A, Sobhany S, Suarez-Uruena A, Swaminathan GJ, Tagari M, Tromm S, Vranken W, Henrick K. (2004). E-MSD: an integrated data resource for bioinformatics. Nucleic Acids Res. 32 Database issue:D211-216.
[18] Westbrook J, Feng ZK, Chen L, Yang HW, Berman HM (2003). Nucleic Acids Res 31: 489-491
[19] Nagarajan N, Yona G. (2004). Automatic prediction of protein domains from sequence information using a hybrid learning system. Bioinformatics. 20:1335-1360.
[20] Kong L, Ranganathan S (2004). Delineation of modular proteins: domain boundary prediction from sequence information. Brief Bioinform. 5:179-92.
[21] Kopp, Schwede (2004) Automated protein structure homology modeling: a progress report. Pharmocogenomics 5:405-416.
[22] Baker D, Sali A (2001) Protein structure prediction and structural genomics. *Science*, **294**, 93-96.
[23] Tizon B, Rodriguez-Torres M, Rodriguez-Belmonte E, Cadahia JL, Cerdan E (1996) Identification of a putative methylenetetrahydrofolate reductase by sequence analysis of a 6.8 kb DNA fragment of yeast chromosome VII. Yeast 12 (10B Suppl):1047-1051.
[24] Cottage A, Clark M, Hawker K, Umrania Y, Wheller D, Bishop M, Elgar G (1999) Three receptor genes for plasminogen related growth factors in the genome of the puffer fish Fugu rubripes. FEBS Lett 443:370-374.
[25] Cottage A, Edwards YJ, Elgar G (2001) SAND, a new protein family:from nucleic acid to protein structure and function prediction. Comp Funct Genom 2:226-235.
[26] Bonangelino CJ, Chavez EM, Bonifacino JS (2002) Genomic screen for vacuolar protein sorting genes in Saccharomyces cerevisiae. Mol Biol Cell 13:2486-2501.
[27] Wang CW, Stromhaug PE, Kauffman EJ, Weisman LS, Klionsky DJ (2003) Yeast homotypic vacuole fusion requires the Ccz1-Mon1 complex during the tethering/docking stage. J Cell Biol 163:973-985.
[28] Edwards YJK, Cottage A (2003) Bioinformatics methods to predict protein structure and function - A practical approach Molecular Biotechnology 23: 139-166
[29] Eisen JA, Fraser CM. (2003) Phylogenomics: intersection of evolution and genomics. Science. 300:1706-1707.
[30] Boeckmann B, Bairoch A, Apweiler R, Blatter MC, Estreicher A, Gasteiger E, Martin MJ, Michoud K, O'Donovan C, Phan I, Pilbout S, Schneider M (2003) The SWISS_PROT protein knowledge base and its supplement TrEMBL in 2003. Nucleic Acids Res 31:365-370
[31] Birney E, Andrews D, Bevan P, Caccamo M, Cameron G, Chen Y, Clarke L, Coates G, Cox T, Cuff J (2004) Ensembl 2004. Nucleic Acids Res, 32:D468-470.
[32] Kulikova T, Aldebert P, Althorpe N, Baker W, Bates K, Browne P, van den Broek A, Cochrane G, Duggan K, Eberhardt R, Faruque N, Garcia-Pastor M, Harte N, Kanz C, Leinonen R, Lin Q, Lombard V, Lopez R, Mancuso R, McHale M, Nardone F, Silventoinen V, Stoehr P, Stoesser G, Tuli MA, Tzouvara K, Vaughan R, Wu D, Zhu W, Apweiler R. (2004) The EMBL Nucleotide Sequence Database. Nucleic Acids Res 32:D27-30.
[33] Garavelli JS, Hou Z, Pattabiraman N, Stephens RM (2001) The RESID Database of protein structure modifications and the NRL-3D Sequence-Structure Database. Nucleic Acids Res 29:199-201.
[34] Claros MG, von Heijne G (1994) TopPred II: an improved software for membrane protein structure predictions. Comput Appl Biosci 10:685-686
[35] Jones DT, Taylor WR, Thornton JM. (1994) A model recognition approach to the prediction of all-helical membrane protein structure and topology. Biochemistry. 33:3038-3049.

[36] Hoffman K, Stoffel W (1993) TMBASE - A database of membrane spanning protein segments. Biol Chem 374:166.
[37] Cserzo M, Bernassau JM, Simon I, Maigret B (1994) New alignment strategy for transmembrane proteins. J Mol Biol 243:388-396.
[38] Thompson JD, Higgins DG, Gibson TJ (1994) CLUSTAL W: improving the sensitivity of progressive multiple sequence alignment through sequence weighting, position-specific gap penalties and weight matrix choice. Nucleic Acids Res 22:4673-4680
[39] Galtier N, Gouy M, Gautier C (1996) SEAVIEW and PHYLO_WIN: two graphic tools for sequence alignment and molecular phylogeny. Comput Appl Biosci. 12:543-548.
[40] Carver TJ, Mullan LJ (2002). Website Update: A new graphical user interface to EMBOSS. Comp Funct Genom 3: 75-78.
[41] Carver T, Bleasby A (2003) The design of JEMBOSS: a graphical user interface to EMBOSS. Bioinformatics 19:1837-1843.
[42] Rice P, Longden, I Bleasby A (2000) EMBOSS: The European molecular biology open software suite. *Trends Genet* **16**, 276-277.
[43] Cuff J. A and Barton, G J (2000) Application of multiple sequence alignment profiles to improve protein secondary structure prediction. *Proteins*. ***40***, 502-511.
[44] Olszewski KA, Yan L, Edwards DJ (1999) SeqFold - fully automated fold recognition and modeling software -validation and application. Theor Chem Acc 11:57.
[45] Kitson DH, Bradretdinov A, Zhu Z-Y,Velikanov M, Edwards DJ, Olszewski K, Szalma S, Yan L (2002) Functional annotation of proteomic sequences based on consensus of sequence and structural analysis. Brief in Bioinform 3:32-44.
[46] Vlahovicek K, Kajan L, Murvai J, Hegedus Z, Pongor S (2003) The SBASE domain sequence library, release 10: domain architecture prediction Nucleic Acids Research 31: 403-405.
[47] Corpet F, Gouzy J, Kahn D (1998) The ProDom database of protein domain families. Nucleic Acids Res 26:323-326.
[48] Henikoff JG, Greene EA, Pietrokovski S, Henikoff S (2000) Increased coverage of protein families with the blocks database servers. Nucleic Acids Res 28:228-230.
[49] Attwood TK, Bradley P, Flower DR, Gaulton A, Maudling N, Mitchell AL, Moulton G, Nordle A, Paine K, Taylor P, Uddin A, Zygouri C (2003). PRINTS and its automatic supplement, prePRINTS. Nucleic Acids Res 31: 400-402.
[50] Sigrist CJ, Cerutti L, Hulo N, Gattiker A, Falquet L, Pagni M, Bairoch A, Bucher P (2002) PROSITE: a documented database using patterns and profiles as motif descriptors. Brief Bioinform. 3:265-274.
[51] Lombard V, Camon E, Parkinson H, Hingamp P, Stoesser G, Redaschi N (2002). EMBL-Align: a new public nucleotide and amino acid multiple sequence alignment database. Bioinformatics 18:763-764.

*Essays in Bioinformatics*
*D.S. Moss et al. (Eds.)*
*IOS Press, 2005*


# Industrial Applications of Genomics, Proteomics and Bioinformatics

Daslav HRANUELI
*Faculty of Food Technology and Biotechnology, University of Zagreb, Zagreb, Croatia*

**Abstract**. Bioinformatics is a general approach underlying current paradigms in the pharmaceutical, agricultural and bio-industrial sectors. The parallel development of genomics, proteomics and informatics has resulted in a number of complex approaches and brought about profound changes within the R & D philosophy of the affected sectors. This chapter aims to provide an overview of how the scientific approach has changed in these three areas.

## Introduction

In his book the "Biotech century", published at the very end of the 20$^{th}$ century, Jeremy Rifkin claimed that never before in the history of humanity, had human beings been faced with such significant new technological and economic challenges as those that lay on the horizon. He believes that by the year 2025, our children and us might be living in a world utterly different from anything human beings have ever experienced before [1].

The analysts of science and technology claim that the "Industrial era" is coming to an end. The industrial era marks the final stage of the age of fire. After thousands of years of putting fire to ore, the age of *pyrotechnology* is slowly burning out. Fire has provided human beings with light, heat and power – the three basic necessities for survival. With fire, human beings can melt down the inanimate world of nature and reshape it into a world of pure utility. However, humankind is now facing three crises simultaneously: a decline of the Earth's non-renewable energy resources, a dangerous build-up of global-warming gases and a steady decrease in biological diversity. After five centuries of fusing, melting and burning inanimate matter to create useful things, we now need a new operational matrix. For the last 20 to 30 years, scientists have been splicing, recombining and mobilising living material into economic utilities. Humanity is, therefore, moving from the age of pyrotechnology to the age of biotechnology. For most of the pyrotechnical age, alchemy - the unsuccessful search for a method by which lead could be transformed into gold – served both as the philosophical framework and as conceptual guide to human beings' technological manipulation of the inanimate matter. However, today the stage is being set for the emergence of a new kind of perception – one that reflects the aspirations and objectives of the new biotechnical age based on *algeny*. Joshua Lederberg's term '*algeny*', refined by Jeremy Rifkin, means the change of the essence of living things and is dedicated to "improvement" of existing organisms and the design of wholly new ones with the intention of "perfecting" their performance. But algeny is much more than that. It is a way of thinking about nature, and it is this new way of thinking that sets the course for the next great era in history. Algeny is likely to emerge as a new philosophical framework and an overarching metaphor for the Biotech Century. Instead of being able to change the inanimate matter, the human race will, for the first time, be in the position to dramatically change living beings by the direct influence on evolution.

People believe that there are many convergent forces coming together to create this powerful new social current. They claim that at an epicentre there is a technological revolution that gives scientists an opportunity to reorganise life at the genetic level. Here are just a few examples of what could happen within the next twenty-five years: (*i*) global corporations and research institutions could hold patents on virtually all genes that make up the blueprints of the human race as well as cells, tissues and organs that human body comprises. They may also own similar patents on tens of thousands of micro-organisms, plants and animals, (*ii*) animal and human cloning could become common, with replication partially replacing reproduction. We could see the creation of a range of new chimeric animals on Earth, including human/animal hybrids that can be used as experimental subjects in medical research and as "donors" for xeno-transplantation. The artificial creation and propagation of cloned, chimeric and transgenic living beings could mean the end of the wild world and its substitution by the bioindustrial world, (*iii*) some parents might choose to have their children conceived in the test tube and incubated in artificial womb outside the human body to avoid unpleasant pregnancy and to ensure safe, transparent environment through which to monitor their unborn child's development. Genetic changes could be made in human foetuses in the womb to correct deadly diseases and disorders and to enhance mood, behaviour, intelligence and physical traits, (*iv*) millions of people could obtain a detailed genetic readout of themselves allowing them to gaze into their own biological futures. The genetic information would give people the power to predict and plan their lives in ways never possible before, (*v*) global agriculture could find itself in the midst of a great transition, with an increasing volume of food grown indoors in tissue culture at a fraction of the price of growing it on land, and (*vi*) tens of thousands of novel transgenic micro-organisms, plants and animals could be released into the Earth's ecosystems for commercial tasks ranging from "bio-remediation" to the production of alternative fuels.

Many people believe that we are at the dawn of one of the great transformations in world history. They claim that in front of us lie the passing of one economic era and the birth of another one. Yet, history has taught us that every new technological revolution brings with it both benefits and costs. The more powerful the technology is in overtaking and controlling the forces of nature, the more demanding the price we will be forced to pay in terms of disruption and destruction of the ecosystems and social systems that sustain life as we know it. The wide-ranging impact that the new genomic technologies provided for the commercial market will have on our lives need to be exhaustively evaluated in the coming years to minimise the risks for future generations and for all other creatures who travel with us on life's journey [2].

A broad range of scientific approaches has currently been applied using genomics, proteomics and bioinformatics in the context of the human genetic blueprint. However, one of the most important and controversial applications of these approaches will be the extension of human life span. Experiments with simple organisms, like the nematode worm *Caenorhabditis elegans* [3] or the fruit fly *Drosophila melanogaster* [4], have already shown that their lives can be extended two to three times by specific genetic modification. People believe that within the next 30 years we will be able to eradicate most of the world's major diseases and that the routine sequencing of individual human genomes will be possible by 2030 extending the average lifespan to 90 years by 2040. Recently, it has been shown that a region on human chromosome 4 might be responsible for the exceptional longevity phenotype [5]. Identification of genes in humans that allow certain individuals to live to extremely old age should lead to insights on cellular pathways that are important for the aging process. If this really happens, birth control on a scale that is unimaginable today, and even expansion of the population to inhabitable regions outside Earth, will have to be seriously considered.

The other component of this development, the information technology, has resulted in equally spectacular advances. The most important result is not so much the appearance of fast computing devices, but rather the linking of computers and databases into one interoperable network that enables researchers to access a wide range of data simultaneously. In the background of this development was a slow paradigm change within biology itself. In fact, molecular biology would never come to existence without laboratory computers, since the complex macromolecular objects cannot be represented and analysed with paper and pencil only. The next step, according to James Watson, happened in the early 1990s when biology turned from data collection towards data processing. This was the advent of sequencing projects, which produced a number of novel tools and services. It became obvious that bioinformatics is an independent, new scientific approach that relies on a number of conventional as well as unconventional elements. Theoretical recognitions, such as the DNA structure, of the theory of molecular evolution were at the core of the new approach. Databases, mostly biological sequence databases, played a highly visible role. New algorithms and computer programs were designed for analysing the databases, and finally a number of dedicated national and international research institutions were created in order to promote the spread of bioinformatics. As a result, bioinformatics is a mature science today, with several large conferences and a number of new textbooks published each year. It is apparent that the new biology and dedicated informatics advance hand in hand.

Therefore, the application of functional genomics can and will be found in three main areas: human health, breeding agricultural plants and domestic animals, and breeding industrial microorganisms. To illustrate them, two examples will be used: the example of brewer's and baker's yeast *Saccharomyces cerevisiae* and the example of industrially important species of *Streptomyces* genus and related genera that produce a large number of pharmacologically important compounds.

## 1. Human Health

The first human gene was cloned in 1975. Fourteen years after the cloning of the first human gene, the Human Genome Project with the acronym HUGO was established at NIH headed by James Watson, later replaced by Francis Collins. British, French, German, Japanese and Chinese scientists joined Americans, more than 1,100 scientists altogether. In spite of that, researchers from Celera Genomics, American company lead by the scientist and entrepreneur Craig Venter, was the first to complete the 'working draft' of human genetic blueprint. Both groups published the first draft of the human genome sequence in February 2001 [6, 7], covering about 95% of the 3 x $10^9$ nucleotides. This work suggested that there were only about 30,000 to 40,000 genes present rather than over 120,000 as had been widely assumed previously. Soon after the completion of the human genome 'working draft', the *Human Proteome Organisation*, with the acronym HUPO, headed by Sam Hanash, was established with the aim to consolidate national and regional proteome organisations into a worldwide organisation. Initial consensus for major objectives included: (*i*) accurate annotation of the human genome sequence with respect to small open reading frames (ORF) by the establishment of a complete list of all distinct proteins (Human Protein Catalogue), (*ii*) production of recombinant proteins from each human ORF, making cDNA clone sets available, (*iii*) production of reporter ligands on the output of each and every ORF product, (*iv*) detailing protein/protein interactions, (*v*) detailing protein/nucleic acid interactions, (*vi*) detailing relative levels of tissue specific protein expression, (*vii*) detailing relative levels of intra-cellular protein expression, (*viii*) establishment of formal links with the structural genomics community, and (*ix*) the status of each from (*i*) to (*viii*) with respect to numerous disease conditions. The overall belief is that HUPO will be much bigger than HUGO with more diversity of niches therein (http://www.hupo.org/).

Thanks to the knowledge of the human genome, in years that are coming medics will be able to predict the diseases each of us is predisposed to. Mankind is on the edge of a new preventive and individual molecular medicine that will be based on pharmacogenomics [8], the use of drugs "tailored" according to the specific genes of an individual. Three years after the birth of the cloned sheep Dolly in Scotland, the Great Britain was the first country in the world where scientists were able to clone human embryos for therapeutic needs. It is believed that the Stem cells of human embryos are hiding keys for curing numerous diseases. The Stem cells will allow the development of tissues that will help in curing Alzheimer and Parkinson diseases, heart diseases, multiple sclerosis, muscle dystrophy, diabetes and many others [9]. With the official publication of the first draft of the human genome and the ensuring rapid progress, a number of important doubts are left open. Some of them are technical, for example, it is one thing knowing the gene but it is another to understand the function of its product. Others are legal, like how much should be known about a gene before a patent can guard it. The third group of doubts are social and ethical. Would we really want to have a diagnosis for an incurable disease 20 to 30 years before its first symptoms appear?

An example of social and ethical doubts comes from the major breakthrough in the fight against malaria that was announced in October 2002 by an international collaboration of scientists from the UK and America. A six-year project to sequence the genome of the *Plasmodium falciparum* parasite, which causes the most deadly form of malaria, was completed. In a separate project, an international consortium of researchers sequenced the genome of the *Anopheles gambiae* mosquito, which is a major vector in the transmission of the parasite to humans. The genomes were published in Nature (*P. falciparum*; [10]) and Science (*A. gambiae*; [11]). Malaria infects at least 500 million people per year and at least 1 million per year die of it. The genome sequences should allow new strategies for combating malaria. As passage through humans is an obligate part of the life cycle of *P. falciparum*, this could well lead to eradication of the species. Many people might view this as a desirable goal, but, as the discussion about destruction of small pox stocks shows, there are potential ethical problems even in this case. Another target is the vector and an approach has been proposed that could drive a mosquito species to extinction [12]. This approach would be to construct an element with a homing endonuclease (HEG) that would be inserted in an essential gene for reproduction. The gene would be chosen so that heterozygotes would not suffer from any disadvantages, which would favour the rapid spread of the HEG element. However, homozygotes would be sterile. Theoretical calculations suggest that the release of enough mosquitoes carrying the HEG element could drive a population to extinction within a short time (e.g. ca. 1 year). Although this would appear attractive from the point of view of reducing malaria infections, mosquitoes play an important role in ecology participating in pollen distribution and being a part of the food chain.

Sequencing of the human genome is important for the understanding of the molecular bases of diseases, as well as for the discovery of new biological targets needed for the development of novel drugs. By looking at proteins of model organisms that are similar to a human protein - deduced from newly sequenced human gene - a lot could be learned about its secondary, tertiary and quarterly structure. This knowledge can then be used to search for chemical compounds that will bind the protein and inactivate it. An early example of the application of such studies in drug discovery comes from a proteomics study, in which protein expression in osteoclasts taken from people with bone tumours was compared with that in osteoclasts taken from healthy people. This revealed that one sequence in particular was over-expressed by osteoclasts from people with bone tumours. The sequence matched a previously identified class of molecules: the cathepsins. With this important lead, researchers from the pharmaceutical industry are trying to find a drug that can bind and inactivate the cathepsin K, as an important target for treating osteoporosis [13]. Microbiologists also use genomics, proteomics and bioinformatics for the comparative phylogenic analysis. Until now (July, 2004), more than 140 bacterial genomes have been sequenced, many of which are human pathogens. The sequencing of many more bacterial genomes is in progress (http://www.tigr.org/). Comparison of small microbial genomes, such as the genome of *Mycoplasma genitalium* having only 517 genes, with the human pathogen *Haemophilus influenzae* containing 1.703 genes, revealed 233 conserved genes reflecting the 'minimal genome' of at most 250 genes important enough to be conserved. This approach is currently being developed further to allow simultaneous analysis of genomes of other pathogens. It was believed that the identification of genes that are highly conserved in these organisms would hopefully deliver a pool of possible targets with the potential for the development of novel anti-infectives [14]. Moreover, the advances in high throughput structural genomics allowed scientists to solve as many structures as possible from a known pathogen genome and then to focus on those that may be useful drug targets [15].

Perhaps it is not exaggerated to state that the genomic paradigm now predominant in the pharmaceutical industry based on a set of complex informatics tools that allow easy handling and mining of genomic information. One group of these tools are related to the access of information, and perhaps the best example is the PubMed system of NCBI, the National Centre of Biotechnology Information, which provides transparent access to molecular as well as literature databases developed at the National Library of Medicine (http://www.ncbi.nlm.nih.gov/). This system was primarily conceived in order to promote spreading of the new biological data within the human health domain, and is accompanied by a number of auxiliary systems designed for practicing physicians, which ensures that the new data will immediately reach the patients. Another publicly available system is Ensemble (http://www.ensembl.org/) developed at the Sanger Centre in Cambridge, UK, which allows researchers to navigate among virtually all-possible types of genomic information. In addition, the pharmaceutical industry develops and uses a wealth of novel informatics tools to handle their proprietary databases. Database management systems such as SRS that allow easy navigation among many data types are typical components of both public and proprietory systems.

Typical players of the pharmaceutical arena are integrating technologies in which a laboratory technology such as microarrays is applied together with a set of specialised computational tools. Transcript profiling technology has reached industry almost immediately after the first scientific reports. Companies such as Incyte and Affymetrix offered cDNA based microarrays and a large number of smaller companies and university associated core facilities provide printed microarray services. Other companies specialised in genome sequencing are offering fully annotated genomic sequences.

## 2. Agricultural Plants and Domestic Animals

DNA sequences of organisms that are important in food production have also been accumulating rapidly. For example, Monsanto has recently produced the first 'working draft' of the rice (*Oryza sativa*) genome sequence [16]. Rice is the world's most important food crop. The International Rice Research Institute in the Philippines estimates that by 2020, four billion people will depend on it. That is one of the reasons why plant geneticists want to sequence its genome – to find unknown genes and gene combinations for better rice quality, yield, and pest protection. Nuclear blueprint of rice is about seven times smaller than that of corn (or human, for that matter) but three times larger than that of the tiny mustard plant *Arabidopsis thaliana*, whose sequence has recently been completed [17]. Monsanto's announcement was good news for the International Rice Genome Sequencing Project, a 10-country consortium led by Japan that has already deposited about seven million bases in the GenBank. The rice genome sequence has been decoded to the level of a 'working draft'. This is the first crop genome to be described in such technical detail, and it will provide a new level of understanding of almost all genes in rice, but it leaves certain details yet to be determined. In the years ahead, rice with better nutritional value, greater yields, and more adaptable to seasons, climates and soils will be developed, both through traditional methods of crop improvement (breeding) and genomic technologies. This research may also lead to the development of rice varieties that require less environmental resources, including land and water, and utilise natural resources more effectively. The availability of detailed information about the rice genome will likely lead to global efforts to improve other major food crops, including wheat, corn, potato, barley, sorghum, millet and others. And indeed, the sequencing of wheat (*Tritium aestivum*) [18], corn (*Zea mays*) [19] and potato (*Solanum tuberosum*) (http://www.tigr.org/tdb/potato/) genomes are in progress.

Similarly, general goal of the Animal Genome Research Program is to determine the genetic makeup of various economically important domestic animals. Committees representing major animal groups - like poultry, swine, sheep and cattle - are developing computer databases similar to that available for mouse genome [20]. These will serve as banks for genomic data representing the entire array of genes of a particular animal. The data will provide a basis for comparative studies among animals, to facilitate correlations between genes and their functions, and also to determine the relative positions of genes in the DNA sequence. The committee responsible for swine genome research has made significant progress in the development of a genetic linkage map. The immediate goals for this committee include continuation of the development of a genetic linkage map and production of swine cells that can be grown independently in a laboratory setting to allow for constant availability. The swine database, USPIGBASE (http://www.genome.iastate.edu/pig), is already available for use. Several genetic linkage maps for cattle have been produced, and these cover approximately 90% of the bovine genome. The "international" map has 201 areas of genetic diversity and is the result of an international collaboration involving ten laboratories in seven countries. A major goal for the immediate future is to develop a consensus linkage map, combining information from all independent maps now available, and to subsequently develop a database from this information. The committee directing the mapping of the poultry genome is striving to develop a consensus genetic linkage map of chickens and to extend this map to other poultry of economic importance. Further, this map will be used to identify genes responsible for specific traits, to work with industries to develop effective applications of this knowledge, and to enhance progress in all of these areas through sharing of information via a database. Researchers in the sheep genome project have been successful in developing genetic linkage maps and work on a consensus genetic linkage map is underway.

In the years ahead, agricultural plants and domestic animals will be developed both through traditional breeding methods and genomic technologies. The primary objective of genome sequencing is to increase our understanding of the structure, organisation, function, expression, and regulation of their genes. Further knowledge in these areas will help to maintain genetic diversity, to improve plant and animal productivity and efficiency, to locate economically important production traits, and finally to provide methods for utilising this information to select desired characteristics in these organisms [21].

Bioinformatics tools used in agricultural research do not markedly differ from those developed within the health sector, and the differences are due to the fact that genetic modifications in plants and animals are legally allowed. There is a wealth of specific databases among which those referring to the regulatory aspects of genetic modifications and biodiversity are of special importance.

**Table 1**. Some industrially important microbial species whose genome sequence has been determined.

| Microorganisms | Biotechnological products | Genome (Mb) | References |
|---|---|---|---|
| *Saccharomyces cerevisiae* | Beer, bread, ethanol, yeast biomass, human recombinant proteins | 12.1 | [23] |
| *Lactococcus lactis* | Cheese and other dairy fermentations | 2.4 | [24] |
| *Lactobacillus plantarum* | Various fermentations | 3.3 | See ref. in [25] |
| *Corynebacterium glutamicum* | L-glutamic acid, L-phenylalanine | 3.3 | [26] |
| *Aspergillus niger* | Citric acid, gluconic acid, glucoamylase | 30.0 | See ref. in [25] |
| *Bacillus subtilis* | Food enzymes | 4.2 | [27] |
| *Escherichia coli* | Human recombinant proteins | 4.6 | [28] |
| *Streptomyces coelicolor* | Model species of antibiotic producers | 8.6 | [29] |
| *Streptomyces avermitilis* | Antiparasitic avermectin | 9.0 | [30] |

## 3. Industrial Microorganisms

Apart from human pathogens, the sequencing of genomes of industrial micro-organisms is also important from the point of view for their application in further breeding. Genome sequencing of industrial micro-organisms, such as brewer's and baker's yeast *S. cerevisiae* or lactococci *lactis* and *plantarum* used in dairy industry, as well as genomes of some micro-organisms whose products are used as food additives or for food processing have also been completed. Secondary metabolites produced by *Streptomyces* species and related genera are also important in industrial production. Among these, antibiotics like Tetracycline and Erythromycin, antiphrastics like Avermectin, coccidiostatics like Monensin, natural insecticides like Spinosyn, animal growth promotants and others are all used in fighting infections in humans, animals, fishes and plants (Table 1) [22]. To explain their application better two examples are used.

### *3.1 Saccharomyces cerevisiae genome sequencing*

Since the industrial strains of the yeast *S. cerevisiae* have been used in a number of biotechnological processes like baking bread as well as the production of beer, industrial ethanol, yeast biomass and human recombinant proteins, and since these species have – as a model eucaryote - a number of other possible applications, the knowledge about their functional genomics will be brifely summarised as an example of industrial micro-organism.

In 1992 a European consortium led by a British scientist Steve Oliver sequenced the first eukaryotic chromosome, chromosome III of the *S. cerevisiae* [31]. This led to the creation of a world wide consortium which, under the leadership of a Belgian scientist André Goffeau, succeeded in deciphering the entire genome of *S. cerevisiae* using a structured, or ordered, approach [23]. The sequence of 12,068 kilobases defines 5,885 potential protein-encoding genes. Approximately 140 of these are ribosomal RNA, 40 genes encode small nuclear RNA molecules, while 275 are transfer RNA genes. In addition, the complete sequence provides information about the higher order organisation of the yeast's 16 chromosomes and allows some insight into their evolutionary history. The major problem to be tackled with during the next stage of the yeast genome project is to elucidate the biological functions of all these genes.

Having the sequence is one thing, but understanding it is quite another. From approximately 6,200 genes of *S. cerevisiae* the function of one-third could be assigned from either previous knowledge or because of a high degree of homology to genes of a known function. Other third could not be unambiguously assigned but has features that at least give some clues to their function. The most surprising discovery was that the last third of genes was of totally unknown function, and was often called orphan genes. This has lead to the world-wide effort to understand the function of all the genes in *S. cerevisiae*, that is European Functional Analysis Network project – the EUROFAN – headed once again by Steve Oliver. This project has grown again into an even bigger project, the so-called Yeast Deletion Project. In one of the published report of the Yeast Deletion Project from five years ago, genomic locations of 1,620 nonessential and 356 essential genes were presented. The distribution of functional classes of essential and nonessential ORFs using the criteria from the Munich information Centre for Protein Sequences was also shown [32].

Completion of the *S. cerevisiae* genome has opened an opportunity for developing new approaches for the evaluation of small molecules and their interaction with living cells in which yeast genome or proteome was used as the unit of function. The Miami conference 'Exploiting Yeast Molecular biology for Therapeutics', summarised by Charles Brenner, has highlighted the latest developments in applied yeast technologies for drug discovery [33]. A number of yeast genes and their corresponding products were identified by classical genetic approaches that began with identification of mutant phenotype and progressed 'forward' to the gene and the product. A number of other genes were discovered by 'reverse' genetic approaches, in which mutants were obtained last. The original reverse genetic experiments were fractionation-based; one purified a protein of interest, sequenced it partially, and then cloned and disrupted the corresponding gene. More recently, reverse genetic approaches have been driven by identification of homologous sequences. The availability of complete genomic information has made possible a new type of reverse genetics based on a novel fractionation schemes. These novel fractionation schemes allowed scientists to start asking questions such as: given a substrate - find the enzyme, given an enzyme – validate it as a drug target, given a target - find a drug, given a drug – find the target and given a pathogenic fungus - find a drug target.

One of the examples of the application of functional genomics in the discovery of novel drugs was the search for 'disinactivators' of human potassium channels. Inactivation of such channels might be associated with sizes and hippocampal ischaemia. For that reason, small molecules that block the association of specific ⌊ and ß subunits were considered to have therapeutic potential. Sequencing of human genome allowed the identification of genes for ⌊ and ß subunits of human potassium channels. The yeast two-hybrid interaction system was constructed to produce growth inhibition, such that drugs that block the interaction restore growth. The new screen for potassium channel disinactivators involved more than 170,000 compounds and has apparently identified a compound with *in vitro* efficacy and specificity for the potassium channels [see references in: 33].

To summarise, although the yeast genome has been completed for more than eight years, the majority of yeast gene functions are still poorly characterised. However, many of the approaches presented in Miami have the potential to assign functions to significant portions of the yeast proteome. As knowledge of yeast cell biology expands, one can expect greater light to be reflected on all eukaryotes and further use of yeast for pharmacological applications.

### *3.2 Genome sequencing of Streptomyces species*

Streptomycetes belong to Gram-positive, mycelial, spore-forming soil bacteria with two important properties. They have an unusual genomic topology of very large linear replicons and synthesise a large number of secondary metabolites, many of which have important pharmacological properties. It is, therefore, not surprising that they have a significant biodiversity potential.

Genetic biodiversity potential of *Streptomyces* species can be illustrated by the following facts. Streptomycetes have linear chromosomes from 8 to 9 Mb, about twice the size of *Escherichia coli* (4.6 Mb [28]) or *Bacillus subtilis* (4.2 Mb [27]) chromosomes containing 4,288 and 4,100 protein-coding genes, respectively. The chromosome of *S. coelicolor*, a model *Streptomyces* species sequenced recently at Sanger Centre [29], is 8,667,507 bp long containing 7,846 protein-coding genes. If one assumes that the known common functions of all saprophytic bacteria for catabolism, metabolism, DNA replication, protein synthesis etc., require at most 4 Mb of coding DNA, the remaining 4 Mb of *Streptomyces* DNA might be species specific. The other question is, what is the structural biodiversity potential of their secondary metabolism? It is well known that out of 19,000 antibiotically active compounds isolated from bacteria to mammals, Streptomycetes synthesise 7,900. Moreover, 75% of all antibiotics important in human and veterinary medicine are produced by *Streptomyces* species [34]. Watve and his collaborators [35] have recently attempted to estimate the number of the yet undiscovered antimicrobials from this genus. The model they developed has shown that the total number of antimicrobial compounds this genus is capable of producing is in the order of 100,000 - less than 10% of what has been discovered so far.

Each *Streptomyces* species is capable of synthesising more than one biologically active secondary metabolite. For example, it has been known that *S. coelicolor* synthesises at least four antibiotics: actinorhodin, undecylprodigiosin, methylenomycin and lipopeptide antibiotic CDA. However, the analysis of *S. coelicolor* genome sequence, and the genome sequence of avermectin producing *S. avermitilis*, suggests that there are many (more than 20) gene-clusters coding for secondary metabolites in each species, which have not yet been analysed [29, 30]. *Streptomyces* secondary metabolites can interact with a number of biological targets such as yet unidentified proteins of different organelles (like ribosomes, membranes, microtubules, chloride ion channels etc.) nucleic acids (both DNA and RNA) and individual proteins (like RNA polymerase, HMG-CoA reductase, FK protein, etc.). It is, therefore, not surprising that *Streptomyces* antibiotics, antifungals, citostatics, immunosuppressants, anticholesterolemics, antiparasitics, coccidiostatics, animal growth promotants and natural insecticides are in commercial use [34]. The main point now is how to use this enormous biodiversity potential?

*Streptomyces* secondary metabolites withstand simple chemical classification, but many best-understood and biologically most active compounds are synthesised by two families of multifunctional enzymes that can assemble unusual carbon and peptide chains, which have important medical, veterinary and agrochemical properties. Polyketide and peptide synthases, abbreviated as PKSs and NRPSs respectively, catalyse condensations of carboxylic acid and amino acid derivatives into polyketide and peptide structures like precursors of Erythromycin or Penicillin, respectively. DNA sequencing of PKS and NRPS gene-clusters showed that they are multi-functional enzymes with modular organisation. Each module is responsible for a single cycle of polyketide and peptide chain extension and contains catalytic domains for necessary ketoreduction, dehydratation and enoylreduction as well as epimerisation, N-methylation and reduction activities. The last modules of both multi-functional enzymes contain thioesterase domains responsible for the release of linear chains from enzymes and their cyclisation. Therefore, there is a one-to-one correlation between the product structure and the active domains in modular PKSs and NRPSs that generate linear polyketide and peptide chains. This allows the prediction of polyketide and peptide backbone structures from DNA sequences. PKSs and NRPSs share considerable DNA homology between them so they must have originated from the same ancestors. Therefore, the creation of directed changes in the backbone structures by genetic manipulation of modules is possible. Major approaches that have been used up until now are targeted manipulations, which are the disruptions, deletions or replacements of certain catalytic domains or whole modules in the existing gene-clusters [see references in: 34 and 36].

There are a number of small biotechnology companies that use these approaches. The approaches of two most important ones, Biotica Technology Ltd. (http://www.biotica.co. uk/) and Kosan Biosciences Inc. (http://www.kosan.com/) are illustrated to show what they do. With the introduction of restriction enzyme sites, scientists at Kosan designed genetic cassettes that allowed relatively easy manipulation of individual active domains by deletion, or insertions of active domains from other modular PKSs. In complex polyketides like macrolides, Biotica and Kosan inactivated specific enzyme active domains or inserted active domains from other clusters to generate novel macrolides with differently reduced oxo groups that possess different stereochemistry. They also deleted whole modules from multi-functional PKSs or used chemo-biosynthesis to generate novel macrolides with smaller or larger polyketide backbones. Moreover, recently the new structural class of polyketides (having 2,4-dioxa-adamantane ring-system) has been isolated from an engineered *Streptomyces* strain, thus supporting the claim that combinatorial biology is capable of producing novel chemotypes. However, all these approaches are labour and time consuming, allowing the development of relatively small libraries of novel polyketides and peptides [see references in: 34].

While studying the genome topology and genetic stability of *S. rimosus*, an industrial producer of the antibiotic oxytetracycline, the frequent interaction between its chromosome and linear plasmid present in the host cells was noticed in a number of strains that had not been in any way selected. Genetic elements like plasmid pPZG103 and the chromosome of the strain MV25W are formed by single crossover between the plasmid pPZG101 and the chromosome end [37]. This suggested a general strategy for obtaining recombinants between two polyketide biosynthesis clusters. The polyketide gene-cluster could be cloned into a linear plasmid vector in between selectable and counter-selectable gene-cassettes [38], introduced in host cells carrying the second gene-cluster cloned within similar gene-cassettes near the end of the chromosome and selected for a single cross over between them (22).

Bioinformatics can also help in accessing the biodiversity in this group of organisms. Until 2000 only 19 polyketide gene-clusters for modular PKSs were cloned and sequenced from a number of different species [34]. However, as it was already mentioned, genome sequencing of *S. coelicolor* and *S. avermitilis* suggests that there are more than 40 presumed gene-clusters in these two species that code for secondary metabolites, which have not yet been analysed [29, 30]. Bioinformatics tools can be used to annotate as yet unanalysed modular polyketide and peptide gene-clusters and to design primers for cloning left and right ends of gene-clusters in order to be able to pull out the entire clusters that are often larger than 100 kb. Computer programmes for modelling recombination between modular polyketide and peptide gene-clusters have been developed. The first programme was written in Turbo PASCAL. The programme outputs a file with the module description of all recombination products. A further programme, written in Java, uses these data to generate a chemical description of products of each module and to give graphical representation of linear polyketide chemical structures [39]. After pre-polyketide and pre-peptide biosynthesis, polyketides and peptides usually undergo cyclisation reactions. It would be very interesting to add a programme to model cyclisation reactions in order to be able to predict the final products in fermentation. To do that a number of PKS and NRPS databases that have been recently developed (Natural Product Gene Database, http://www.npbiogene. com/; A Database of Modular Polyketide Synthases, http://www.nii.res.in/pksdb.html; A knowledge based resource for analysis of Non-ribosomal Peptide Synthetases and Polyketide Synthases, http://www.nii.res.in/nrps-pks.html) can be used.

Additional biodiversity can be envisaged from recent findings that have shown natural existence of mixed complexes. For example, PKS-like modules, responsible for the incorporation of a polyketide moiety within the peptide chains, have been found. One such example is the biosynthesis of Bleomycin from *S. verticillus*. And indeed, DNA sequencing of the Bleomycin gene-cluster showed that the $7^{th}$ module in this enzyme is indeed the PKS module. NRPS-like modules have also been found within PKSs. In such situations, NRPS-like modules are responsible for the incorporation of an amino acid moiety within the polyketide chains as occurs in the biosynthesis of the antibiotic Rifamycin from *Amycolatopsis mediterranei*. Consequently, it is very much likely that PKSs and NRPSs could also be recombined, both *in silico* and in the laboratory [see references in: 22].

To summarise, *Streptomyces* species and related genera undoubtedly have very significant genetic biodiversity potential. The structural biodiversity of their secondary metabolites is also very significant. It has been shown that combinatorial biosynthesis in *Streptomyces* can be used to generate novel chemical entities, so there is an obvious need for further work with Streptomycetes and their secondary metabolites. The bioinformatic programmes mentioned would be useful tools for predicting novel polyketide, non-ribosomal peptide and/or mixed structures *in silico* that might then be produced by an appropriate genetic manipulation in the laboratory.

## 4. Conclusions

Genomics, proteomics and bioinformatics have brought about fundamental changes, and one cannot exist without the other. Development of bioinformatics will shift many experiments from the laboratory to the computers, although it is obvious that predictions cannot go without experimental confirmations. The race between companies will be won by those who will be able to mine databases best, and finally it is also about *brainware*, not only hardware or software. All these factors seem to point towards an unprecedented concentration of technological means within the pharmaceutical industry as well as agriculture. According to a survey of Time magazine, data mining and bioinformatics will be within the 10 hottest jobs of the 21-century [40]. It is obvious that all biologists have to be educated in this field.

## Acknowledgments

This work was supported by the grant 0058008 from the Ministry of Science, Education and Sports, Republic of Croatia.

## References

[1] Rifkin, J. The Biotech Century: Harnessing the Gene and Remaking the World. Phoenix, London, 1999.
[2] Hranueli, D. New technologic, economic and social challenges. Perspectives, PLIVA Global Review, **1:** 6-9, 2002.
[3] Murakami, S., P.M. Tedesco, J.R. Cypser & T.E. Johnson. Molecular genetic mechanisms of life span manipulation in Caenorhabditis elegans. Ann. NY Acad. Sci., **908:** 40-49, 2000.
[4] Leips, J. & T.F. Mackay. Quantitative trait loci for life span in Drosophila melanogaster: interactions with genetic background and larval density. Genetics, **155:** 1773-1788, 2000.
[5] Puca, A.A., M.J. Daly, S.J. Brewster, T.C. Matise, J. Barrett, et al. A genome-wide scan for linkage to human exceptional longevity identifies a locus on chromosome 4. Proc. Natl. Acad. Sci. USA, **98:** 10505-10508, 2001.
[6] The Human Genome, Nature, **409:** 2001.
[7] The Human Genome, Science, **291:** 2001.
[8] McLeod, H.L. Pharmacogenetics: more than skin deep. Nature Genet., **29:** 247-248, 2001.
[9] Asahara, T., C. Kalka & J.M. Isner. Stem cell therapy and gene transfer for regeneration. Gene Therapy, **7:** 451-457, 2000.
[10] Gardner, M.J., N. Hall, E. Fung, O. White, M. Berriman et al., Genome sequence of the human malaria parasite Plasmodium falciparum, Nature, **419:** 498-511, 2002.
[11] Holt, R.A., G.M. Subramanian, A. Halpern, G.G. Sutton, R. Charlab et al., The genome sequence of the malaria mosquito Anopheles gambiae, Science, **298:** 129-149, 2002.
[12] Burt, A. Site-specific selfish genes as tools for the control and genetic engineering of natural populations, Proc. R. Soc. Lond. B, Published online, 2002.
[13] The business of the human genome. Supplement to Scientific American, July 38-57, 2000.
[14] Allsop, A.E. New antibiotic discovery, novel screens, novel targets and impact of microbial genomics. Curr. Opin. Microbiol., **1:** 530-534, 1998.
[15] Sharff, A. & H. Jhoti. High-throughput crystallography to enhance drug discovery. Curr. Opin. Chem. Biol., **7:** 340-345, 2003.
[16] Delseny, M., J. Salses, R. Cooke, C. Sallaud, F. Regad et al., Rice genomics: Present and future. Plant Physiol. Biochem., **39:** 323-334, 2001.[17] The Arabidopsis Genome Initiative. Analysis of the genome sequence of the flowering plant Arabidopsis thaliana. Nature, **408:** 796-815, 2000.
[18] Lagudah, E.S., J. Dubcovsky & W. Powell. Wheat genomics. Plant Physiol. Biochem., **39:** 335-344, 2001.
[19] Brendel, V., S. Kurtz & V. Walbot. Comparative genomics of Arabidopsis and maize: prospects and limitations. Genome Biol., **3:** Reviews 1005, 2002.[20] Mouse genome. Nature, **420:** 2002.
[21] Hranueli, D. Where functional genomics can be applied. Perspectives, PLIVA Global Review, **1:** 28-33, 2002 (http://www.pliva.com/perspectives).
[22] Hranueli, D. & J. Cullum. Bioinformatics of Streptomycs species and food production, pp. 333-340. **In:** Z. Kniewald et al. (Eds.), Current Studies of Biotechnology - Vol. III. Food. Croatian Society of Biotechnology, Zagreb, Croatia, 2003.
[23] Goffeau, A., B.G. Barrell, H. Bussey, R.W. Davis, B. Dujon, et al. Life with 6000 genes. Science, **274:** 546, 563-567, 1996.
[24] Bolotin, A., P. Wincker, S. Mauger, O. Jaillon, K. Malarme et al. The complete genome sequence of the lactic acid bacterium Lactococcus lactis ssp. lactis IL1403. Genome Res., **11:** 731-753, 2001.
[25] de Vos, W.M. Advances in genomics for microbial food fermentations and safety. Curr. Opin. Biotechnol., **12:** 493-498, 2001.
[26] Hayashi, M., H. Mizoguchi, N. Shiraishi, M. Obayashi, S. Nakagawa, J. Imai, S. Watanabe, T. Ota, M. Ikeda. Transcriptome analysis of acetate metabolism in Corynebacterium glutamicum using a newly developed metabolic array. Biosci. Biotechnol. Biochem., **66:** 1337-1344, 2002.
[27] Kunst, F., N. Ogasawara, I. Moszer, A.M. Albertini, G. Alloni, et al. The complete genome sequence of the gram-positive bacterium Bacillus subtilis. Nature, **390:** 249-256, 1997.
[28] Blattner, F.R., G. Plunkett, C.A. Bloch, N.T. Perna, V. Burland, et al. The complete genome sequence of Escherichia coli K-12. Science, **277:** 1453-1474, 1997.

[29] Bentley, S.D., K.F. Chater, A.M. Cerdeno-Tarraga, G.L. Challis, N.R. Thomson, et al. Complete genome sequence of the model actinomycete Streptomyces coelicolor A3(2). Nature, **417:** 141-147, 2002.

[30] Ikeda, H., J. Ishikawa, A. Hanamoto, M. Shinose, H. Kikuchi, et al. Complete genome sequence and comparative analysis of the industrial microorganism Streptomyces avermitilis. Nature Biotechnology, **21:** 526-531, 2003.

[31] Oliver, S.G., Q.J. van der Aart, M.L. Agostoni-Carbone, M. Aigle, L. Alberghina, et al. The complete DNA sequence of yeast chromosome III. Nature, **357:** 38-46, 1992.

[32] Winzeler, E.A., D.D. Shoemaker, A. Astromoff, H. Liang, K. Anderson, et al. Functional characterization of the S. cerevisiae genome by gene deletion and parallel analysis. Science, **285:** 901-906, 1999.

[33] Brenner, C. A cultivated taste for yeast. Genome Biol., **1:** Reviews 103, 2000.

[34] Hranueli, D., N. Perić, B. Borovička, S. Bogdan, J. Cullum, P.G. Waterman & I.S. Hunter. Molecular biology of polyketide biosynthesis. Food Technol. Biotechnol., **39:** 203-213, 2001.

[35] Watve M.G., R. Tickoo, M.M. Jog & B.D. Bhole. How many antibiotics are produced by the genus Streptomyces? Arch. Microbiol., **176:** 386-390, 2001.

[36] Mootz, H.D., D. Schwarzer & M.A. Marahiel. Ways of assembling complex natural products on modular nonribosomal peptide synthetases. Chembiochem, **3:** 490-504, 2002.

[37] Pandza, S., G. Biuković, A. Paravić, A. Dadbin, J. Cullum & D. Hranueli. Recombination between the linear plasmid pPZG101 and the linear chromosome of Streptomyces rimosus can lead to exchange of ends. Mol. Microbiol., **28:** 1165-1176, 1998.

[38] Cullum, J., M. Aikawa, D. Hranueli, R. Lal, G. Padilla, A. Paravić & K. Vongerichten. Genetic methods for the manipulation of polyketide-producing actinomycetes pp. 167-174. **In:** Z. Kniewald et al. (Eds.), Current Studies of Biotechnology - Vol. II. Environment. Croatian Society of Biotechnology, Zagreb, Croatia, 2001.

[39] Tupath, H., J. Pfeiffer, I. Pfeifer, D. Deckbar, T. Fleige, H. Peitz & J. Cullum. A computer program to model recombination products between modular polyketide clusters, pp. 291-295. **In:** V. Lelas et al. (Eds.), Proceedings of the 4th Croatian Congress of Food Technologists, Biotechnologists and Nutritionists (Central European Meeting), Zagreb, 2002.

[40] Peters, T. What will we do for work. Time, **155:** 48-53, 2000.

# Appendix

# Student papers

*Essays in Bioinformatics*
*D.S. Moss et al. (Eds.)*
*IOS Press, 2005*



# ß-Spectrins and their Homologues – Comparative Studies and Consensus Sequence Construction

Anna FOGTMAN*
*Institute of Biochemistry and Molecular Biology, University of Wroclaw, Poland*

**Abstract.** The β-spectrin family of proteins was the subject of the analysis of amino acid replacenents at aligned positions. The homologous and non-homologous positions were subjected to an analysis of the interrelations among occurring residues and the mechanism of variability using the algorithm of genetic semihomology [6]. 67 β-spectrin sequences were collected and 55 of them were subjected to an comparative analysis. After in-depth studies of the global multiple alingnment, a consensus sequence was construscted. It was the base of the detailed analysis of genetic relations among all the amino acid residues occuring the same positions of homologous sequences. Such examination shows a detailed picture of the relations among the representatives of the β-spectrin family and gives a possibility of following the evolutionary paths of the protein family arising, what is the base of further analytic examinations of the β-spectrin family.

## Introduction

Spectrin was first identified as a major component of the erythrocyte membrane cytoskeleton, controlling its organization, stability and shape. Nowadays it is known, that spectrins are common in cells of all types of tissues of *Vertebrates* and *Invertebrates*; they take part in many processes essential for normal functioning of a cell. Spectrin is a cytoskeletal protein important for keeping the right shape of the cell, resilience of a membrane during a mechanical stress. It determines the distribution of transmembrane proteins and the organization of organelles in the cytoplasm. Spectrin, through its interactions with the hydrophobic part of the cell membrane, is a factor taking part in building the system of actin filaments. Recently a new role of this protein has been found as a participant in secretic pathways of cells [Beck and Nelson, 1998; De Matties and Morrow, 1998]. Diverse isoforms of spectrin and spectrin binding proteins – ankyrins are present on the surface of Golgi structures, transported intermediates and on membranes of endocythotic pathway. Spectrin plays a crucial role for stabilization of the cell membrane, organization of domains of integral proteins, controling mobility of the membrane receptors, cell adhesion, nerve impulse transduction, synthesis of the secretic vesicles and their transport among organelles, also for development and morphogenesis of the embrion cells. Changes in the primary structure that are caused by genetic mutations, cause injuries or absence of the membrane cytoskeleton proteins, which disturb the interactions: cytoskeleton - cell membrane. Those disturbations lead to deformations, loss of elasticity and diminish of the cell surface. An example of such disturbances are hereditary hemolytic anaemias.

*e-mail: *fogi@grid.icm.edu.pl*

A spectrin molecule is a heterodimer composed of domains: α and β connected by noncovalent bonds with each other. Chains of those domains are winded round the same axis, anti-parallel. The tetramer is 200 nm long with the diameter 3 nm and it is the basic functional unit of spectrin molecule.

β-Spectrins concern 19 segments; the first of them is a high conservative N-terminal domain composed of two neighbouring domains that are calponin homologues. The centre of the molecule is occupied by 17 triple helical repeats and the terminal part – by a domain concerning PH – motif – a region homologic to pleckstrin. The actin binding sites are localized in the middle part of the tetramer (segment 15).

In this paper the β-spectrin family of proteins was subjected to a theoretical analysis of the variability mechanisms in their primary structure, through examinations of similarities and differences within the family. The correlation of the variability of several positions and functions of the proteins was also analysed. On the basis of the global multiple alignment and consensus sequence that was constructed, a level of semihomology and identity was estimated.

## 1. Materials and methods

The initial material used it this research were β-spectrin amino acid sequences obtained from Swiss Prot[1] database and through BLAST[2] programme. The first step was to find a model sequence of human erythroid β–spectrin (access number: P11277) and then using BLAST programme – to find its homologues. 67 sequences were colleted; 12 of them were rejected – they were recurrent sequences, incomplete or totally distinct in their amino acid composition than other sequences. The algorithm of genetic semihomology [6] was used for the analysis of the correlation between amino acids at semihomologous and non-homologous positions, mechanism of variability, location of gaps, the multiple alignment and the consensus sequence construction. The multiple alignment was constructed tentatively by the Test[3] programme aligning two sequences, based on the algorithm of genetic semihomology. The graphic figure of the multiple alignment was created using Protein Calculator[4] (v0.901 Beta version) programme, colouring the conservative positions and constructing tentatively the consensus sequence.

## 2. Results and discussion

### *2.1 General characteristics of algorithm of genetic semihomology*

The algorithm of genetic semihomology assumes, that the basic (but not the only) mechanism of evolutionary diversity of proteins is single point mutation that may lead to the replacement one amino acid residue by another in one or more positions of homologous sequences. It is based on simple, clearly defined rules; connections between codons of several amino acid residues make it's sense. The main part of the algorithm is a three-dimensional diagram showing a network of genetic relations among amino acids (Fig. 1.).

---

[1] www.expasy.ch/sprot/

[2] www.ncbi.nlm.nih.gov/blast/Blast.cgi

[3] Programmes are properties of Interdysciplinary Centre for Mathematical and Computational Modelling, Warsaw University.

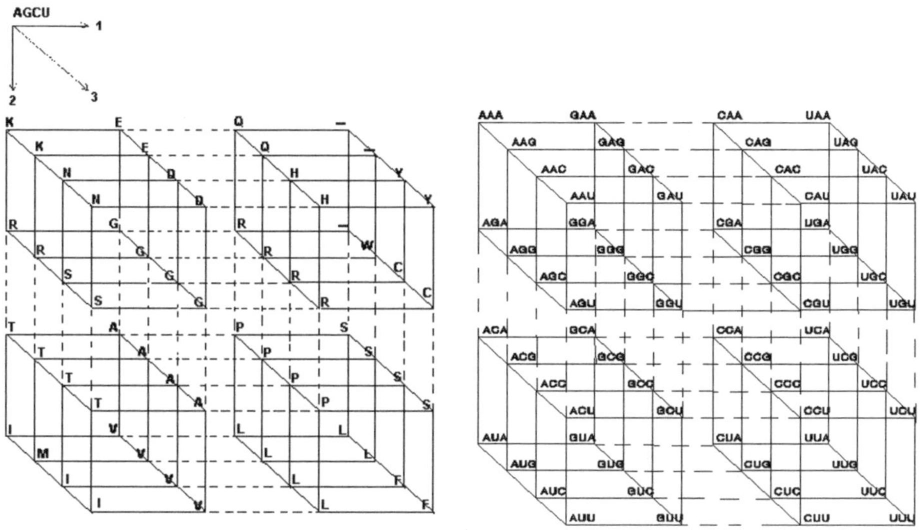


Diagram A Diagram B

**Figure 1.** Diagrams of semihomologous relations among amino acids (Diagram A) and their codons (Diagram B). The codons of residues along each axis differ by only one nucleotide. The diagram setting shows the codon changes at first (axis 1), second (axis 2) or third (axis 3) position in order: A→G→C→U. Those diagrams form the basis of the non-statistic genetic senmihomology algorithm. Full lines mean transitions, dashed lines – transversions.

A single nucleotide replacement concerns three types of mutations: transition – within the same group (purine to purine or pyrimidine to pyrimidine), transversion – between two groups (purine to pyrimidine or conversely) and cryptic mutation – replacement one nucleotide by another within the trird residue of the codon, without changing it's sense. This mutation type and the presence of six-codon amino acids determine a better variety of proteins. The cryptic mutation process itself isn't subjected to selection, but it increases the spectrum of amino acid diversity. That is why serine is able to replace 12 different amino acid residues (including a cryptic mutation) – two times more frequently than methionine (one codon). This is a proof why six-codon amino acids play a crucial role in increasing the field of diversity in occupied positions (especially serine).

The algorithm of genetic semihomology breaks the rule of using the Markov model as a tool for comparing two protein sequences. Opposite to this model, the algorithm predicts the amino acid residue occupying this position in the future with taking into consideration the fact, what kind of residue was occupying this position in the past. The algorithm of genetic semihomology assumes close relations among amino acids residues and their codons; the same residues in different positions in the respect of their ability to a substitution are not equal. It treats every amino acid residue as individuality, taking into consideration what kind of amino acid occurred the position in the past. It is based on a theory, not statistics. Its sense is the minimalisation of assumptions and the influence of the user to the analytic process.

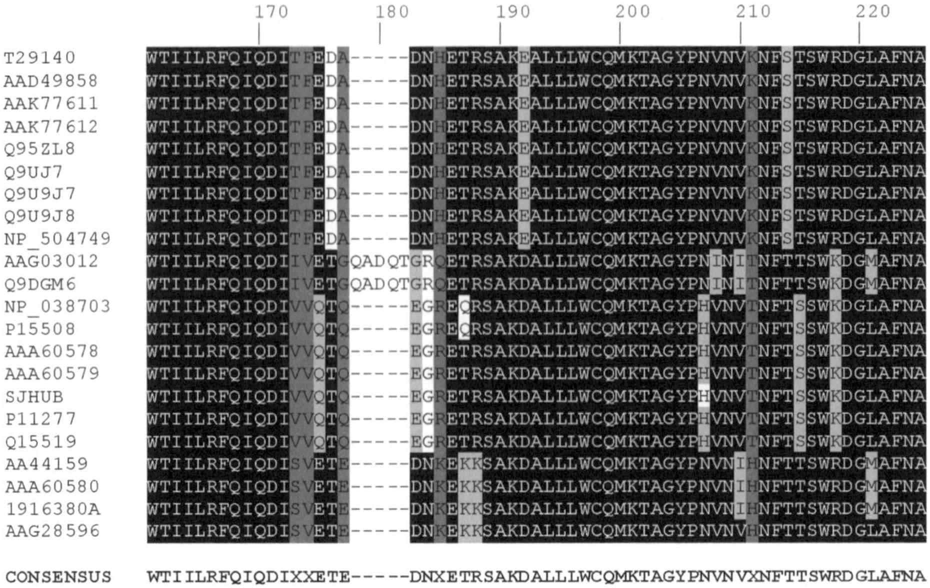


**Figure 2.** A fragment of multiple alignment of β-spectrin sequences with the consensus sequence.

*2.2 The global multiple alignment*

The aim of the multiple alignment construction is to create a picture of identities, similarities and differences of primary structures being compared. It is made by aligning sequences in the way that several amino acids in the same positions are relate with each other as close as possible. It is a three-colour combination; conservatives positions are negatives, residues semihomologous to them are light grey and residues semihomologous to each other – dark grey (Fig. 2.).

The global multiple alignment of β-spectrin protein family and the consensus sequence concern 2668 positions (34 pages A4), that is why showing the whole alignment in this paper is impossible[5].

The rate of homology within the β-spectrin family is estimated as moderated, c.a. 50% (according to the consensus sequence). There are different parts of the alignment to be noticed: parts with a very high level of identity – these are whole fragments of conservative positions and parts with a low rate of homology. The N-terminal fragment (positions 45-695) has a high level of identity – there are continuous fragments of conservative residues, locally interrupted by single semihomologous positions. This conservative nature of this fragment is probably determined by the protein's functions. A domain homologous to calponine (CH domain) is located in this area – responsible for binding action. The CH domain plays the main role of this protein, crucial for the existence of the cell.

There is a gap in positions: 697-706 (9 positions) – only 12 sequences has amino acid residues in those positions. These are very characteristic sequences – rich in glycine and alanine residues. The middle area of the global multiple alignment is rich in

[5]The whole multiple alignment is accessible in B.S. Thesis: "®-Spectrins and their homologues – comparative studies and consensus sequence construction" - A. Fogtman, Institute of Biochemistry and Molecular Biology, University of Wroclaw, Poland, 2003.

tryptophane residues. This amino acid is a large molecule, that does not fit to a very ordered structure like α-helise forming the middle of the spectrin molecule (triple helical repeats). It probably forms the links between the consecutive repeats in the triple helical repeat chain. The fragment including residues: 2160-2490 concerns 100% non-homologous and non-identical positions. The rate of variety in this area is very high - this fragment (330 amino acid residues) is dominated by deletions. The diversity of this area suggests that it is not a crucial fragment for the main protein (β-spectrin) functions and the amino acid composition of this segment depends on the localization and functions characteristic for an individual protein. The C-terminal part of the β-spectrin molecule begins with the residue 2490. The homology within this fragment is moderate; definitely lower than the homology within the N-terminal fragment. It concerns plenty of deletions, the rate of identity is minimal.

Apart from changes of similarity of comparing sequences characteristic for whole regions, there are sporadic, single point mutations noticed within the sequences. Probably these are results of defensive processes because of lethal mutations caused by appearing of STOP codons. Those codons must be removed – by putting a single deletion. This situation takes place especially in positions rich in amino acid residues, that have codons similar to STOP codons: Leu and Ser (position 342 – those codons are semihomologous to codons: UAA, UAG and UGA). Presence of such a type of deletions in proteins suggests their taxonomic similarity.

### *2.3 The consensus sequence*

The consensus sequence5 is a peculiar summary all the positions in the global multiple alignment of β-spectrin protein family, concerns a piece of information about general structure of a given family. The consensus sequence (Fig. 3.) is composed of three types of signs: letter amino acid symbols (in conservative positions) – in the β-spectrin family consensus sequence, when a position was occupied by one type of amino acid residue in at least 54,55%. The "X" sign means a position with indefinite amino acid residue – there are genetic relations among amino acids occupying this position, but there is no amino acid residue, that would occur in this place very often. The rate of semihomologous residues in this position must be at least 29,09%. The "-" sign means a deletion – in case of exceeding the limit of 49,09% deletions in occurring position.

```
MXXX---XDXX-XXXXXX--YX-XXXXXXXXD-XXXXXX----XSSARLFERSRIKALADEREXVQKKTFTKWVNSHLAR
VXCXIXDLYXDLRDGXMLXXLLEVLSGEXLPKPTKGXMRIHCLENVDKALQFLKEQRVHLENXGSHDIVDGNHRLTLGLI
WTIILRFQIQDIXXETE-----DNXETRSAKDALLLWCQMKTAGYPNVNVXNFTTSWRDGLAFNALIHKHRPDLXDFXKL
XKS-N-AXYNLQXAFXXAEXXLGLXKLLDPEDVXVXXPDEKSIITYVVTYYHYFSKMKXLAVEGKRIGKV-XDXAXEXXK
MIEXYEXLASDLLXWIEXTIXXLNXRXFANSLXGVQXQLQAFXXYRTXEKPPKFXEKGNLEVLLFTIQSKMRANNQKXYX
PREGKLISDINKAWEXLEKAEHERELALRXELIRQEKLEQLAXRFDRKAAMRETWLXE-NQRLVSQDNFGXDLXAVEAAX
```

**Figure 3.** A fragment of β-spectrin consensus sequence (positions: 1-480). The residues in bold occupy their positions in at least 89,09% (extremely conservative).

After construction the β-spectrin consensus sequence, basic parameters that characterize the whole β-spectrin family were calculated (Tab. 1.).

**Table 1.** Identity rate, as a percent of of identical residues in the whole pool of amino acids.

| Positions | Total | Percentage content |
|---|---|---|
| Identic | 1313 | 49,21% |
| 89,09% in total pool among identic | 507 | 19% identic pool<br>38,61% in total pool |
| Semihomologous | 980 | 36,73% |
| Deletions | 375 | 14% |

On the basis of the consensus sequence, estimating the general parameters characterizing the whole family of proteins (β-spectrins) is possible. The consensus sequence is a perfect average of results, it decreases considerably their amount and gives a reliable piece of information about probable location of amino acid residues, that are responsible for forming very important structural and functional units of proteins.

## 3. Summary

The B.S. Thesis, that was the basis of writing this paper, concerns also a detailed analysis of amino acid composition of the β-spectrin protein family, as well as the analysis of genetic relations among amino acid residues occurring several positions. Using the algorithm of genetic semihomology makes the analysis of the primary structure of proteins easier and more reliable. Such analysis of the protein primary structure is only an introduction to complete examinations of structure and functions of proteins; mechanism of variability of proteins, location of gaps, mutational correlations at particular positions and their contact with each other, evolutionary pathways and future evolutionary changes of protein structure.

Currently, the research within the primary structure of members of the β-spectrin protein family is continued. The research concerns improving the features of the algorithm of genetic semihomology (ascribing concrete values of probability to particular changes of amino acid residues within several positions) and testing it using the β-spectrin protein family. The studies will also probably concern detailed examinations of the evolutionary pathways (in the past and in the future) within β-spectrins.

## References

[1] Bennett V. and Baines A.J.; "Spectrin and Ankyrin-Based Pathways: Metazoan Inventions for Integrating Cells Into Tissues", Physiological Reviews, Vol. 81, No. 3, 1353-1391, July 2001.
[2] Broderick M.J.F, Winder S.J.; "Towards a Complete Atomic Structure of Spectrin Family Proteins", Journal of Structural Biology 137, 184-193, 2002.
[3] Djinovic-Carugo K., Gautel M., Ylanne J., Young P.; "The spectrin repeat: a structural platform for cytoskeletal protein assemblies", FEBS Letters 513, 119-123, 2002.
[4] Gimona M., Djinovic-Carugo K., Kranewitter W.J., Winder S.J.; "Functional plasticity of CH domains", FEBS Letters, Vol. 513, Issue 1, 98-106, 2002.
[5] Hanus-Lorenz B., Hryniewicz A., Lorenz M., Sikorski A.F.; „Spektryny – roznorodnosc form i funkcji powszechnie wystepujących bialek cytoszkieletowych", Kosmos, Tom 50, Nr 3, 243-262, 2001.
[6] Leluk J.; "A New Algorithm for Analysis of the Homology in Protein Primary Structure", Computers and Chemistry, Vol. 22, No. 1, 123-131, 1998.
[7] Regularities in mutational variability in selected protein families and the Markovian model of amino acid replacement", Computers and Chemistry 24, 659-672, 2000.
[8] Leluk J.; "A non-statistical approach to protein mutational variability", BioSystems 56, 83-93, 2000.
[9] Leluk J., Konieczny L., Roterman I.; "Search for structural similarity in proteins", Bioinformatics, Vol. 19, No. 1, 2003.

[10] Leluk J., Hanus-Lorenz B., Sikorski A.F.; "Application of genetic semihomology algorithm to theoretical studies on various protein families", Acta Biochimica Polonica, Vol. 48, No. 1/2001.

[11] Matteis M.A., Morrow J.S.; "Spectrin tethers and mesh in the biosynthetic pathway", Journal of Cell Science 113, 2331-2343, 2000.

[12] Meglicz A., M.S. Thesis: "Bialkowe inhibitory kinaz - analiza pokrewienstwa, zmiennosci, mechanizmow roznicowania oraz relacji genetyczno-strukturalnych", Institute of Biochemistry and Molecular Biology, University of Wroclaw, Poland, 2003.

[13] Thomas G.H., Newbern E.C., Korte C., Bales M.A., Muse S.V., Clark A.G., Kiehart D.P.; „Intragenic Duplication and Divergence in the Spectrin Superfamily of Proteins", Molecular Biology and Evolution 14(12), 1285-1295, 1997.

[14] Zdyb A., B.S. Thesis: "Kinazy bialkowe-semihomologiczne zestawienie sekwencji, relacje genetyczne, konstrukcja i analiza sekwencji konsensusowej", Institute of Biochemistry and Molecular Biology, University of Wroclaw, Poland, 2002.

*Essays in Bioinformatics*
*D.S. Moss et al. (Eds.)*
*IOS Press, 2005*


# Bioinformatics - Computational Support for Genome Analysis

Fahri Salih KOCABAS
*Middle East Technical University (METU) Computer Engineering Department*
*06530 Ankara, Turkey. (e-mail: fkocabas@tsk.mil.tr)*

**Abstract.** The major goal of bioinformatics is the analysis of sequence, structure and function relationships. In these studies, lab experiments and computational work must validate and consolidate each other, and findings of both initiatives expedite each other's improvement. This process requires experts who can both work at lab bench and in computer applications. This chapter summarises a computer scientist's views on the diverse fields of bioinformatics.

## Introduction

The work of bioinformatics requires to orchestrate different disciplines like molecular biology, math, computer science, statistics etc. to have a united focus on its objectives in a team oriented work environment. It is easy to state but difficult to implement. The existence of double major scientists, appealing grants and the enthusiastic nature and the challenge of the subject may well be organised and utilised to start and maintain such a bioinformatics study. Therefore, being a computer scientist, the author values the information contained in this article even if most of the content is known by related disciplines. It is so because the bringing the related information together under the supervision and experience of a computer scientist is valuable. The major concepts largely focused on sequence analysis are visited in the second part whereas the concluding remarks and tips for future studies are given in the last part.

In gene and aminoacid sequence analyses, the sequences of related ones were observed to be similar; thus, corresponding portions matched in their alignments. It is known that strong similarity indicates the homology where homology means a common evolutionary history whereas similarity emerges for some other criteria, not for a common ancestor [1]. Alignment by utilising basic computer science techniques presents solutions to the question of the relatedness of sequences. The genetic, functional and structural relations are under examination in this regard.

Other than comparison analyses, the computational requirements of molecular biology could mainly be listed as: set of tools powered by integrated knowledge bases; solid, complete methodologies; computation techniques enriched with introduction of probability, uncertainty, fuzziness, learning mechanisms, heuristics, approximation, knowledge discovery and the like. One important aspect is to decide over the trade off between the sensitive, exact solution and exponential computational running times. The bottom line is that the environment in which the bioinformatics problem resides must well be reflected in designing the optimum data structures and algorithms. It is also the main course of ongoing advances that bioinformatics graduate professionals will employ bioinformatics - specific computational frameworks in line with the advances in related disciplines in coming years.

## 1. Analysis of current work

The basis for comparison of protein and gene sequences for similarity is to examine if they are related by evolution (they have a common ancestor). However, random mutations in the sequences with common ancestor develop over time as well as similar portions come up for different structures and functions and this should be considered in studies. In parts of the sequence that are critical for the function of the protein, hardly any mutations will be accepted; nearly all changes in such regions will destroy the function [2].

One important algorithm used in sequence analysis is Dynamic Programming (DP). In DP, large tables are built with all known previous results. The solution of the problem then depends on the solutions of smaller ones in the table. A recursive structure for computing optimal score in DP algorithm is designed and interdependent sub solutions are filled in the table using the recurrence rule. The tables are created iteratively based on an optimal recurrence function and result is computed in a bottom up fashion. The construction of this table should be made efficiently since scanning of the table leads to quadratic running times. What if (a) combining the solutions of smaller problems of the same kind to form the solution of a larger one is not be possible, (b) the number of small problems to solve are unacceptably large (c) the costs are fractional in which the efficiency of DP is limited? The reduction in search space and employing other techniques like Top Down DP, Divide and Conquer, Greedy Approach and Progressive Sequence Alignment, by accompanying and replacing the procedure might help in that matter. The bottom line is that DP is applicable when the subproblems are not independent and, the problem must be an optimisation problem.

Assumptions and inferences made are based on the evolutionary change and constitute the context in which the alignment process takes place. An optimal alignment is the one with maximum number of matches and minimum number of mismatches and gaps. The score of an alignment is the sum of position scores. The gap penalty used in scoring scheme is important. It helps deciding whether or not to accept a gap or insertion in an alignment when it is possible to achieve a good alignment at some other neighbouring points in the sequence. One can not let gaps and insertions occur without penalty, otherwise an unreasonable alignment with gaps would result. Biologically, it should be natural for a protein to accept a different residue in a position, rather than having parts of the sequence deleted or inserted. Gaps and insertions should therefore be more rare than point mutations/substitutions [2].

In pairwise alignments, there is a two-dimensional matrix with the sequences on each axis, and the elements in the matrix are initially the substitution coefficients, which are then operated on to locate the best path through the matrix. The number of operations required to do this is approximately proportional to the product of the lengths of the two sequences. Dot plot as a graphical tool can help in aligning two sequences. Pairwise sequence alignment is basis for the other analyses even for experimental design of PCR primer design. But, there are some problems with pairwise alignments. For example, when many sequences that are significantly similar to the query sequence are obtained, comparing each sequence to every other may become impractical as the number of sequences increases. Then, multiple sequence alignment, where all similar sequences can be compared in one single figure or table is employed. The basic idea is that the sequences are aligned on top of each other, so that a co-ordinate system is set up, where each row is the sequence for one protein, and each column is the same position in each sequence. Each column corresponds to a specific residue in the prototypical protein. One may have to introduce gaps in sequences at positions where there were no gaps in the corresponding pairwise alignment; thus, multiple alignments typically contain more gaps than any given pair of aligned sequences.

In multiple sequence alignment, similar sequence motifs are identified and protein families are analysed. The general method of multiple alignment has been to extend the pairwise alignment method into a simultaneous n-wise alignment by using a DP algorithm in n dimensions. One can implement and visualise this algorithm easily in the case of three sequences by setting up a three-dimensional matrix instead of the two-dimensional one. Then, the same procedure as for the two-sequence case is followed. The result is a path that goes diagonally through the cube-shaped matrix from one corner to the opposite. The problem here is that the time to compute this n-wise alignment becomes expensive as the number of sequences grows. The algorithmic complexity is something like $O(c^{2n})$, where c is a constant, and n is the number of sequences. This is not an acceptable performance. Rather than doing a simultaneous n-wise alignment, using so-called progressive alignment method could be preferred. Then, the alignment is built up in stages where a new sequence is added to an existing alignment using some rules to determine in which order and how the sequences should be added [2]. However, there are supporting approaches such as approximation algorithms, heuristics and pruning the search space based on the contextual information. Scoring criteria for multiple alignment could be *entropy methods* scoring each column based on the probability distribution of the characters in it; *tree alignment metrics* assuming knowledge of an existing phylogenic tree and weight differences between closely related sequence pairs as more important than distant pairs; *sum of pairs metric*, which is the most popular, summing up the cost of the k(k-1)/2 pairs of symbols in each column as an upper bound. Finding the optimal sum of pairs alignment is non polynomial (NP Complete). However, by exploiting the lower bounds given by each pairwise DP matrix, one can heuristically reduce the number of states in the multiple DP matrix and hope to find the optimal alignment of say 6-7 sequences of e.g. 200 characters in a reasonable amount of time. Since the applications of multiple alignment fall beyond the range of exact solution algorithms, we must employ heuristic methods. For example, randomly picking a sequence for deletion from the alignment and then reinserting it at the position, which either maximises the score or with probabilities biased toward the maximum score, a successful but not necessarily optimal result is reached.

Alignment of aminoacid sequences differs from the nucleotide sequences. For nucleotide sequences, a mismatch between sequences is usually scored as 1 whereas for aminoacids, the possible pathways in which one aminoacid may be replaced by another need to be considered. For example, Cysteine (TGT) and Tyrosine (TAT) have single but Cysteine and Methionine (ATG) have 3 changes. The alignment of Cysteine with Tyrosine is less costly than the alignment it with Methionine. For more than three average sized protein sequences, Heuristic Progressive Alignment gives better results rather than DP approach, but not guaranteed to find the optimal alignment. In Progressive Alignment procedure; n-1 pairwise alignments calculating distance matrix; neighbour-joining tree is formed based on the similarity values in the distance matrix; Progressive alignment following the neighbour-joining tree is performed where the most closely related sequences are first aligned. However, there is no objective function and if an error is introduced early in the alignment, it becomes impossible to correct it later in this procedure.

## 2. Conclusion and further work

For a complete method to support genome analysis (RNA, DNA, Proteins), the environment in which the genome data resides must be examined precisely. Then, the assumptions, criteria, limits, patterns, profiles, scoring schemes, attributes, associations, design and application of the algorithms and the like are defined and initialised accordingly. So that, combining the math and statistics with computational support and biological data come into the process. Manual analysis and editing the attributes given above over the colour coded alignments are revisited and the values are inserted into the work where necessary. In some cases, even when one has a lot of information about the proteins, such as active site residues, secondary structure, 3D structure, mutations, etc, it may still be necessary to make a manual alignment to fit all the data. [2].

There is exponential growth in known sequences, sequence and structure alignments. The analysis data of those studies should be geared to the needs of bioinformaticians. For example, the outcome of the decision whether it is similar or homologous affects the whole process. It must again be considered that certain regions (structure and function) are of more crucial residues. When two protein sequences have more than 25 % identical residues aligned, the corresponding 3D structures are said to be very similar implying similar functionality. Therefore, the sequence alignment of proteins remains to be an approximate predictor of the underlying 3D structural alignment. However, experimental findings for evolutionary background should consolidate these studies [3].

The operations like match, mismatch, insertion, deletion and introduction of gaps with varying numbers, definitions even with different scoring subschemes can be utilised in scoring schemes. Depending on the context, some changes are more plausible than others and probabilistic interpretation of how likely one alignment versus another is performed. The success depends not only on the parameters such as insertion and deletion penalties, substitution coefficients but also on the order in which sequences are added to the multiple alignment process. A number of rules are used to increase the success rate of the procedure like each sequence is weighted according to how different it is from the other sequences. Of many different possible scoring schemes, one can employ position-specific scores. For example, if one knows from other sources like its 3D structure that a gap should not be allowed in a certain part of a sequence, then higher gap penalty values could be determined in relevant calculation.

In overall calculation, the employment of local and global alignments or combination of them where better fits should be considered. Local alignments in which the regions with high degree of similarity in two sequences rather than globally aligning them from head to toe may be preferred and done to support the global alignment. Sort and search techniques may be borrowed in running alignment procedure based on the contextual information. A Context Sensitive grammar may be formed to model the contextual information within the enacted environment of the related process. Clustering of large multiple alignments supported with alternative representations could well be performed. How can we represent a pattern of residues as found in a multiple alignment? And how can we use such a pattern to search for it in other protein sequences? The formalism devised to describe the kind of patterns we need: is regular expressions to describe particular languages in restricted cases.

The selection and employment of algorithms constitute the major issue when we are searching large databases. For example, a database of size $10^9$, one can not run DP algorithm to query a string of length up to 500, because of exponential running times. However, this problem can be handled in different ways: (a) Implementing the DP algorithms in hardware, thus executing them much faster. The disadvantage is its high cost. Furthermore, by using parallel hardware, the problem can be distributed efficiently to a couple of thousands of processors, and the results can be integrated later. This approach is costly, too. (b) Using heuristics that work much faster than the original DP algorithms and exact algorithms. Here are some measures to take: due to the huge DB size, Preprocessing of the rather stable portions of database is done; Substitutions are much more likely than insertions and deletions; We expect homologous sequences to contain a lot of segments with matches or substitutions, but without insertions and deletions and gaps. These segments can be used as starting points for further searching. [4]. Learning algorithms of artificial neural networks supported with uncertainty, probabilities, fuzziness, heuristics could be utilised. So that learning mechanism can steer the running of the algorithm in guidance of contextual information. A multithreaded parallel implementation of sequence comparison by a DP algorithm could well be employed. The algorithmic steps and data space of the problem can be designed specifically for parallel implementation. The problem might be solved in different set-ups to validate, consolidate and further improve the result. Time and space complexity must well be balanced in the followed procedure, though.

Bioinformatics work is multi-disciplined in nature but, not complex. The way ahead, a road map for a computer science expert in the field of bioinformatics might be to attend (computational) molecular biology classes and workshops; to review genome data and existing supporting computational tools; to examine the analysis requirements for genome data including sequence, pattern, association structures and concepts; to study the latest work and literature to determine the present technology; to join and exchange views with a team of different expertise. These issues equally apply for other disciplines that are inherent in bioinformatics, as well. This initiative lays the required groundwork to identify and solve the bioinformatics problems with bioinformatics-specific frameworks.

One major goal of bioinformatics is the analysis of sequence, structure and function relationships. In those studies, lab experiments and computational work must validate and consolidate each other. The findings of both initiatives expedite each other's improvement. This process requires experts who can both work at lab bench and in computer applications. Better algorithms, improved scoring tables, solid semantic models will all emerge with better understanding of huge experimental data residing in large annotated databases. This remains to be the major challenge of our time.

## References

[1] Huelsenbeck, J. Ogihara, M. (2001). Lecture 7. *CS 120.*

[2] Kraulis, P. (2000). Structural Biochemistry and Bioinformatics Lecture Notes. *Stockholm Bioinformatics Center.*

[3] Sabbiah, S. An Overview of the Computational Analysis of Biological Sequences. *Stanford Univ. Bioinformatics Center, Singapore.*

[4] Shamir, R. (2001). Algorithms for Molecular Biology. Lecture Notes. *Tel Aviv University School of Computer Science.*

*Essays in Bioinformatics*
*D.S. Moss et al. (Eds.)*
*IOS Press, 2005*



# Prediction of Signal Peptides and Signal Anchors of Cytochrome c Nitrite Reductase from *Desulfovibrio desulfuricans* ATCC 27774 Using Bioinformatic Tools

*Luisa L. GONÇALVES[1,2,3], Maria Gabriela ALMEIDA[1,2], Jorge LAMPREIA[1], José J.G. MOURA[1] and Isabel MOURA[1].*
*[1]REQUIMTE, CQFB, Departamento de Química, Faculdade de Ciências e Tecnologia, Universidade Nova de Lisboa, 2829-516 Monte de Caparica, Portugal.*
*[2]Instituto Superior de Ciências da Saúde-Sul, Campus Universitário – Quinta da Granja, 2825-511 Caparica, Portugal.*
*[3]Present Adress: Faculty of Pharmacy, Room #514, 19 Russel Street, Toronto, Ontario M5S 2S2, Canada.*

**Abstract.** The cytocrome c nitrite reductase (ccNir) isolated from the sulphate-reducing bacterium Desulfovibrio desulfuricans ATCC 27774 is a hetero-oligomeric complex composed by two subunits (61 KDa and 19 KDa), encoded by genes nrfA and nrfH, respectively. We report the use of bioinformatic predictive models in order to access of ccNir most relevant topological characteristics, namely signal peptides and signal anchors. We made used of a combined method of SignalP V2.0 (SignalP-HMM and Signal-NN) in association with TMHMM 2.0 for the prediction of the presence and location of signal peptide cleavage sites, to discriminate between cleavable signal peptides and N-terminal transmembrane anchors segments and, to predict of transmembrane helices.

## Introduction

Sub cellular protein sorting, i.e. the processes through which proteins are routed to their final destination within a cell, is a fundamental attribute of cellular life. In general, sorting depends on "signals" that can already be identified by looking at the primary structure of a protein. N-terminal signal peptides (also referred to as signal sequences or leader sequences) target proteins to the secretory pathway in eukaryotic cells and for translocation across the cytoplasmatic membrane in bacteria [1]. They have a conserved three-region design with a positively charged amino-terminal segment (n-region), a central hydrophobic segment (h-region) and, a more polar c-terminal segment (c-region) that is recognised by the membrane bound signal peptidase enzyme. The general signal peptide structure is conserved among different proteins and also across different species [2]. Although general physiochemical properties are conserved among proteins within the same cellular localisation, the primary structure is low conserved. Signal peptides (SP) are often cleaved off of the mature proteins upon arrival at the sub cellular destination site. Otherwise, the remaining signal peptide anchors the protein to the membrane and is referred to as a "signal anchor" (SA) [3]. Signal anchors have both an n- and h- region, and no cleavage site.

In bacteria, three types of signal peptidases are known so far [for a review see 4 and references therein]. The type II signal peptidases (Spase II; EC 3.4.23.36), or proliprotein signal peptidases (Lsp), cleave lipoproteins when a large hydrophobic residue is present at the –3 position and a modified cysteine is present at the +1 position, being the consensus cleavage site "A|(G/A)|C" [5].

The most currently used method to identify the presence and location of signal peptides cleavage sites, in amino acid sequences from different organisms, is the neural network-based signalP predictor. Signal IP combines two different neural networks, one that discriminates between residues that belong and do not belong to a signal peptide (S-core) and, one that was conceived to recognise signal peptidase cleavages sites (C-score) [1]. The cleavage site is predicted by multiplying together the C-score and the negative "derivative" of the S-score, while the discrimination between proteins that have and do not have a signal peptide is based on the mean S-score evaluated from the N-terminus to the predicted cleavage site.

SignalP V2.0 comprises two signal peptide prediction methods, Signal-NN (based on neural networks) and, SignalP-HMM (based on hidden Markov models). According to SignalP server (http://www.cbs.dtu.dk/services/SignalP-2.0) SignalP-H provides not only a prediction of the presence of a signal peptide and the position of cleavage site, but also an approximate assignment of n-, h- and c-regions within the signal peptide. Additionally, for eukaryotic data, the HMM version has an improved discrimination between signal peptides and uncleaved signal anchors, but has a lower accuracy in predicting the precise location of the cleavage site [1].

Some proteins have sequences that initiate translocation in the same way as SPs do, but are not cleaved by signal peptidase. As the rest of the polypeptide chain is translocated trough the membrane, the resulting protein remains anchored to the membrane by the hydrophobic region, with a short N-terminal cytoplasmic domain. The uncleaved signal peptide is kwon as a signal anchor SA, and the resulting protein is known as a type II membrane protein. SAs differ from SPs in other respects than the cleavage sites: they have longer hydrophobic starches and the N-terminal region of the hydrophobic stretches can also be much longer [6].

Multiheme cytocrome c nitrite reductase (ccNiR) isolated from the sulphate-reducing bacteria Desulfovibrio desulfuricans ATCC 27774 is a membrane bound enzyme that catalyses the dissimilatory nitrite reduction to ammonia in a six-electron step. It is a key enzyme involved in the second and terminal step of the dissimilatory nitrate reduction pathway of the nitrogen cycle and plays an important role on bacterial respiratory energy conservation [7,8]. It has recently been shown that ccNiR is a hetero-oligomeric complex composed by two subunits (63 KDa and 19 KDa) both containing c-type hemes, encoded by genes nrfA and nrfA, respectively [9].

Based on the primary sequence determined by chemical and DNA sequencing (described on references 9 and 10) we used a combined method of SignalP V2.0 (SignalP-HMM and Signal-NN) in association with TMHMM 2.0 for the assessment of ccNir most relevant topological characteristics.

## 1. Primary Structure

Chemical Sequencing. The N-terminal amino-acid sequence of D. desulfuricans ATCC 27774 ccNiR subunits and their internal peptides were determined by automated Edman degradation on a Procise™ Protein Sequencer (model 491, Applied Biosystem) as described, in detail, in the literature [9].

The internal peptide sequences obtained by enzymatic cleavage, as well as the nrfA and nrfH sequences have been submitted to the EMBL database under the accession number AJ316232. The data on the alignment and homology of both nrfA and nrfH have already been discussed in references 9 and 10.

We used SignalP V2.0 (SignalP-HMM and Signal-NN) [11,12] for the prediction of the presence and location of signal peptide cleavage sites for gram-negative bacteria and the program TMHMM 2.0 to predict of transmembrane helices [13,14]. SignalP and TMHMM are available under the prediction server page of the Center for Biological Sequence Analysis at www.cbs.dtu.dk/services/tmhmm2.0.html. The signal sequence of lipoproteins was examined using the program Lipop accessed at PSORT WW Server (http://psort.nibb.dtu.dk )[15].

The N-terminal sequences of both NrfA and NrfH are the following:

NrfA, 24 XQDVSTELKAPKYKTGIAETETKMSAFKGFPQQYASYMKNNE

NrfH, 1 GTPRNGPWLKWLLGGVAAGVVLMGVLAYAMTTTDQRP

## 2. Results and Discussion

### *2.1 Primary Structure*

NrfA. As already been described in previous papers [9,10], the deduced amino acid sequence of NrfA (518 aa) contains four classical c-type heme-binding motifs CXXCH and a fifth heme-binding site CWXCK, where the proximal histidine residue was replaced by a lysine. Excluding the cleaved 23 N-terminal aminoacids and the heme prostetic groups, it has a molecular mass of 56768 Da. The addition of five hemes gives 59848 Da.

The sequence of nrfA encodes for a precursor signal peptide, which shows the "LA(G/A)|C" consensus motif recognised by signal peptidase II [10].The prediction given by SignalP, using the HMM version, gives a maximum cleavage site probability between Gly23 and Cys24 (Fig.1). Interestingly, the NN version, although referred in the literature to have a better performance in predicting the cleavage site location in gram-negative bacteria [1], gave slightly different results, being the maximum cleavage site probability between Ser28 and Thr29. Nevertheless, the results are positive for the presence of a signal peptide based on S-score (output form signal peptide networks) and mean S-score values (Fig. 1B). The above mentioned cleavage site was experimentally confirmed by N-terminal sequencing of the mature protein. Accordingly to the HMM prediction, the N-terminal sequence of NrfA starts at the $24^{th}$ residue. It shall be stressed out that the chemical sequencing by Edman degradation doesn't recognise cysteines. Additionally, signal peptidase II cuts upstream of a cysteine residue to which a gliceride-fatty acid lipid is attached [16]. For this reason, the signal sequence of lipoproteins, i.e., proteins with a covalently attached lipid molecule in their mature N-terminus, was examined using the program Lipop. This program also predicted a lipid attachment to Cys24 with a sequence consensus motif of CQDV, which gave us an additional evidence for the correct cleavage site position predicted by the HMM version. Curiously, none of the nrfA from other organisms published in the literature [see 9 and 10 and references there in] shows this consensus motive. Thereby, the presence of a lipidic component attached to Cys24 may be a particular feature of NrfA from D. desulfuricans.

### *2.2 NrfH*

As previously reported [9], the deduced amino acid sequence of NrfH subunit (154 aa) shows four CXXCH consensus sequences. It has a predicted molecular mass of 16764 Da, excluding the heme groups. The attachment of four hemes leads to a total molecular mass of 19228 Da.

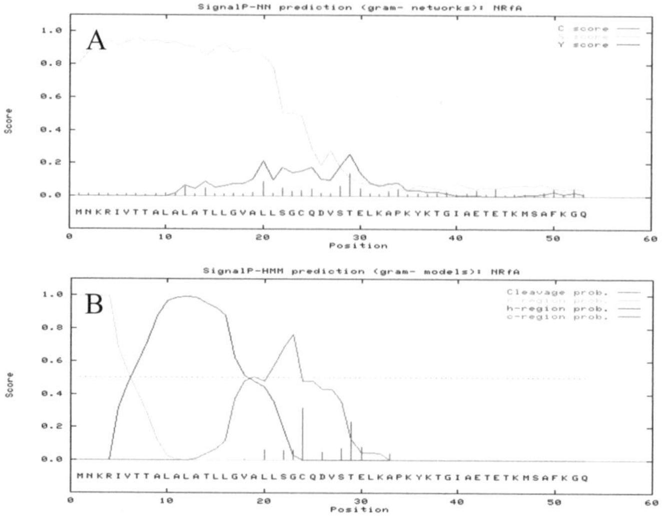


**Figure 1.** Prediction of signal peptides for gram-negative bacteria, using NN- and HMM versions of SignalP. C, S and Y-scores represent, respectively, the output from cleavage site networks, the output from signal peptide networks, and the output of the combined cleavage site score, given by $Y_i = \sqrt{Ci\Delta dSi}$ . n-region: positively charged amino-terminal segment; h-region: central hydrophobic segment; and c-region: polar c-terminal segment. NrfA (SignalP-NN) – Positive results; Cut-off between Ser28 and Thr29. NrfA (SignalP-HMM) – Positive results; Cut-off between Gly23 and Cys24.

Both versions of Signal P predict that nrfH encodes for a signal peptide (based on y- and S-score), with a maximum cleavage site probability between Ala29 and Met30 (Fig.2). However, conflicting results were obtained when TMHMM 2.0 was run for the search of transmembrane helices. Apparently, this subunit is expected to be a transmembrane protein, with the bulk of the protein facing the periplasm. The N-terminus (residues1-6) remains in the cytosol, while residue 7-29 is predicted to form a transmembrane helix, which most likely acts as a membrane anchor (Fig.3). Similar profiles were obtained with JPRED – a consensus method for protein secondary structure prediction - available at http://www.expasy.org. Considering that a similar topological behaviour is often observed among c-type cytocromes from bacteria [9 and references therein], these results suggest that NrfA is devoided of a periplasm signal. Although not surprising for SignalP-NN, where the discrimination between SAs and SPs has proved to be poor (according to S-score, 50% of the SAs are predicted as SPs), critical reviews on the prediction of organellar targeting signals [6] recommend the combination of SignalP with one of the available prediction methods for transmembrane helices, as for example, PHDhtm and/or TMHMM. This problem seems to be reasonably overstepped with the TMHMM program, which was developed by E. L. Sonnhammer and co-workers [13] with an integrated architecture based on SignalP-HMM and an HMM-based transmembrane helices prediction method. This suggestion has revealed to be of particular usefulness with type II membrane proteins where false positive results are often found.

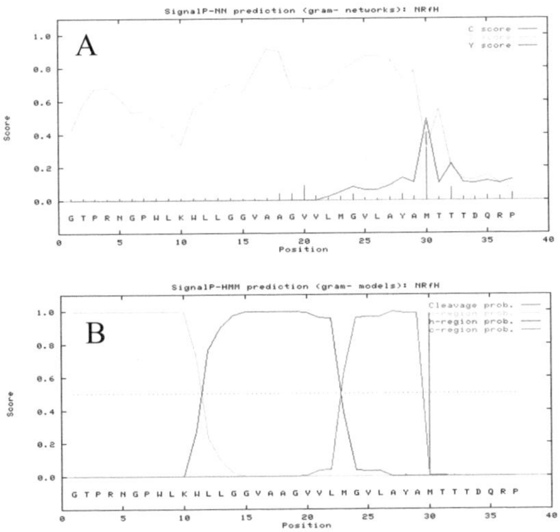


**Figure 2.** Prediction of signal peptides for gram-negative bacteria, using NN- and HMM versions of SignalP. C, S- and Y-scores represent, respectively, the output from cleavage site networks, the output from signal peptide networks, and the output of the combined cleavage site score, given by $Y_i = \sqrt{Ci\Delta dSi}$ . *n-region*: positively charged amino-terminal segment; h-region: central hydrophobic segment; and *c-region*: polar c-terminal segment. *NrfH* (SignalP-NN) – Positive results; Cut-off between Ala29 and Met30. NrfH (SignalP-HMM) – Positive results; Cut-off between Ala29 and Met30.

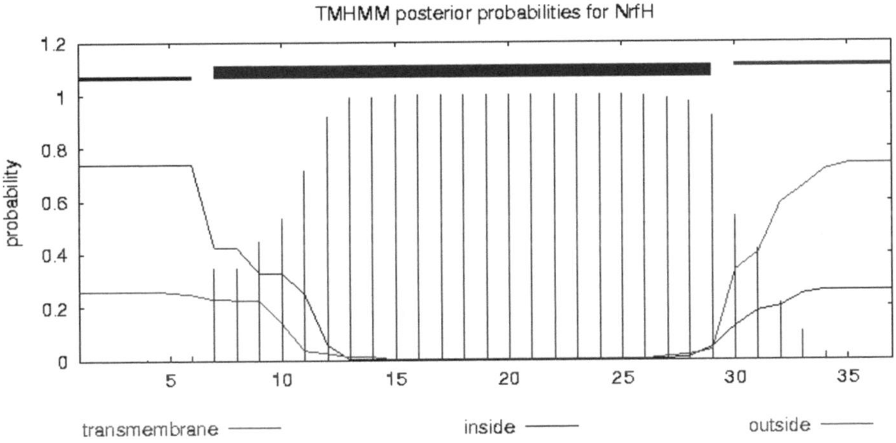


**Figure 3.** Transmembrane helix prediction for NrfH subunit by TMHMM 2.0. NrfH – transmembrane helix: segment 7-29; inside: segment 1-6; and, outside: segment 30-37.

## 3. Conclusions

The first issue to be considered when predicting of signal peptides and other protein sorting signals is which program shall be used. The profusion and the continuous rising of protein databases make it an uneasy task, considering that any given application will require careful consideration regarding the best balance between sensitivity and specificity. Otherwise, many of these methods use different approaches and focus on specific cell types and/or signal sorting pathways. Some of the most popular tools for predicting protein signals use Neural Networks (e.g. SignalP-NN and Predotar at http://www.inra.fr/predotar), Hidden Markov models (e.g. SignalP-MM), Weight Matrices (e.g. Emboss Pscan at http://www.hgmp.mrc.ac.uk) and/or Integrated Methods (e.g. TargetP and Psort), being the last one an integrated approach of the methods mentioned above.

Our results suggest that it is advisable to compare the output of several programs to increase the reliability of the overall data and to make a final decision. As reported earlier, some programs, like SignalP, are useful for initial detection of signal peptides, but this initial approach may have some shortcomings, as for example when predicting signal peptides of type II membrane proteins, that prompt for further analysis with other prediction methods/programs. However, the sequence analysis methods described above, when used in meaningful combinations, can generally provide reliable predictions.

## References

[1] Emanuelsson, O. and von Heijne, G. (2001) Prediction of oganellar targeting signals. Biochimica et Biophy. Acta 154, 114-119.

[2] Nakai, K. (2000) Protein sorting signals and prediction of subcellular localisation. Advances in protein chemistry 54, 277-344.

[3] Nadershahi, A. (2002) Prediction of cell localisation, November 4, 1-11. http://www.micab.umn.edu/ 8006/litreviews/afshin.pdf.

[4] Prágai, Z., Tjalsma, H., Bolhuis, A., Maarten van Dijl, J., Venema, G. and Bron, S. (1997) The signal peptidase II (lsp) gene of Bacillus subtilis. Microbiology 143, 1327-1333.

[5] von Heijne, G. (1985) Signal sequences: the limits of variation. J. Mol. Biol. 184, 99-105.

[6] Nielsen, H., Brunak, S. and von Heijne, G. (1999) Machine learning approaches for the prediction of signal peptides and other protein sorting signals. Protein Engineering. 12, Nº.1, 3-9.

[7] Zumft, W.G. (1997) Cell biology and molecular basis of denitrification. Microbiol. Mol. Biol. Rev. 61, 533-616.

[8] Bercks, B.C., Fergunson, S.J., Moir, J.W.B. and Richardson, D.J. (1995) Enzymes and associated electron transport systems that catalyse the respiratory reduction of nitrogen oxides and oxyanions. Biochem. Biophys. Acta 1232, 97-173.

[9] Almeida, M.G., Macieira, S., Gonçalves, L.L., Huber, R., Cunha, C.A., Romão, M.J., Costa, C., Lampreia, J., Moura, J.J.G. and Moura, I. (2003) Isolation and characterization of cytocrome c nitrite reductase subunits (NrfA and NrfH) from Desulfovibrio desulfuricans ATCC 27774. Re-evaluation of the spectroscopic data and redox properties. Eur. J. of Biochem. 270,1-12.

[10] Cunha, C.A., Macieira, S., Dias, J.M., Almeida, G., Gonçalves, L.L., Costa, C., Lampreia, J., Huber, R., Moura, J.J.G., Moura, I. and Romão, M.J.(2003) Cytochrome c nitrite reductase from Desulfovibrio desulfuricans ATCC 27774. The relevance of the two calcium sites in the structure of the catalytic subunit (NrfA). J. Biol. Chem. 278(19), 17455-65.

[11] Nielsen, H., Engelbrecht, J., Brunak, and von Heijne, G. (1997) Identification of prokaryotic and eukaryotic signal peptides and prediction of their cleavages sites. Protein Engineering 10, 1-6.

[12] Nielsen, H. and Krogh, A. (1998) Prediction of signal peptides and signal anchors by a hidden Markov model. In Proceedings of the Sixth International Conference on Intelligent Systems for Molecular Biology (ISMB 6), AAAI Press, Menlo Park, California, pp.122-30.

[13] Sonnhammer, E.L.L., Heijne, G.V. and Krogh, A. (1998) A hidden Markov model for predicting transmembrane helices in protein sequences. In Proceedings of the Sixth International Conference on Intelligent Systems for Molecular Biology (Glasgow, J., Little-John, T. Major, F., Lathrop, R., Sankoff, D. and Sensen, C., Menlo Park eds.) pp.175-182, AAAI Press, CA, USA.

[14] Krogh, A. Larsson, B., von Heijne, G. and Sonnhammer, E. L. L. (2001) Predicting transmembrane protein topology with a hidden Markov model: Application to complete genomes. Journal of Molecular Biology 305 (3), 567-580.

[15] Nakai, K. and Kanehisa, M. (1991). Expert system for predicting protein localisation sites in Gram-negative bacteria. PROTEINS: Structure, Function, and Genetics 11, 95-110.

[16] Hayashi, S. and Wu, H.C. (1990) Lipoproteins in bacteria. J. Bioenerg. Biomembr. 22, 451-471.

*Essays in Bioinformatics*
*D.S. Moss et al. (Eds.)*
*IOS Press, 2005*



# Graph Representations of Oxidative Folding Pathways

Vilmos ÁGOSTON[1], Masa CEMAZAR[2,3] and Sándor PONGOR[2]
[1]*Bioinformatics Group, Biological Research Center, Hungarian Academy of Sciences, Temesvári krt. 626726 Szeged, Hungary*
[2]*Protein Structure and Bioinformatics Group, International Centre for Genetic Engineering and Biotechnology, Area Science Park, 34012 Trieste, Italy*
[3]*Current address: Institute for Molecular Bioscience, University of Queensland, St. Lucia 4072, QLD, Australia*

**Abstract.** Oxidative folding combines the formation of native disulfide bond with the conformational folding resulting in the native three-dimensional fold. Oxidative folding pathways can be described in terms of disulfide intermediate species (DIS) containing a varying number of disulfide bonds and free cysteine residues, which can also be – as opposed to the majority of protein folding states –isolated and experimentally studied. Each DIS corresponds to a family of folding states (conformations) that the given DIS can adopt in three dimensions. The oxidative folding space can be represented as a network of DIS states interconnected by disulfide interchange reactions reactions that can either create/abolish or rearrange disulfide bridges. Such networks can be used to visualize folding pathways in terms of the experimentally observed intermediates. In a number of experimentally studied cases, the observed intermediates appear as part of contiguous oxidative folding pathways.

## Introduction

Levinthal's paradox, introduced in 1968 [1], stated that the folding of a protein would last more than the age of the universe, if it went through looking for the native conformation by adapting every single conformation possible. There have been many propositions regarding how the conformational space is restricted so that the folding time is reduced to the experimental range. We know today that most single domain proteins are able to fold effectively *in vitro* to their native folds within seconds. The obvious flaw in stating the paradox itself is actually the fact that the search for the native conformation is unbiased with no stabilisation of particular conformations. Today it has been widely accepted that the native state is the energetically most favourable one on the *potential energy surface*. Actually, each conformational state of the protein assumes a certain position on this surface, which means that not all states are equal in free energy and hence the search for the native fold cannot be unbiased. The way this view has evolved to form theories about folding pathways is the following. Already Levinthal stated that there exist specific pathways for folding. By restricting the molecules to those pathways the polypeptide chain does not need to undergo an extensive search of all the conformational space. In 1973 Anfinsen proposed that the information coded in the amino acid sequence of a protein completely determines its folded structure and that the native state is the global minimum of the free energy [2]. Later, a variety of theories emerged, for example the framework model, the diffusion-collision model, the nucleation model, the hydrophobic-collapse model and the jigsaw model. The hydrophobic-collapse and the framework models were favoured over the nucleation model, because they imply the existence of folding intermediates, which were discovered soon after. All proposed mechanisms and models were able to explain particular pieces of experimental data, but none provided a clear explanation of the folding principles or a solution to Levinthal's paradox (for a collection of reviews see: [3]).

The current, unified view of protein folding presented in some highly cited reviews by Dobson and co-workers [4,5], underlies the fact that protein folding is a progression in which both native and non-native contacts stabilise native-like structural features. The folding either proceeds through a hydrophobic collapse to a compact globule that has stabilising interactions or through a slow formation of a folding core (nucleus), which then rapidly proceeds towards the native state. Folding is thus seen as a step-wise behaviour, sampling regions of the landscape that are downhill in energy. An important element in the "new view" of protein folding is the folding funnel, which was first introduced by Onuchic and associates [6]. This is one way of representing the folding landscape with the free energy (enthalpy and entropy) as a function of folding progress variable, also known as the fraction of the native contacts. In the light of this simple surface (see Figure 1), it is possible to understand a number of features of the folding process. There are three kinds of states that can be easily distinguished in the folding funnel.

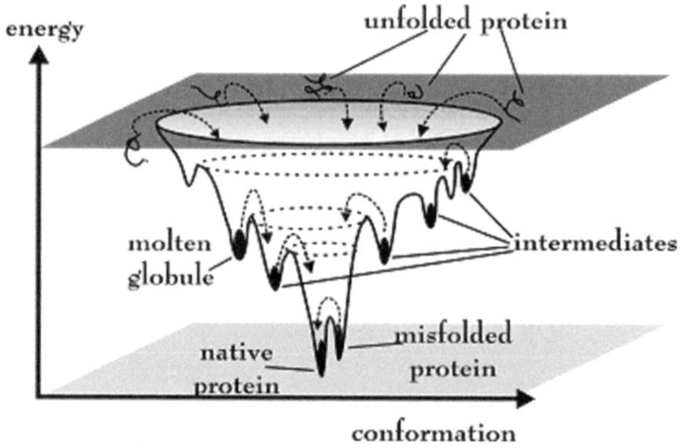


**Figure 1.** Schematic representation of the energy landscape of protein folding. The energy of a protein is displayed as a function of the topological arrangement of atoms. Adapted from Cemazar [7].

The initial state from which the folding proceeds is extremely heterogeneous and encompasses a large conformational space of rapidly inter-converting states. It seems generally accepted that the unfolded or denatured states are not completely random as one would expect for a theoretical polymer. On the contrary, it has intrinsic propensities for native and non-native like interactions, which funnel the folding process either through global or local conformational preferences. Compact denatured states, commonly known as molten globules, are lower in energy in the folding funnel. These have been in the past defined with a set of well-defined features such as a set of secondary structural elements in the absence of tertiary structure. In contrast, at the bottom of the funnel we find a highly compact state, where the close packing of the side chains is essential for a well-defined conformation. This is the so-called native state [4,5].

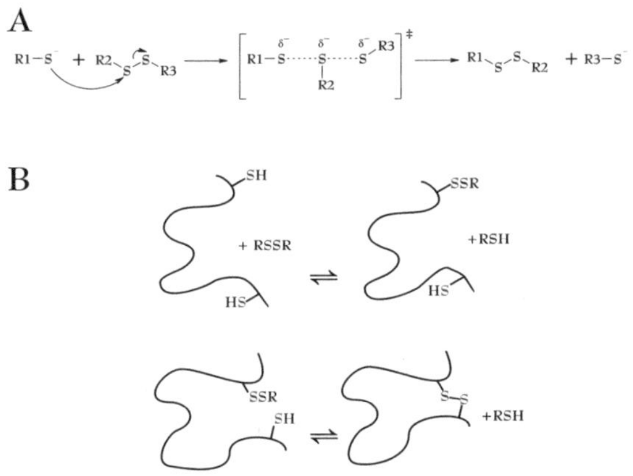


**Figure 2.** A. Thiol-disulfide exchange mechanism: in the pH range above 8, cysteine thiols are readily converted to thiolate anions ($RS^-$), which are potent nucleophiles. $RS^-$ anions attack a disulfide bond, displacing one sulfur atom and forming a new bond with the other sulfur atom (nucleophilic substitution). The rate-determining step of this concerted process is the formation of a transition state with a partial transfer of the negative charge ($\delta^-$) over the three sulfur atoms. B. The formation of a disulfide bond on the polypeptide chain (solid curve) with the help of a small molecule reagent (thiol form: RSH, disulfide form: RSSR). The two steps both proceed via a thiol-disulfide exchange reaction. The first step shown is intermolecular and the second intramolecular. The rate of the intramolecular step is relevant to protein folding, since it also involves conformational changes.

The particular kind of folding that this article is concerned with is oxidative folding, which is the fusion of native disulfide bond formation with conformational folding. This complex process is guided by two types of interactions: first, non-covalent interactions giving rise to secondary and tertiary protein structure, and second, covalent interactions between cysteine residues, which transform into native disulfide bridges. The process of disulfide formation is a simple chemical reaction in which two SH groups join to form a disulfide link (Figure 2A). If the SH groups are on a polypeptide chain, the *in vitro* reaction can be promoted by an external redox system such as a mixture of oxidized and reduced glutathione, or cysteine and cystine, respectively. *In vivo*, the oxidative power comes from specific agents such as the molecular chaperones protein disulfide isomerases. The underlying mechanism is disulfide interchange (Figure 2B). There are two kinds of reactions: in a *redox reaction* a protein disulfide bond is created (or abolished), i.e. the oxidative state of the polypeptide is changed. This is the case when one of the participants of the reaction (say RSH) is not part of the protein. In a *shuffling reaction* both participants of the disulfide interchange are protein-bound, so the oxidative state of the polypeptide does not change. In view of these possibilities it becomes obvious that there are a great many ways in which disulfide bridges can form and rearrange during the folding process. Today it is generally accepted that non-covalent interactions guide the process of folding and formation of disulfide bridges will lock the protein into the right conformation. The advantage of oxidative folding as opposed to general protein folding is that disulfide intermediates can be chemically isolated and studied using such techniques as acid trapping of the intermediates and analysis of the disulfide bridges using a combination of enzymatic cleavage and mass spectrometry. There is a body of literature in describing the pathways of oxidative folding in terms of disulfide intermediates [8-10], and our goal is show how graph theory can be used for this purpose.

Graph theory has been applied to many aspects of protein research (for a review see [11]). Applications to protein folding followed two broad approaches. First, protein structure itself can be considered as a graph consisting of various interactions (such as covalent bonds, hydrogen bonds, spatial vicinities, contacts etc.) as edges, the nodes being atoms or residues of the protein. It was found, among others, that the so-called contact order, i.e. the average sequence distance between residues in atomic contact, seems to be a key determinant of folding speed [12]. Another line of research concentrates on characteristic networks of interatomic contacts that may form stabilization centres in protein structures and can be the reason of the stability of various proteins [13,14]. It was found that populated conformations seen in molecular dynamics simulations contain characteristic networks of residues [15,16].

Another line of research was triggered by the finding that the robustness and stability of networks may be the result of simple topological properties that are invariant throughout various technical as well as biological systems including social organization, electrical networks, road networks and the Internet [17]. In the following years the network topology of a large number of systems have been described, and it was found that some topology classes, like those characterized by a scale-free distribution of the degree (number of links at each node), or the so called "small world models" that are characterized by densely connected subnetworks loosely linked between each other, are indeed found in various systems within and without biology (for a review see e.g. [18]). The various network types were described in terms of a number of simple measures borrowed from graph theory, such as the clustering coefficient, the diameter of the graph etc. This approach was later extended to descriptions of the entire folding space, using the folding states as nodes, and transitions as links between them. As the folding states of native systems cannot be readily studied by physical methods, the investigations were first directed to model systems. Scala and associates [19] described the folding states of short peptides using Monte Carlo simulation on lattice models. They found that that the geometric properties of this network are similar to those of small-world networks, i.e. the diameter of the conformation space increases for large networks as the logarithm of the number of conformations, while locally the network appears to have low dimensionality. Shahnovitch and co-workers analysed the folding states of proteins during molecular dynamics simulations. It was found that the folding space is reminiscent of scale-free network, characterized by a majority of less populated states as well as some highly populated states reminiscent of "hubs" seen in other systems [20].

Our purpose is to describe the folding space of the oxidative folding process using graph theory. This is an intriguing task since the number of folding states defined in terms of disulfide links is relatively small, as compared to "ordinary" folding. We will approach the problem in two steps: i) using graph theory to describe the disulfide intermediates, and to enumerate the states of the folding space. ii) using a graph-like representation of the folding space to visualize the experimentally studied folding pathways.

## 1. Graph representation of oxidative folding intermediates

In proteins containing disulfide bonds, usually all cysteines form part of disulfide bridges, and the disulfide topology can be unequivocally described by defining which cysteines are connected. For example, a topology 1-3, 2-4 means that a protein with 4 cysteines has two disulfide bridges that connect cysteines (1,3) and cysteines (2,4) respectively. Cysteines can be labelled by their sequence position, or – as in the previous example – in a serial order from the N-terminus (Figure 3).

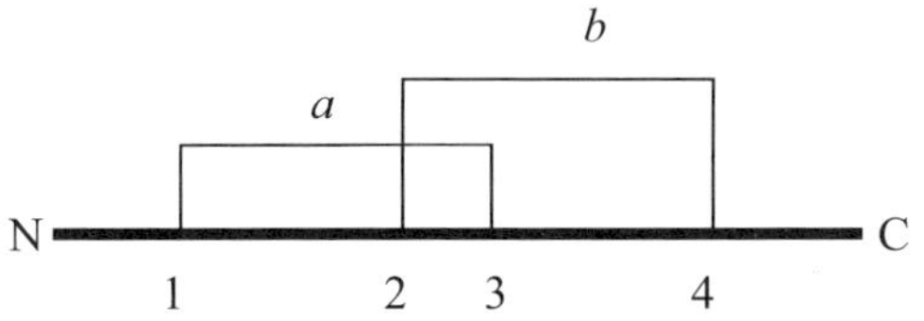


1-3, 2-4 or *abab* topology

**Figure 3.** Nomenclature for disulfide topologies. Disulfides can be labeled by the sequence positions, or simply by the sequential number of the cysteine residues they connect (*1-3, 2-4* topology). Alternatively, it is customary to alphabetically label the disulfide bridges, and describe the topology by assigning the bridge label to the cysteines, starting from the N terminus (*abab* topology).

The number of fully connected (disulfide bonded) isomers in a protein chain with $n$ disulfide bonds ($2n$ cysteines) can be deduced from simple combinatorial considerations as $(2n)!/(n!*2^n)$. According to this formula proteins with two disulfide bridges have 3 fully oxidized isomers, 3-disulfide proteins have 15 and 4-disulfide proteins have 105. In other words, the number of intermediates increases very fast as a function of the number of constituent cysteines, and it has been hypothesized that the reason why the number of cysteines in autonomously folding protein domains is not very large is because the too high number of possible intermediates would slow down the folding process.

| | 1 | 2 | 3 | 4 |
|---|---|---|---|---|
| 1 | | | 1 | |
| 2 | | | | 1 |
| 3 | | | | |
| 4 | | | | |

→

| | 1 | 2 | 3 | 4 |
|---|---|---|---|---|
| 1 | | 1 | | |
| 2 | | | | |
| 3 | | | | 1 |
| 4 | | | | |

1-3, 2-4 1-2, 3-4

**Figure 4** Adjacency matrices of two disulfide topologies of a peptide with two disulfide bridges

For a complete description of the folding process we have to consider both fully oxidized intermediates and the ones with free cysteine residues. For this purpose we will use a formal description of the intermediates as (undirected) graphs, with cysteines as nodes and disulfide bridges as edges (the main chain will not be represented). For the majority of naturally occurring protein structures the resulting graphs will be extremely simple especially if described as an adjacency matrix. Such an adjacency matrix is symmetrical, and contains 1 if two cysteines form a disulfide bonds and zero otherwise. As one cysteine can form only one disufide bridge, each column and each row of the resulting matrix will have atmost one value of 1. The adjacency matrix of two disulfide topologies of a 2-disulfide proteins are shown in Figure 4.

## 2. Description of the oxidative folding space as graphs

The graph descriptions introduced above can be applied both to fully and to partially oxidized intermediates, and the transitions between them can be conveniently described by comparing the adjacency matrices of the two states. The sum of the elements in the i-th column plus the i-th row ($S_i = \Sigma_j A_{ji} + \Sigma_j A_{ij}$) shows if the i-th cysteine forms a bridge. The sum of the differences calculated between these measures of two adjacency matrices describing two intermediates, ($SD = \Sigma_i \Delta S_i$) shows how many cyesteins gained or lost a pair. If two states are connected by a disulfide interchange reaction, the number of disulfide bridges NB remains the same by definition, and it is easy to show that SD will differ exactly by 2. For redox steps in which one disulfide bridge is established or lost, NB and SD will increase or decrease by one and two, respectively. On the above basis one can easily enumerate, for a protein with any number of cysteine residues, a) the oxidative folding states and b) the possible transition steps between them. In other words one can draw a network of all possible oxidative folding pathways. The characteristics of a few systems are summarized in Table 1.

**Table 1.** Number of possible intermediates in and graph parameters of oxidative folding networks.

| N of cysteines | N of intermediates (nodes) | Redox transitions | Shuffling transitions | Total no of transitions (edges) | Clustering coefficient C | Average path length |
|---|---|---|---|---|---|---|
| 1 | 1 | 0 | 0 | 0 | 1.000 | 0.000 |
| 2 | 2 | 1 | 0 | 1 | 1.000 | 1.000 |
| 3 | 4 | 3 | 3 | 6 | 1.000 | 1.000 |
| 4 | 10 | 12 | 12 | 24 | 0.400 | 1.467 |
| 5 | 26 | 40 | 60 | 100 | 0.410 | 1.810 |
| 6 | 76 | 150 | 240 | 390 | 0.247 | 2.293 |
| 7 | 232 | 546 | 1050 | 1596 | 0.253 | 2.640 |
| 8 | 764 | 2128 | 8736 | 10864 | 0.181 | 3.149 |
| 9 | 2620 | 8352 | 19152 | 27504 | 0.182 | 3.550 |
| 10 | 9496 | 34380 | 83520 | 117900 | 0.142 | 3.977 |

The results show that on one hand, the clustering coefficient of the system decreases while on the other, the average path length increases with the number of cysteines. Both findings are consistent with the view that the folding space of peptides with many cysteines may be too complex and thus the systems may be unable to fold fast enough.

The pathways can also be graphically represented, and in order to simplify the resulting picture, we chose a 3D representation wherein the states having the same number of disulfide bridges are placed on separate planes. In this representation, the shuffling transitions are within the planes, and the redox edges connect adjacent planes.

It is noted that the experimental methods do not reveal all possible intermediates; some of them may be too short-lived or not abundant enough so as to be noticed an isolated. In spite of these limitations, the folding pathways appear as connected subgraphs within the network of all possible intermediates, showing that the experimental techniques actually identified states that can interconvert into one another. Only in EGF do we see an "isolated" intermediate which suggest that some intermediates of the pathway were not observed experimentally.

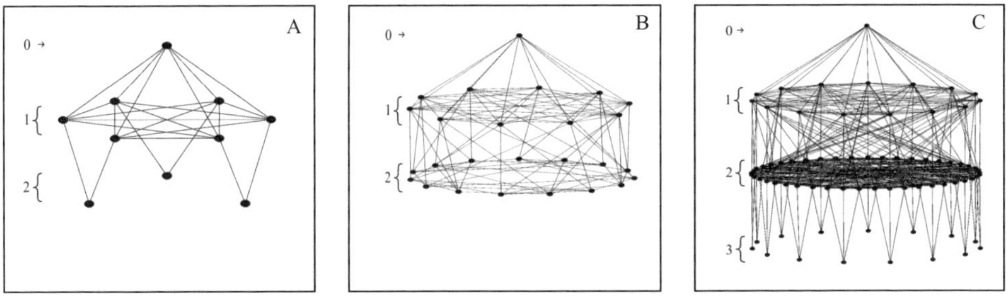


**Figure 5.** Three dimensional representation of the oxidative folding space of polypeptides with 4,5 and 6 cysteine residues (A, B and C, respectively). The nodes represent intermediates, the number of disulfide bridges is indicated with numbers on the left of each panel. The edges indicate disulfide exchange transitions. Zero indicates the fully reduced state, nodes in the lowest plane are the fully oxidized intermediates, one of which is the native state. Edges within the same plane indicate shuffling reactions (interchange between two protein-bound disulfides), edges between planes are redox transitions in which a disulfide bridge is created or abolished.

The network representations shown in Figure 3 are three-dimensional representation of the entire oxidative folding space described in terms of chemically well-defined disulfide intermediates. Species with the same number of disulfide bridges are placed on the same plane, so shuffling reactions, which do not change the number of disulfide bridges are represented as edges within the same plane. On the contrary, reactions in which a disulfide bridge is gained or lost, are represented as edges between two neighbouring planes. The fully reduced state (zero disulfide bridges) is on top, the fully oxidized species, on of which is the native state, is on the bottom. Panel B shows a peptide with 5 cysteines, such as granulocyte-colony stimulating-factor [21, 22] in which the native state contains one free cysteine residue that is not part of a disulfide bridge. In this case the native state can in principle rearrange into other species, so there are shuffling edges also in the lowest plane in the figure. In most of the known cases, the number of cysteines is an even number, so the fully oxidized DISs cannot readily interconvert into each other. In some cases this might be an obstacle: the propeptide of BPTI contains an additional free cysteine that seems to facilitate the folding of the molecule. The propeptide is subsequently cleaved and in this way the structure is locked into the native disulfide configuration [23]. The oxidative folding pathways can be pictured as routes within the full network, starting at the fully reduced species and ending at the native state. In the literature there are a few well-studied examples in which folding intermediates have been determined. Three examples, bovine pancreatic trypsin inhibitor, insulin-like growth factor and epidermal growth factor are shown in Table 2 and Figure 6.

BPTI's folding pathway was the subject of an intense dispute in the early 1990's, but later resulted in one of the most extensively studied oxidative folding pathways and a major protein folding model. With some differences, BPTI's pathway was characterised with the predominance of only a limited number of folding intermediates that adopt mainly native disulfide bridges and native-like structures. It is important to remember that 1- and 2-disulfide intermediates were present, but no 3-disulfide species apart from the native protein was detected on this pathway. One of the most abundant intermediates is a two disulfide species with two native disulfide bonds and a native-like structure. Formation of the third disulfide (Cys14-Cys38) is the last step of the folding process. A prevalence of the native-like structures and native disulfide bridges points to the conclusion that non-covalent

**Table 2**. Disulfied intermediates experimentally observed in the oxidative folding of various proteins

| Protein | Disulfide intermediates[1] | Ref. |
|---|---|---|
| Bovine pancreatic trypsin inhibitor (BPTI) | 3-5; 1-6; 3-5, 1-2;<br>3-5, 1-4; 3-5, 2-4;<br>1-6, 2-4; 3-5, 1-6;<br>**1-6, 3-5, 2-4;** | [24,25] |
| Insulin-like growth factor (IGF) | 2-6; 2-6, 3-5; 2-6, 1-4;<br>2-6, 4-5; 2-6, 1-3;<br>2-6, 1-3, 4-5; **1-4, 2-6, 3-5;** | [26-28]8] |
| Epidermal growth factor (EGF) | 2-3; 1-2; 4-6; 5-6;<br>3-4; 2-4, 5-6; 2-5, 3-4;<br>1-6, 2-5, 3-4; 1-2, 3-4, 5-6;<br>**1-3, 2-4, 5-6;** | [29] |

[1]The intermediates are described with the notation given in Figure 3. The native disulfide connectivity is given in bold, the fully reduced species is not explicitly included.

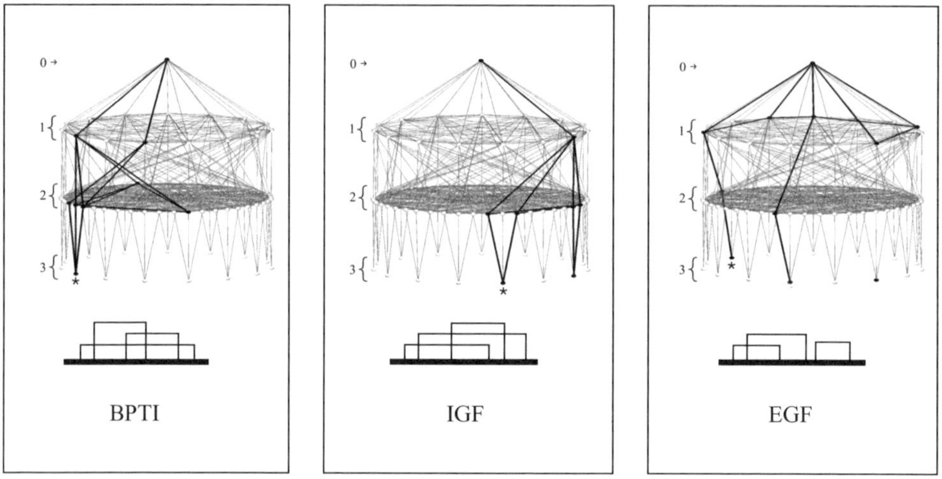


**Figure 6.** The oxidative folding pathways of bovine pancreatic trypsin inhibitor (BPTI), insulin-like growth factor (IGF) and epidermal growth factor (EGF). The native state is marked by asterisk.

interactions that are specific to the amino acid sequence can guide the initial stages of the folding process and hence admit a very limited number of disulfide species on the pathway.

Oxidative folding of the fully reduced EGF [29] proceeds through 1-disulfide intermediates and accumulates rapidly as a single stable 2-disulfide intermediate (designated as EGF-II), which represents up to more than 85% of the total protein along the folding pathway. Among the five 1-disulfide intermediates that have been structurally characterized, only one is native, and nearly all of them are bridges by neighbouring cysteines. Extensive accumulation of EGF-II indicates that it accounts for the major kinetic trap of EGF folding. EGF-II contains two of the three native disulfide bonds of EGF, Cys(14)-Cys(31) and Cys(33)-Cys(42). However, formation of the third native disulfide (Cys(6)-Cys(20)) for EGF-II is slow and does not occur directly. Kinetic analysis reveals that an important route for EGF-II to reach the native structure is via rearrangement pathway through 3-disulfide scrambled isomers. Epidermal growth factor (EGF) [29] forms both non-native three-disulfide isomers as well as a predominant species with two native disulfides (EGF-II).

## 3. Conclusions, scope and limitations

The graph representations described here provide a simple method to visualise folding pathways as studied by experimental methods.

The picture emerging from these representations confirm that the folding pathways of oxidative folding are contiguous routes that connect the fully reduced state to the native state. If we try to reconcile this picture with the three-dimensional energy landscape of oxidative folding, the energy of the protein will be a function of which disulfide bonds are present and of the extent of conformational folding (Figure 7). The protein molecules will have folded successfully when they reach the lowest energy point, which represents the native species, both in terms of disulfides and conformation. The non-native disulfide intermediates lie in the local energy minima, from where they have to be re-activated to reach the native state. This picture suggests a qualitative explanation for the observation that non-native disulfide intermediates can be the necessary steps of the folding pathways. The cited case of pro-BPTI is an indirect proof for that. In the analysis of a small 3-disulfide peptide, AAI we found that a disulfide intermediate with no native disulfide bonds is in fact the most abundant species [30,31].

The current approach is limited by the fact that the 3D images of oxidative folding pathways cannot be generated fully automatically. (A drawing application that uses the Tulip package (www.tulip-software.org) is available from VA, vilagos@nucleus.szbk.u-szeged.hu). A further plausible improvement would include colouring of the folding states by quantitative properties and look for correlations between the coloured areas of the network and the experimentally determined folding pathways.

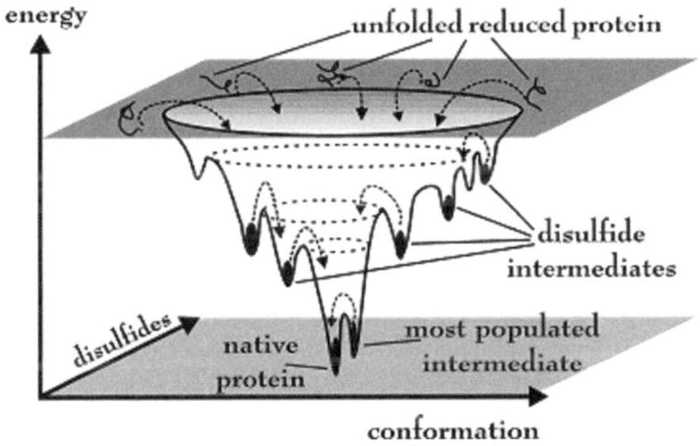


**Figure 7.** The energy landscape of oxidative protein folding [7]. The energy of the protein is displayed as a function of which disulfide bonds are present and the extent of conformational folding. The local minima represent non-native disulfide intermediates, which are kinetic traps.

## References

[1] Levinthal C. Are there pathways in protein folding? J. Chim. Phys. 1968;65:44-5.
[2] Anfinsen CB. Principles that govern the folding of protein chains. Science 1973;181:223-30.
[3] Pain RH. Mechanisms of Protein Folding. Oxford, New York: Oxford University Press, 2000.
[4] Dobson CM, Karplus M. The fundamentals of protein folding: bringing together theory and experiment. Curr Opin Struct Biol 1999;9 (1):92-101.
[5] Dinner AR, Sali A, Smith LJ, Dobson CM, Karplus M. Understanding protein folding via free-energy surfaces from theory and experiment. Trends Biochem Sci 2000;25 (7):331-9.
[6] Onuchic JN, Socci ND, Luthey-Schulten Z, Wolynes PG. Protein folding funnels: the nature of the transition state ensemble. Fold Des 1996;1 (6):441-50.
[7] Cemazar M. Oxidative folding of a cystine knot protein: the Amarathus alpha-amylase inhibitor. International Centre for Genetic Engineering and Biotechnology. Trieste, Italy: Open University, UK, 2003. pp. 130.
[8] Chang JY. Evidence for the underlying cause of diversity of the disulfide folding pathway. Biochemistry 2004;43 (15):4522-9.
[9] Wedemeyer WJ, Welker E, Scheraga HA. Proline cis-trans isomerization and protein folding. Biochemistry 2002;41 (50):14637-44.
[10] Welker E, Wedemeyer WJ, Narayan M, Scheraga HA. Coupling of conformational folding and disulfide-bond reactions in oxidative folding of proteins. Biochemistry 2001;40 (31):9059-64.
[11] Vishveshwara S, Brinda KV, Kannan N. Protein Structure: Insights from Graph Theory. Journal of Theoretical and Computational Chemistry 2002;1 (1):187-211.
[12] Plaxco KW, Simons KT, Baker D. Contact order, transition state placement and the refolding rates of single domain proteins. J Mol Biol 1998;277 (4):985-94.
[13] Magyar C, Tudos E, Simon I. Functionally and structurally relevant residues of enzymes: are they segregated or overlapping? FEBS Lett 2004;567 (2-3):239-42.
[14] Selvaraj S, Gromiha MM. Importance of hydrophobic cluster formation through long-range contacts in the folding transition state of two-state proteins. Proteins 2004;55 (4):1023-35.
[15] Vendruscolo M, Paci E, Karplus M, Dobson CM. Structures and relative free energies of partially folded states of proteins. Proc Natl Acad Sci U S A 2003;100 (25):14817-21.
[16] Vendruscolo M, Paci E, Dobson CM, Karplus M. Three key residues form a critical contact network in a protein folding transition state. Nature 2001;409 (6820):641-5.
[17] Albert R, Jeong H, Barabasi AL. Error and attack tolerance of complex networks. Nature 2000;406 (6794):378-82.
[18] Dorogovtsev SN, Mendes JFF. Evolution of Networks: From Biological Nets to the Internet and Www (Physics). Oxford, New York: Oxford University Press, 2003.
[19] Scala A, Amaral LAN, Barthelemy M. Small-world networks and the conformation space of a short lattice polymer chain. Europhysics Letters 2001;55 (4):594-600.
[20] Dokholyan NV, Shakhnovich B, Shakhnovich EI. Expanding protein universe and its origin from the biological Big Bang. Proc Natl Acad Sci U S A 2002;99 (22):14132-6.
[21] Cantrell MA, Anderson D, Cerretti DP, Price V, McKereghan K, Tushinski RJ, Mochizuki DY, Larsen A, Grabstein K, Gillis S, et al. Cloning, sequence, and expression of a human granulocyte/macrophage colony-stimulating factor. Proc Natl Acad Sci U S A 1985;82 (18):6250-4.
[22] Werner JM, Breeze AL, Kara B, Rosenbrock G, Boyd J, Soffe N, Campbell ID. Secondary structure and backbone dynamics of human granulocyte colony-stimulating factor in solution. Biochemistry 1994;33 (23):7184-92.
[23] Weissman JS, Kim PS. The pro region of BPTI facilitates folding. Cell 1992;71 (5):841-51.
[24] Creighton TE. The disulfide folding pathway of BPTI. Science 1992;256 (5053):111-4.
[25] Weissman JS, Kim PS. Reexamination of the folding of BPTI: predominance of native intermediates. Science 1991;253 (5026):1386-93.
[26] Hober S, Uhlen M, Nilsson B. Disulfide exchange folding of disulfide mutants of insulin-like growth factor I in vitro. Biochemistry 1997;36 (15):4616-22.
[27] Milner SJ, Carver JA, Ballard FJ, Francis GL. Probing the disulfide folding pathway of insulin-like growth factor-I. Biotechnol Bioeng 1999;62 (6):693-703.
[28] Yang Y, Wu J, Watson JT. Probing the folding pathways of long R(3) insulin-like growth factor-I (LR(3)IGF-I) and IGF-I via capture and identification of disulfide intermediates by cyanylation methodology and mass spectrometry. J Biol Chem 1999;274 (53):37598-604.
[29] Chang JY, Li L, Lai PH. A major kinetic trap for the oxidative folding of human epidermal growth factor. J Biol Chem 2001;276 (7):4845-52.

[30] Cemazar M, Zahariev S, Lopez JJ, Carugo O, Jones JA, Hore PJ, Pongor S. Oxidative folding intermediates with nonnative disulfide bridges between adjacent cysteine residues. Proc Natl Acad Sci U S A 2003;100 (10):5754-9.

[31] Cemazar M, Zahariev S, Pongor S, Hore PJ. Oxidative folding of Amaranthus alpha-amylase inhibitor: disulfide bond formation and conformational folding. J Biol Chem 2004;279 (16):16697-705.

*Essays in Bioinformatics*
*D.S. Moss et al. (Eds.)*
*IOS Press, 2005*


# The Application of Bioinformatics Techniques in Genetic Identification and Profiling of Rare Grape Varieties Indigenous to Croatia

Jasenka PILJAC
*Laboratory for plant tissue culture, Department of molecular genetics, Institute 'Ruđer Boškovich', Bijenička c. 54, 10 002 Zagreb, Croatia*

**Abstract** Genetic profiling using microsatellite markers provides a highly efficient method for characterizing and identifying grape varieties. This work describes the use of genetic markers, including single sequence repeat markers, in the discovery of genetic relatedness of the American cultivar Zinfandel and autochthonous Croatian grape varieties (*Vitis vinifera L.*)

Identification of vine varieties and cultivars dates as far back as ancient Greeks and Romans. This branch of viticulture belongs to ampelography (*ampelos* –vine, *grafos* – to write), and while there are no precise figures on the number of grape cultivars in exploitation, it is estimated that about 5,000 exist in different collections worldwide (Alleweldt 1988). The number of cultivar names in use is even larger. Due to undocumented trade and mislabelling, genetically synonymous cultivars often carry completely different names, depending on where they are grown. Sometimes, two separate names may be used for cultivars grown in the same country, on different islands (Bulić 1949). Identification of original cultivars is important, not only because of the accurate count and preservation of *Vitis vinifera* L. genotypes present in the world today, but also because of legal requirements in many countries whereby wine is identified by the variety name. The two major concerns for viticulturists and wine producers are therefore: (i) to find an objective method of identification of grapevine cultivars and eliminate the naming errors, (ii) to understand the origin of cultivars in use today and determine which genetic events lead to production of some famous and economically important cultivars (Bowers and Meredith 1997, Meredith et al. 1999).

The classic wine grape varieties are all *Vitis vinifera* L., which is native to Europe and Western Asia. However, the exact origin of Zinfandel, one of the most important red wine cultivars of the United States ('the spirit of American pioneers captured in a bottle' – Prof. Charles Sullivan, VOA interview with Jagoda Bush), has been a mystery to Americans ever since its arrival on the continent. As a part of the collaborative research project undertaken between the University of California at Davis, Department of Viticulture and Enology (Genetics/Biotechnology laboratory ran by Professor Carole Meredith) and Department of Viticulture and Enology, Faculty of Agronomy, University of Zagreb, I participated in the search for the original genetic match for Zinfandel, that has long been believed to exists in Dalmatia - coastal Croatia.

In 1998, under the guidance of Professor Carole Meredith and with the help of Zagreb scientists, we performed extensive sampling of native Croatian grapevine varieties in southern Dalmatian and on the islands that morphologically resembled Zinfandel and could thus be potential Croatian Zinfandel counterparts. I performed genetic profiling of all the varieties using microsatellite markers and compared the profiles with that of Zinfandel. I compiled the results of this research in my Ph.D (which resulted in the finding of Zinfandel in Croatia). dissertation defended at the University of Zagreb and entitled '*Investigation of relatedness between Zinfandel and autochthonous Croatian grape varieties (Vitis vinifera L.)*'.

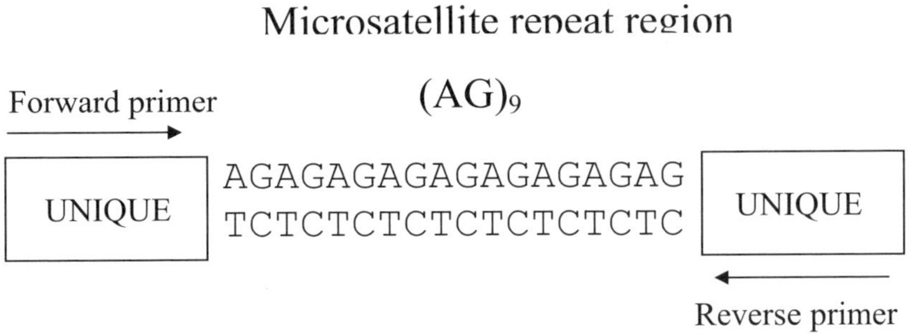


**Figure 1.** A schematic representation of a microsatellite repeat.

Delseney et al. (1983) first demonstrated the existence of simple sequence motifs in plant nuclear DNA. **Simple sequence repeat** (SSR) markers are di-, tri- or tetra-nucleotide repeats numbering in thousands in every eukaryotic genome (Zietkiewicz et al. 1994, Goldstein and Schlötterer 1999). They are locus-specific and codominant, making complicated parentage relationships resolvable (Bowers and Meredith 1997, Meredith et al. 1999) and enabling the reconstruction of grapevine pedigrees (Sefc et al. 1998b). They also proved to be useful in resolving dilemmas regarding induced crosses – e.g. Müller Thurgau (Regner et al. 1996, Dettweiler et al. 2000a) and confirming synonyms (Cipriani et al. 1994, Botta et al. 1995, Bowers et al. 1996, Lopes et al. 1999, Maletić et al. 1999, Lefort et al. 2000). Further advantages of SSR markers in identification and parentage analyses are: (i) their reproducibility – consistent microsatellite profiles were obtained for the same cultivars in different years (Botta et al. 1995) and different laboratories (Grando and Frisinghelli 1998, Lefort et al. 2000), (ii) high degree of polymorphism observed (between 5-10 alleles per marker) enabling differentiation of clones (Vignani et al. 1996, Regner et al. 2000), and (iii) objectiveness in comparison to ampelographic or isozyme methods alone. Each simple sequence repeat region is flanked by unique sequences, and PCR primers complementary to the flanking sequences can uniquely detect each SSR. The PCR primers that are used to detect an SSR locus are so specific that they recognize only a single location in the plant DNA. All the fragments amplified by a pair of primers represent alleles of a single locus. Two bands at the same position represent identical DNA sequences. The banding patterns of each cultivar are easy to distinguish, and the information (expressed as the length, in nucleotides, of each band) may be communicated to other research groups for comparison. This makes SSRs the method of choice in cultivar identification.

Since Thomas and Scott (1993) published the first grape microsatellite markers, many additional polymorphic markers have been developed (Bowers et al. 1996, 1999a, Sefc et al. 1999). Because international cooperators can easily use the same SSR markers and because results obtained can easily be shared electronically, it is possible to compare the SSR profiles of varieties grown in different countries without exchanging gel images or importing cuttings or DNA. Six markers are in most cases sufficient to differentiate cultivars, but when resolving complicated relationships or defining parentage trees between closely related cultivars, many more are needed. For the purpose of my research, SSR markers were employed to identify and resolve complicated relationships between a group of closely related Croatian cultivars and Zinfandel. A schematic linear representation of a microsatellite repeat is shown in Figure 1.

Genetic results obtained by SSR molecular markers presented in my doctoral dissertation support the hypothesis of a Croatian origin of Zinfandel, as well as a high degree of relatedness between Zinfandel and autochthonous Croatian cultivars. In fact, we discovered that Zinfandel is an offspring in the cross between two autochthonous Croatian wine grape varieties, Plavac mali (the predominant red cultivar on the Pelješac peninsula) and Dobričić (a neglected red variety on the island of Šolta). Since Zinfandel is in the center of complex genetic relationships with native Croatian cultivars, and even plays a parental role in one, we hypothesize that this variety in the past served as the pollinator of local varieties. It was most probably eradicated and discontinued from further exploitation in Croatia due to several reasons: pests and diseases at the turn of the $20^{th}$ century that resulted in a major loss of native Croatian varieties, modern production demands, and a shift of viticultural emphasis in Croatia towards introduced varieties such Chardonnay, Rhine riesling, etc.

Based on allele frequency analyses of Greek, Italian and Croatian *Vitis vinifera* L. gene pools, and discovered cultivar relationships, I concluded that the Croatian gene pool is the most likely source of Zinfandel. Subsequently to the completion of my Ph.D. research, several (only 9!) Zinfandels (under the local name of Crljenak kaštelanski, Figure 2) were discovered in an old vineyard in Kaštel Novi, near the major Dalmatian port of Split. Their genetic match with Californian Zinfandel was confirmed in Professor Meredith's laboratory at UC Davis.

The observed gene diversity of 77.7%, which I calculated for the Croatian varieties, reveals a substantial level of genetic variation in the Croatian population of cultivars and points to the significance of preservation of more than 100 unique genotypes found in Croatia today.

I reckon that the bioinformatics tools to which I was introduced in the NATO sponsored course held in May of 2003 in Dubrovnik, Croatia, will help me in expanding the analyses and confirming the findings of my research. The online available databases of DNA sequences of various organisms keep increasing and, hopefully, they will soon include the genetic profiles of major grapevine cultivars exploited in Europe.

## References

[1] Alleweldt G. (1988) The genetic resources of *Vitis*. Genetic and geographic origin of grape cultivars, their prime names and synonyms. Second ed. Federal Research Center for Grape Breeding, Geilweilerhof, Germany.

[2] Bulić S. (1949) Dalmatinska ampelografija, Poljoprivredni nakladni zavod, Zagreb.

[3] Botta R., Scott N. S., Eynard I., Thomas M. R. (1995) Evaluation of microsatellite sequence-tagged site markers for characterizing *Vitis vinifera* cultivars, *Vitis*, 34(2):99-102.

[4] Bowers J. E., Meredith C. P. (1997) The parentage of a classic wine grape, Cabernet Sauvignon, *Nature genetics*, 16:84-87.

[5] Bowers J. E., Dangl G. S., Vignani R., Meredith C. P. (1996) Isolation and characterization of new polymorphic simple sequence repeat loci in grape (*Vitis vinifera* L.), *Genome*, 39:628-633.

[6] Cipriani G., Frazza G., Peterlunger E., Testollin R. (1994) Grapevine fingerprinting using microsatellite repeats, *Vitis*, 33:211-215.

[7] Delseny M., Laroche M., Penon P. (1983) Detection of sequences with Z-DNA forming potential in higher plants, *Biochem Bioph. Res. Commun.*, 116:113-20.

[8] Dettweiler E., Jung A., Ziprian E., Töpfer R. (2000a) Grapevine cultivar Müller-Thurgau and its true to type descent, *Vitis*, 39(2):63-65.

**Figure 2.** Kaštelanski crljenak, the Croatian genetic counterpart of Zinfandel. (Photo by Jasenka Piljac)

[9] Goldstein D. B., Schlötterer C. (1999) Microsatellites: Evolution and applications, Oxford University Press, Oxford.

10] Grando M. S., Frisinghelli C. (1998) Grape microsatellite markers: Sizing of DNA alleles and genotype analysis of some grapevine cultivars, *Vitis,* 37:79-82.

[11] Lefort F., Anzidei M., Roubelakis-Angelakis K. A., Vedramin G. G. (2000) Microsatellite profiling of the Greek Muscat cultivars with nuclear and chloroplast SSRs markers. *Quaderni della Scuola di Specializzazione in Scienze Viticole ed Enologiche*, 23:56-80.

[12] Lopes M. S., Sefc K. M., Eiras Dias E., Steinkellner H., Laimer da Camara Machado M., Da Camara Machado A. (1999) The use of microsatellites for germplasm management in a Portuguese grapevine collection, *Theor. Appl. Genet*., 99:733-739.

[13] Maletić E., Sefc K. M., Steinkellner H., Kontić J. K., Pejić I. (1999) Microsatellite variability in grapevine cultivars from different European regions and evaluation of assignment testing to assess the geographic origin of cultivars, *Theor. Appl. Genet*., 100:498-505.

[14] Meredith C. P., Bowers J. E., Riaz S., Handley V., Bandman E. B., Dangl G. S. (1999) The identity and parentage of the variety known in California as Petite Sirah, *Am. J. Enol. Vitic*., 50(3):236-242.

[15] Regner F., Steinkellner H., Turetschek E., Stadhulber A., Glössl J. (1996) Genetische Charakterisierungvon Rebsorten (*Vitis vinifera*) durch Mikrosatelliten – Analyse, *Mitteilungen Klosterneuburg*, 46:52-60.

[16] Regner F., Wiedeck E., Stadlbauer A. (2000) Differentiation and identification of White Riesling clones by genetic markers, *Vitis*, 39(3):103-107.

[17] Sefc M. K., Guggenberger S., Regner F., Lexer C., Glössl J., Steinkellner H. (1998a) Genetic analysis of grape berries and raisins using microsatellite markers, *Vitis,* 37:123-125.

[18] Sefc M. K., Regner F., Turetschek E., GlÖSSl J., Steinkellner H. (1999) Identification of microsatellite sequences in *Vitis riparia* and their applicability for genotyping of different *Vitis* species, *Genome,* 42:367-373.

[19] Thomas M. R., Scott N. S. (1993) Microsatellite repeats in grapevine reveal DNA polymorphisms when analyzed as sequence-tagged sites (STSs), *Theor. Appl. Genet*., 86:985-990.

[20] Vignani R., Bowers J. E., Meredith C. P. (1996) Microsatellite DNA polymorphism analysis of clones of *Vitis vinifera* Sangiovese, *Sci. Hortic.* 65:163-169.

[21] Zietkiewicz E., Rafalski A., Labuda D. (1994) Genome fingerprinting by simple sequence repeat (SSR)-anchored polymerase chain reaction amplification, *Genomics*, 20:176-183.

*Essays in Bioinformatics*
*D.S. Moss et al. (Eds.)*
*IOS Press, 2005*


# Papaya (*Carica papaya*) Fruit Ripening I - Pectinmethylesterase (PME) cDNA Cloning and Expression during Fruit Development and Ripening

Aladje BALDÉ[1]; Manuela M.C. GOUVEIA.[2] and Maria Salomé PAIS[1]
[1]*Laboratory of Plant Biotechnology, ICAT; Ed. ICAT, Campo Grande, 1749-016 Lisboa, Portugal*
[2]*Departamento de Biologia, Univ. Madeira, Largo do Município, P-9050, Funchal, Portugal*

**Abstract.** Pectinmethylesterase, an enzyme involved in cell wall softening of papaya fruit was isolated. The structure of this cDNA and its expression during development and ripening of the fruit was analysed. Northern Blotting and was used to determine the expression of pectinmethylesterase genes along fruit development and ripening. PME is differentially expressed in the inner and outer mesocarp. The levels of PME activity increase gradually with maturation until day 7 of ripening. The pectinmethylesterase activity increases differentially from the outer mesocarp to the inner mesocarp along ripening. These values are similar for 7 days ripened fruits which corresponds to 70% ripening. After that ripening stage there are no significant differences between PME in inner and outer mesocarp and the PME activity is reduced of about 10%. The phylogram generated using an alignment of the deduced amino acid sequences of PME and of 10 PMEs homologues from other plant species revealed that pectinmethyl esterase from papaya fruits presents higher similarity with tomato PME sequences than with the other PMEs sequences available. The amount of total RNA in the mature ripe fruit duplicated the amount of total RNA in the green fruit. All the cDNAs were expressed at similar levels at the inner and outer mesocarp tissues during the different stages of fruit ripening. However, its expression was highest for ripening stages 1, 3, 5 and 7 decreasing thereafter to lower levels of expression. These results show that increase in mRNA translation parallels the increase in PME activity until day 7 of ripening.

## Introduction

Papaya exportation has problems associated to long distance transport, due to adulteration of the pulp whenever the fruits are caught as soon maturation is achieved. To solve this problem, the current practice is to collect fruits, for exportation, in very precocious phases of maturation, which brings, as consequence, the adulteration of the organoleptic characteristics of this fruit. The process of fruit softening is commercially important because it often dictates early harvest of fruit to avoid damage in subsequent handling, which can result in failure to develop optimum flavour and colour. Finally, excessive softening and the associated enhancement in pathogen susceptibility limit the post-harvest life of papaya.

Papaya fruit is susceptible to chilling injury with critical temperatures ranging between 10-15 C. In papaya, the symptoms of chilling injury are more evident upon returning the fruits to higher ripening temperatures (Chan et al., 1985; Lyons & Breidenbach, 1987).

Pectinmethylesterase (PME) activity has been reported to increase during the development of banana (Brady, 1976), apple (Knee, 1978), avocado (Awad et al., 1979) and papaya (Paul & Chen., 1983) fruits. The exact role of PME in *Carica papaya* fruit development and ripening is yet to be determined. However, it has been hypothesised that destherefication of pectin by PME and further depolymerisation by polygalacturonase (PG) are involved in fruit softening. This hypothesis is based on the observation that demethylation of pectin by PE causes a several fold increase in cell wall solubilisation by polygalacturonase (Pressey and Avants, 1982). PME, in addition to other pectolytic enzymes, has been implicated in fruit ripening (Basic et al., 1988). This cell wall metabolising enzyme is responsible for the demethylation of galacturonic acid residues in high molecular weight pectin, each methyl group being converted to a proton and methanol (Hall et al. 1993). According to Ali et al. (1993), PG, PME and β-galactosidade may collectively play significant roles in the development of the chilling injury symptom of increased – susceptibility – to disease commonly observed in papayas upon returning chill-stored fruits to warmer environments.

The aim of this study was to investigate the significance of PME to differential softening and to characterise the PME expression during papaya fruit development and ripening at the biochemical level and molecular mRNA translation.

## 1. Material and Methods

### *1.1 Plant material and sampling*

After harvest, mature green papayas were brought from Guinea-Bissau to the Laboratory and were allowed to ripe at 25 °C. The fruits were sampled at different ripening stages (1, 3, 5, 7, 9 and 11 days), cut transversally in two parts and seeds were removed. Inner mesocarp was separated from the outer mesocarp and was homogenised each in liquid nitrogen using warring blander. The homogenised pulp was instantly frozen at –80 °C.

### *1.2 RNA isolation*

The tissues were ground to a fine powder in liquid nitrogen using a warring blender. Using a metal spatula, chilled in liquid nitrogen, the powder was quickly transferred to tubes containing mixture (1:1) of extraction buffer (sodium acetate, EDTA and SDS) and phenol (pH 4.3), preheated at 65 °C for 5 min. After homogenisation by vortexing for 5 min, ½ x volume of Chloroform: Isoamyl alcohol (24:1) was added. After vortexing for 5 min, the homogenate was spined at 10000 rpm for 10 min at 4 °C. Using a sterile glass pipette, the upper aqueous phase was transferred to polypropylene tubes and equal volume of chloroform isoamyl alcohol (24:1) was added. Vortexing for 5 min and centrifugation for 10 min at 10000 rpm at 4 °C were repeated. Using a sterile glass pipette, the upper aqueous phase was again transferred to polypropylene tubes and equal volume of chloroform: isoamyl alcohol was added and vortexed for 5 min. The sample was transferred to Corex tubes and spined at 10000 rpm for 10 min at 4 °C. The upper aqueous phase was transferred to Corex tubes and 1/3 volume of 8 M LiCl was added and precipitation took place overnight at 4 °C. A new centrifugation at 10000 rpm for 10 min at 4 °C. Pellet was dissolved with 2 M LiCl by vortexing and centrifuged 10 min at 10000 rpm at 4 °C. The two previous steps were repeated twice. The pellet was dissolved with 3 M Sodium acetate by vortexing and centrifuged 10 min at 10000 rpm at 4 °C. The two previous steps were repeated. The pellet was washed twice with 70% ethanol, air dried and dissolved into 100 µl of water (DEPC treated).

### *1.3 Oligonucletide Design and RT-PCR*

Degenerated oligonucleotides were designed based on regions of high homology between aligned PME-deduced amino acid sequences from *Lycopersicon esculentum* (Bird, et al., 1993-1994; Pear et al., 1993; Bridges, et al. 1988; Ray, et al 1988), *Phaseolus. vulgaris* (Recourt et al., 1992; 1995), *Petunia inflata* (Um et al., 1994) and were synthesised.

First-strand cDNA was synthesised from 2 µg of total RNA from mature fruit of papaya. RNA was incubated in 20 µl of 1 x first-strand buffer (50 mM Tris-HCl, pH 8,3, 75 mM KCl, and 3 mM MgC12), 0.5 mM each dNTP and 100 ng of oligo(dt)17, 10 mM DTT and 20 units of RNAs in at 65 °C for 10 min and then placed on ice. 1 µl of MMLV-RT (200 units/µl) was added and the reaction was incubated at 37 °C for 1 h. Reaction was then heated to 95 °C for 5 min, and then placed on ice or stored at -20 °C until further use.

1 µl of first-strand reaction was used as a template in PCR. The reaction mixture was composed of 10 Tris-HCl, pH8.3 50 mM KCl 1 mM MgC12, 0.2 mM dNTPs, 100 pM each PE1 and PE2 primers, and 0.2 µl of Taq polymerase. The conditions for amplification were 94° C for 4 min and 35 cycles of 94 °C for 1 min, 55 °C 1 min, 72° C for 1 min and then 72 °C for 7 min. Product was gel purified using Qiaex (Qiagen) and the product was cloned in Bluscripts KSII. Cloned PCR product was sequenced and analysis was carried out using the DNASTAR software.

### *1.4 Northern blotting analysis*

Twenty micrograms of total RNA from papaya fruit was separated by glyoxal denaturation agarose gel electrophoresis and transferred to nylon membrane (Hybond-N, Amersham), according to the manufacturer's instructions. Membrane was probed with [$\alpha$-$^{32}$P]dCTP-labelled insert DNA from CpPME (partial-length PCR clone). The probe was labelled by Ridiprimer DNA labelled kit (Amersham) labelled the probe. The hybridisation was carried out overnight at 65 °C in 7% (w/v) SDS, 0.5 M phosphate buffer, 2 % (w/v), blocking reagent (Boheringer) with approximately 50 ng of labelled probe. The blot was washed twice in 2xSSC and 0.1 % (w/v) SDS at 65 °C, twice in 0.1xSSC and 0.1 % (w/v) SDS at 65 °C. Blot was exposed to film at -80 °C overnight.

### *1.5 PME extraction and purification*

Extraction of PME was as described in Fayyaz et al. (1994). Briefly, after thawing at 4 °C 100 g of papaya pulp, previously frozen at –80 °C were homogenised with 200 ml of 2 M NaCl solution pH 8.0. After adjusting the pH to 8.0, the homogenate was incubated in a cold room at 4 °C for 5 hours under stirring condition. During the incubation period, the pH of the homogenate was maintained at pH 8.0 by adding either 2 M NaOH or 2 M HCl. The homogenate was centrifuged at 24000xg for 30 min at 4 °C. Solid ammonium sulphate sufficient to give 30 % precipitation was added to the extract with continuous stirring. The extract was centrifuged at 24000xg for 30 min at 4 °C. The precipitate was discarded and solid ammonium sulphate was added to the supernatant to give 90 % saturation and was allowed to stand for 4 hours at 4 °C. The precipitate was centrifuged at 24000xg for 30 min and pellet dissolved in 0.02 M, sodium phosphate buffer at pH 7.5. The enzyme solution was dialysed for 36 hours against several changes of 15 volumes of 0.02 M, sodium phosphate buffer solution, pH7.5. The dialysed solution was clarified by centrifugation at 24000xg for 30 min and pellet was discarded. The enzyme solution was concentrated by ultrafiltration using an Amicon system. The concentrated enzyme was applied to a CM-Sephadex C-50 column (2x6x37 cm) which had been previously equilibrated with 0.02 M pH 7.5 sodium phosphate buffer. The column was washed with the equilibration buffer and the enzyme was eluted by using 500 ml of linear gradient of 0.1 M NaCl in 0.02 M, pH 7.5 sodium phosphate buffer. Fraction were collected and assayed from protein and PE activity. The active fractions obtained from the previous step were combined and concentrated using Amicon filter. The concentrate sample were applied to a of Sephadex G-100 (2x6x65 cm ) column, equilibrated with 0.02 M sodium phosphate buffer pH 7.5 containing 0,2 M NaCl and 0.02 sodium aside. The enzyme was eluted with the same buffer, until the absorbency at 280 nm of the effluent was negligible. Active enzyme fractions were pooled, concentrated as above. Pectinmethylesterase activity was assayed using the method of H. Fayyaz et al. (1993).

### *1.6 Southern Blotting*

Total genomic DNA was isolated by CTAB method (Auto e data...) 10 μg of DNA were digested with restriction enzymes BamHI, EcoRI and HindIII (Boheringer), separated on 0.8% agarose gels, and transferred to Nitro-cellulose membranes according to the manufacturer's (Amersham) instructions. Membranes were probed with gel-purified, [$\alpha$-$^{32}$P]dCTP-labeled insert DNA from cpPME1 (partial-length RT-PCR clones), under the conditions described above for RNA-blot hybridisation, washed in 5x SSC and 0.1% SDS at 65 °C and 0.2x SSC and 0.1 SDS at 65 °C and exposed to film with one intensifying screen at –80 °C for overnight.

### *1.7 Phylogenetic Analysis*

The deduced amino acid sequence of cpPME1 and pPME2 were aligned to 10 amino acids sequences of pectinmethylesterase gene. Homologies between the deduced amino acid sequences of PME were determined using Clustal V multiple-sequence alignment software. The sequences were: 6 from tomato fruit PME, 3 from *Phaseolus. vulgaris* PME and 1 from *Petunia inflata* PME. The PME phylogenetic tree was inferred from the aligned sequences using the maximum parsimony algorithm of the DNASTAR software.

## 2. Results and Discussion

A characteristic feature during the ripening of papaya fruit is softening. Softening is the result of the structural changes in the cell wall caused by the activity of hydrolases (Hubert 1983). Pectinmethylesterase, an enzyme that catalyses demethylation of the $C_6$ carboxyl group of galacturonosyl residues, may play an important role in determining the extent to which pectin is accessible to degradation by polygalacturonase. Indeed, it has been suggested that the increased susceptibility of tomato fruit cell walls to polygalacturonase action during ripening is due to the action of pectinmethylesterase (Koch et al., 1989).

In *Carica papaya,* fruit softens differentially in relation to the position of the tissue. Based on carotenoid development, Paul and Chen (1983) considered that papayas ripen from the endocarp towards the outer mesocarp, the same being suggested on the base of fruit softening (Chan et al., 1981; Lazan et al., 1989) The inner mesocarp tissue is softer and its firmness decreases more rapidly during ripening than that of the outer mesocarp tissue. The levels of PE activity increase gradually with maturation (fig.1). The pectinmethylesterase activity increases differentially from the outer mesocarp to the inner mesocarp along ripening (fig.1). These values are similar for 7 days-ripened fruits, which corresponds to 70% ripening. After that ripening stage there are no significant differences between PME in inner and outer mesocarp and the PME activity is reduced of about 10%. The reduction of activity does not parallel the total protein values that increase with ripening. The highest values of PME activity, have been found for the same ripening stage as Polygalacturonase activity is highest (data not shown) which is in agreement with results from other authors according which PME, an enzyme that catalyses demethylation of the $C_6$ carboxyl group of galacturonosyl residues, may play an important role in determining the extent to which pectin is accessible to degradation by polygalacturonase (Koch et al., 1989). According to Harriman et al. (1991) during tomato ripening the levels of PME protein continue to increase beyond the turning stage while the PME activity begins to decline. As the fruit ripens, pectin solubility and depolymerisation increase (Lazan et al., 1995). According to these authors tissue softening is more closely related to changes in β-galactosidase activity than to PG or PME activity. Similar results have been obtained by Harrimann et al. (1991) for tomato PME activity. Differential ripening has also been reported for papaya ACC activity by Chan (1991). In *Persea americana* fruits, PME decline from its maximum at the time of picking to a low level early in the climateric (Awad and Young, 1979).

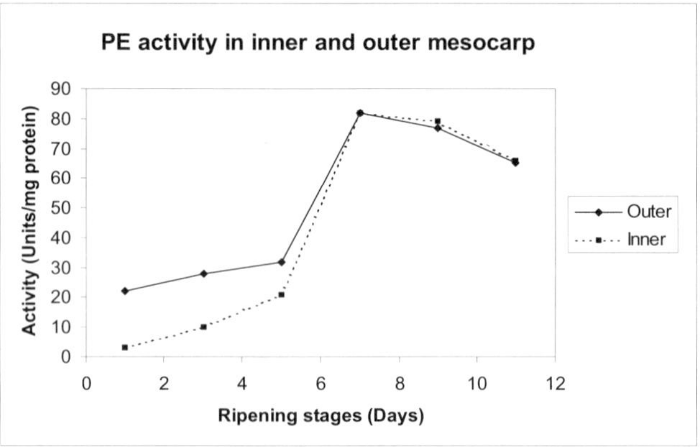


**Figure 1.** Changes in PME activity along ripening of papaya fruit mesocarp. Dot line represents the activity at the inner mesocarp and the continuous line at the outer mesocarp.

Fruits at maturation stages (1, 3, 5, 7, 9 and 11 days respectively) prepared for sampling and RNA extraction produced yields of the purified RNA in the range of 300-450 µg RNA per g/fw of tissue of ripening fruit and 150- 200 mg per g/fw of non ripe fruit tissue. In all cases $A_{260}/A_{230}$ values were about 1.9-2.0, which indicates high purity of the RNA preparation. The quality of RNA was evaluated by RT-PCR applications, Northern Blotting analysis, mRNA isolation and cDNA library construction.

The construction of cDNA libraries is a basic step in most molecular biological techniques.

Degenerated oligonucleotides were designed based on regions of high homology between aligned PE-deduced amino acid sequences from *Lycopersicon esculentum* were used to amplify partial-length of papaya cDNA from reverse-transcribed RNA of mature ripe (70 % yellow) fruit mesocarp of papaya. The amplified product was 649 bp, as predicted from the sequences of known Pectinesterases. RT-PCR clones were cloned in Bluscripts vector (Strategene) and were sequenced.

The cDNA library constructed from papaya fruit mRNA was screened using, as probe, an insert from partial-length cDNA obtained by RT-PCR. The resulting cDNA clone was 1620 bp in length and contained a complete open reading frame. The length of cDNA clones correspond to the size of the most abundant corresponding mRNA, and it is assumed that they represent full-length mRNAs. All the cDNAs contained complete open reading frames

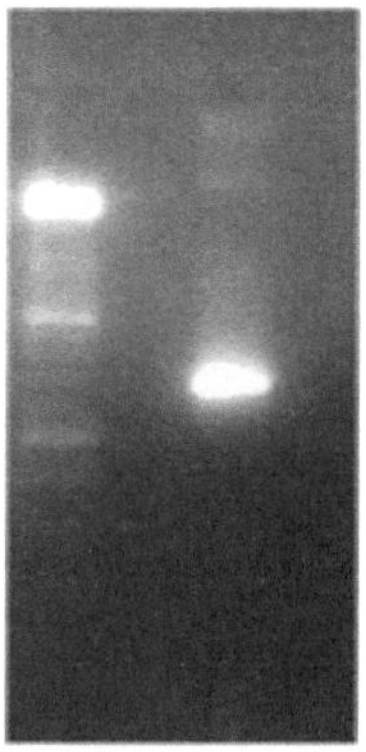

**Figure 2.** PCR amplification of PME using degenerated primer

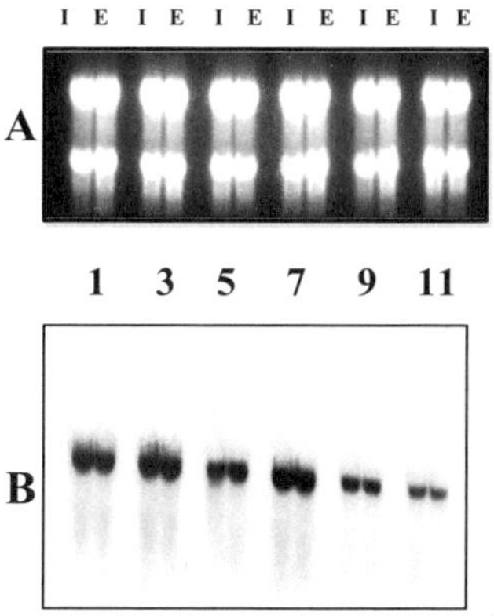


**Figure 3.** Northern Blotting of Carica papaya total RNA using papaya PME partial gene

The phylogram generated using an alignment of the deduced amino acid sequences of papaya PME and of other 10 PMEs published primary aminoacid sequences from other plant species is presented in fig.6. From this phylogram it can be observed that pectinmethyl esterase from papaya fruits presents higher similarity (between 35 and 82%) with tomato PME sequences than with the other PME sequences available.

The amount of total RNA in the mature ripe fruit (stages 9 and 11) duplicated the amount of total RNA in the green fruit (stages 1 and 3) (fig.4). In all the stages of fruit ripening, the amount of total RNA in the inner and in the outer mesocarp was similar (data not shown).

Northern blot analysis was carried out to examine the level of mRNA expression in fruit at different ripening stages. All the cDNAs were expressed at similar levels at the inner and outer mesocarp tissues during the different stages of fruit ripening. However, its expression was highest for ripening stages 1, 3, 5 and 7 decreasing thereafter (stages 9 and 11) to very low levels of expression (fig.4). These results show that increase in mRNA translation parallels the increase in PME activity.

Cloning of a tomato fruit PME cDNA clone allowed the characterisation of mRNA levels that are highest in immature green fruit and then decline throughout maturation and ripening (Ray et al., 1988). This pattern of mRNA accumulation did not parallel the increase in PME enzyme activity previously reported by Seymon et al. (1987), Tucker et al. (1982).

## References

[1] Awad M, Young R E (1979). Postharvest variation in cellulase, polygalacturonase and pectin methylesterase in avocado (Persea americana) fruit in relation to respiration and ethylene production. Plant Physiol. 64:306-308.
[2] Bird,C.R. (1993) Direct Submission. JOURNAL Submitted (17-AUG-1993) to the EMBL/GenBank/DDBJ databases. C.R.
[3] Brady C J (1976). The Pentinesterase of pulp banana fruit. Aust. J. Plant Phygiol 3:163-172
[4] Bridges,I.G., Schuch,W.W. and Grierson,D (1988) Anti-sense regulation of plant gene expression.
[5] Patent: EP 0271988-A 3 22-JUN-;
[6] Fayyaz A , Asbi B. A ; Ghazali H. M., Che Men Y. B. & Jiap (1994) Purification and Molecular Properties of Papaya Pectinesterase.Food Chemistry 49:373-378
[7] Gouveia,M.M.C., Balde,A., Pais,M.S., Mooibroek,A. and Recourt, K. (.... )Characterisation of pectinase cDNAs in fruit of Carica papaya L. Unpublished.
[8] Ripeness and tissue depth effects on heat inactivation of papaya ethylene-forming enzyme (Chan, H. Jr., J. Food Sci. 56(4): 996-998.
[9] Hubert et al. (1983): The role of the cell wall hydrolases in fruit softening. Hortic. Ver. 5:169-219.
[10] Knee M. (1979) Metabolism of polygalacturonase in apple fruit cortical tissue during ripening. Phytochemestry 17:1262-1264.
[11] Koch, J. L. et al. (1989): Tomato fruit cell wall I. Use of purified tomato polygalacturonase and pectinesterase to identify developmental changes in pectin. Plant physiology 91: 91:816-822.
[12] Lazan H. Selmat M. K. and Ali Z. M. (1995) B-galactosidase, polygalacturonase and pectinesterase in differential softening and cell wall modification during papaya fruit ripening. Physiol. Plant 95:106-112.
[13] Mu,J.-H., Stains,J. and Kao,T.-h. (1994)Characterization of a pollen-expressed gene encoding a putative pectin esterase of Petunia inflata JOURNAL Plant Mol. Biol. 25, 539-544.
[14] Paull R E, Chen N. J. (1983) Postharvest variation in cell wall degrading enzymes of papaya (Carica papaya L.) during fruit ripening. Plant physiol. 72:382-385.
[15] Pear,J.R., Sanders,R.A., Summerfelt,K.R., Martineau,B. and Hiatt,W.R. (1993). Simultaneous inhibition of two tomato fruit cell wall hydrolases, pectinmethylesterase and polygalacturonase, with antisense gene constructs. Antisense Res. Dev. 3 (2), 181-190.
[16] Pear,J.R., Sanders,R.A., Summerfelt,K.R., Martineau,B. and Hiatt,W.R. (19..) Simultaneous inhibition of two tomato fruit cell wall hydrolases, pectinmethylesterase and polygalacturonase, with antisense gene.
[17] Pressey R. Avants J. K. (1982) Pectin enzymes in “long keeper” tomatos. HortScience 17:398-406.
[18] Pressey R. Avants J. K. (1982) Solubilization of cell wall by tomato polygalacturonase effects of pectinesterase. J. Food Biochem. 6:57-74.
[19] Ray,J., Knapp,J., Grierson,D., Bird,C. and Schuch,W. (1988). Identification and sequence determination of a cDNA clone for tomato pectin esterase JOURNAL Eur. J. Biochem. 174 (1), 119-124.
[20] Recourt,K. (1992). Direct Submission. Submitted (05-AUG-1992) to the EMBL/GenBank/DDBJ databases. K.
[21] Recourt,K., Laats,J.M., Stolle-Smits,T., Wichers,H.J., Van Dijk,C. and Ebbelaar,C.E.M. (1990). Molecular characterisation of bean pectin esterases and the expression during pod development. Unpublished.
[22] Seymour G B. Lasslett Y, Tuker G A (1987) Differential effects of pectolytic enzymes on tomato polyuronides *in vivo* and *in vitro* Phytochemistry 26:3137-3139.

*Essays in Bioinformatics*
*D.S. Moss et al. (Eds.)*
*IOS Press, 2005*




# Organogenic Nodule Formation in Hop (*Humulus lupulus* var. Nugget)

Ana Margarida FORTES, Maria Salomé PAIS
*Lab. of Plant Biotechnology, ICAT, FCUL, Campo Grande, 1749-016 Lisboa Portugal*

**Abstract.** This work aimed to study some of the processes involved in organogenic nodule formation in *Humulus lupulus* var. Nugget. Organogenesis and *in vitro* somatic embryogenesis from differentiated plant cells are complex morphogenic processes involving physiological, biochemical, molecular and elemental tissue and cell changes. These morphogenic processes play pivotal roles in plant biotechnology. Knowledge on the signals involved in their induction, formation and development will enable in the future a controlled induction of morphogenesis.

In a first approach, it was studied the sequence of histological and histochemical events occurring from internodes inoculation until the development of shoot buds [1]. Cell division was observed in both cambial and cortical regions during the first week of culture establishment. Divisions of cortical cells led to the formation of an incipient callus tissue. Prenodular structures of cambial origin appeared surrounded by these calluses and gave rise to nodules from which shoot buds were formed. Nodules kept separating into "daughter nodules" from which arose an increasing number of shoot buds. Iodide staining showed a strong starch accumulation in callus tissue and in prenodular structures. During shoot bud primordia formation starch content decreased in nodules and was probably mobilized for organ initiation and development. Control explants, which never gave rise to organogenic nodules nor regenerated plantlets, accumulated starch at a much lower extent than explants cultured on media with growth regulators. This suggested that a differential pool of sugars could play an inductive role in organogenic nodule formation.

Previous studies carried out during induction of somatic embryogenesis in other plant species suggested callose and cutin deposition as a way to isolate cells from the surroundings, which might cause metabolic changes leading to embryo formation. With the purpose of investigating if such a deposition occurred along morphogenesis induction in hop, callose and cutin accumulation was followed by staining with Aniline Blue and Nile Red and by immunolocalization using antibodies raised against callose [2]. A cutin layer showing bright yellow autofluorescence appeared surrounding cells or groups of cells committed to express morphogenic competence and enter mitosis. This cutin layer that evolved to a randomly-organized network appeared underneath a callose layer and may create a specific cellular environment with altered permeability and altered receptors providing conditions for entering cell cycle. The fact that only an incipient callose accumulation was observed in control explants suggested the involvement of callose in the initiation of the morphogenic program leading to nodule formation. A scanning electron microscopic study during the organogenic process showed that before shoot bud regeneration, the cutin layer increased in thickness and acquires a smooth texture (Fig. 1).

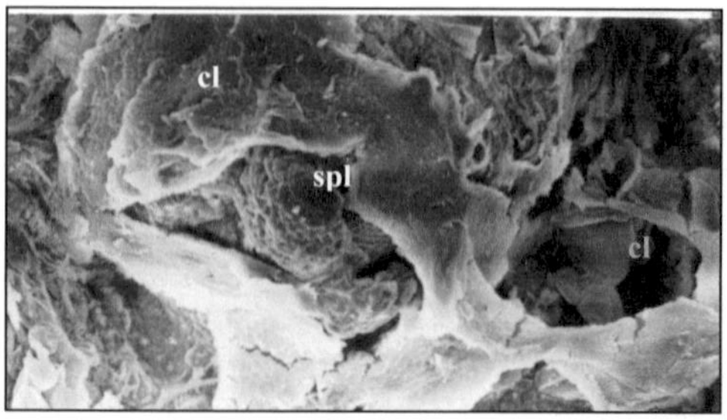


**Figure.** Scanning electron microscopy image showing a cutin layer (cl) of smooth texture over organogenic nodules regenerating shoot primordia (sp).

This cutin layer was specific to nodular organogenic regions and disappeared with plantlet regeneration. This layer was suggested to control permeability to water and solute transfer throughout plantlet regeneration.

Lipoxygenases have been related to several processes of growth and development as well as stress response. Studies of lipoxygenases during organogenic nodule formation in hop showed that they are developmentally regulated throughout the process [3]. Lipoxygenase activity and lipid peroxides presented a huge increase during the first week of culture, which could indicate a role for lipoxygenase and lipoxygenase products in response to wounding in hop, as reported for other systems. Western blotting analysis showed a de novo synthesis of lipoxygenase (LOX) isoenzymes in response to wounding. The antibody used detected two different isoenzymes with molecular masses of approximately 74, and 98 kDa (Fig. 2). A partial cDNA fragment (1000 bp) coding for a lipoxygenase was cloned through a Reverse Transcriptase- Polimerase Chain Reaction based approach and may correspond to the most expressed isoenzyme during this period. As shown using Blast-n (NCBI Database BLAST program) [4] this fragment shares 79 % identity with Prunus dulcis LOX mRNA.

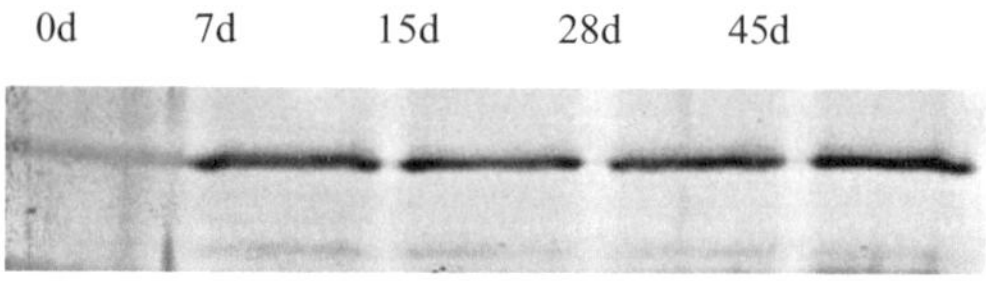


**Figure 2.** Proteins from different culture periods (d-days) after SDS-PAGE and immunoblotted with polyclonal antisera for LOX. The upper band corresponds approximately to a 98 kDa isoenzyme whereas the lower one which is less intense corresponds to a 74 kDa isoenzyme.

Confocal analysis of lipoxygenase immunofluorescence revealed the presence of the enzyme in cortical cells of induced internodes and in prenodular cells, mostly appearing as cytoplasmic spots. Some of them were identified as lipid bodies by cytochemical and double immunofluorescence assays, suggesting the involvement of a lipid-body lipoxygenase during nodule formation. Immunogold labeling detected lipoxygenase in peroxisomes, lipid bodies and plastids of nodular cells. The quantification of the labeling density provided statistical significance to the localization of lipoxygenase (three different isoenzymes) in the three compartments, which suggested a possible involvement of

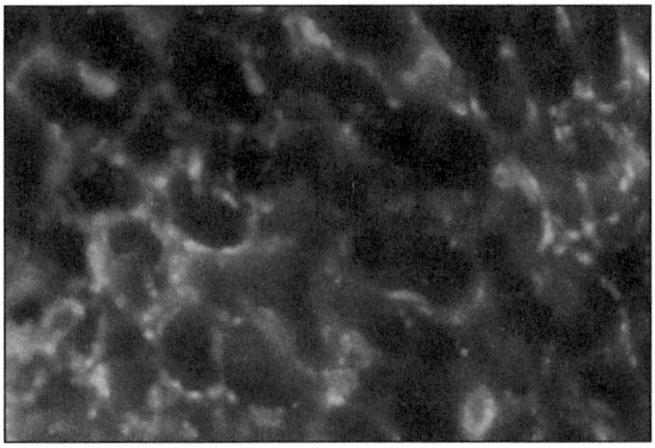

**Figure.** Confocal image of a nodular cell showing AOC in chloroplasts. Plant tissue was incubated with a polyclonal antibody anti-AOC and with the secondary antibody anti-rabbit-Alexa Fluor 488, and further stained with DAPI to confirm the absence of immunofluorescence signal in the nucleus.

lipoxygenase in metabolic functions of these organelles during organogenic nodule formation and plantlet regeneration.

In order to enable the study of the transcript pattern during organogenic nodule formation and to isolate genes related to morphogenesis and wounding response, a cDNA library was constructed using RNA extracted from induced internodes along 24 h and from organogenic nodules. The amplified library titter was $1.25 \times 10^{10}$ pfu/ ml. A random in vivo excision was performed and 60 clones sequenced which showed that full-length cDNAs could be obtained by screening of this library. Amplification by PCR of cDNA inserts from the plasmid library revealed that their size was ranged between 400 bp and 2000 bp. The library redundancy was approximately 40 %.

Sequences clones were checked for identities using the NCBI Database BLAST program. Homologues of peroxidase, cytochrome P450, metalothionein, and pectinesterase were isolated among the selected clones. The correspondent mRNAs may be differentially expressed throughout organogenic nodule formation since they play roles on wound response, cell division and differentiation processes as well as on biosynthesis of growth regulators.

A crucial step in the biosynthesis of jasmonic acid is the formation of its stereoisomeric precursor, cis(+) 12-oxophytodienoic acid, which is catalyzed by allene oxide cyclase. Study of allene oxide cyclase expression during organogenic nodule formation revealed that this enzyme is involved in the response of internodes to wounding, nodule formation and plantlet regeneration from these nodules [5]. A Reverse Transcriptase- Polimerase Chain Reaction based approach using degenerated primers is being undertaken in order to clone the AOC homologue from hop. Western blotting analysis using an antibody raised against allene oxide cyclase from tomato showed elevated levels of allene oxide cyclase in response to wounding peaking at 24 h after internode inoculation. Jasmonic acid levels increased at this time (62,2 nmol/ g FW) whereas 12-oxophytodienoic acid levels reached the highest levels 12 h after wounding (1440,3 nmol/ g FW). Allene oxide cyclase is mostly present in vascular bundles of inoculated internodes which may be a first indication that the systemin signalling pathway is operating also in hop. During prenodule and nodule formation, allene oxide cyclase levels were still high. Jasmonic acid and 12-oxophytodienoic acid levels decreased respectively to 10 and 118 pmol/ g FW during nodule formation. Levels of 12-oxophytodienoic acid expressed per mg of protein presented a five-fold increase during plantlet regeneration (a higher increase than that detected for jasmonic acid) suggesting that it may play a different role from jasmonic acid in the process. Double immunolocalization experiments with an antibody raised against Rubisco together with Lugol staining showed that allene oxide cyclase (AOC) is present in amyloplasts of prenodular cells and in chloroplasts of vacuolated nodular cells (Fig. 3) whereas meristematic cells showed little allene oxide cyclase accumulation. The presence of allene oxide cyclase in non-photosynthetic tissues may be related to the ability of jasmonic acid to stimulate carbon and nitrogen accumulation that will be used later in developmental processes.

## References


[1] Fortes AM, Pais MS (2000). Am. J. Bot. 87 (7), 971-979.

[2] Fortes AM, Testillano P, Risueño MC, Pais MS (2002). Physiol. Plant. 116, 113-120.

[3] Fortes AM, Coronado MJ, Testillano P, Risueño MC, Pais MS (2003). J. Histochem. Citochem. 52 (2), 227-241.

[4] Altschul SF, Madden TL, et al. (1997). Nucleic Acids Res. 25, 3389-3402.

[5] Fortes AM, Miersch O, et al. (2003). DGF symposium on Plant Oxylipins, Goettingen, Germany, pp. 40.

*Essays in Bioinformatics*
*D.S. Moss et al. (Eds.)*
*IOS Press, 2005*



# Single Nucleotide Polymorphism in Xenobiotic and Estrogen Metabolizing Genes and Breast Cancer Susceptibilty in Turkish Population

Neslihan AYGÜN KOCABAS

*Gazi University, Faculty of Pharmacy, Department of Toxicology*

*06330 Etiler, Ankara - Turkey*

**Abstract** The relationship among human genetic polymorphism, cancer susceptibility is increasingly important for risk assessment, early diagnosis and prevention, of clinical disease and cancer. This work analyses single nucleotide polymorphism (SNP) in human xenobiotic and estrogen metabolising genes and it is suggested that combinations of polymorphic enzymes may be better predictors of cancer risk than polymorphisms in one or two genes alone.

Many of the low penetrance susceptibility genes involved in xenobiotic and estrogen metabolism are polymorphically distributed within the human population. Single nucleotide polymorphisms (SNPs) in these genes are great deal of attention has been paid to the role in cancer epidemiology. Inherited alterations in the activity of cytochrome P450 *1B1* (*CYP1B1),* Catechol *O*-methyltransferase (*COMT),* Manganese superoxide dismutase (*MnSOD)* hold the potential to define differences in estrogen metabolism and, thereby, possibly explain inter-individual differences in cancer susceptibility associated with estrogen-mediated carcinogenesis The *CYP1B1* (L432V), *COMT* (V158M), *MnSOD* (Ala-9Val) genotypes to examine estrogen metabolism and influence of age of menarche / menopause, and N-acetlytransferase; *NAT2* (**4, *12A, *5A, *5B, 5C, 6, 7*) genotypes to detect environmental exposure were determined by using different polymerase chain reaction-restriction fragment length polymorphisms (PCR-RFLP) based genotyping assays in breast cancer patients and healthy women.

Sites in the DNA sequence where individuals differ at a single DNA base are called single nucleotide polymorphisms (SNPs). Single nucleotide polymorphisms (SNPs) are the most common genetic variations and occur once every 100 to 300 bases. Genetic variation also plays a role in whether a person has a higher or lower risk for getting particular diseases. Single gene differences in individuals account for some traits and diseases. More complex interrelationships among multiple genes and the environment are responsible for many common diseases, such as diabetes, cancer. In the postgenomicsera many more discoveries will begin with elucidation of genetic polymorphisms in candidate genes (e.g. those known to be involved in the metabolism, transport, or targets of the candidate medication) [1]. The analysis of SNPs within gene participating in the metabolism of various xenobiotics including carcinogens, influences the individual risk of cancer will help to understand of the gene-gene and gene-environment interactions in the process of human carcinogenesis, to identify individuals / populations who are at a very high risk because of their increased genetic susceptibility and to change the approach of therapeutics and health risk assessment [2]. Striking ethnic dissimilarities, as well as inter-individual differences, in genes involved in drug metabolism are well known. This refers to the enzymes participating in carcinogen metabolism phase I and to those participating in phase II [3]. Many xenobiotic agents are activated or detoxified by these important metabolizing enzymes. Given the fact that genetic polymorphisms in these enzymes may cause inter-individual variability in the genotoxic damage induced by xenobiotics, individual risk assessment has to be evaluated by taking into account individual genetics. The genetic principles of the polymorphisms of enzymes participating in the metabolism of environmental carcinogens have already been quite well explained at the DNA level. However, it is still difficult to determine precisely the role of genetic diversity, and the associated variations of enzyme functions in individual's susceptibility to the carcinogenic action of the chemicals present in the occupational and communal environment [4]. Gene environmental interactions can explain why some individuals develop cancer and others do not, for same level and quality of exposure. Also, it can explain why some people are particularly sensitive to low levels of carcinogenic exposures. Except for some occupational exposures, most human exposures to carcinogens are through mixtures where single carcinogens have very low concentrations. In cancer epidemiology, great deal of attention has been paid to the role of common population polymorphisms in genes controlling carcinogen metabolism [5].

As observed in drug and chemical metabolism, there is considerable interindividual genetic variability in the metabolic and biosynthetic pathways in steroidogenesis. These person to person differences might define subpopulations of women with higher lifetime exposures to hormone dependent growth promotion or to cellular damage from particular estrogens and estrogen metabolites. Such variation could explain a portion of the cancer susceptibility associated with reproductive events and hormone exposure [6]. Many of the low penetrance susceptibility genes involved in estrogen metabolism are polymorphically distributed within the human population. cytochrome P450 *1B1 (CYP1B1),* Catechol *O*-methyltransferase *(COMT),* Manganese superoxide dismutase *(MnSOD)* genes are known common genetic polymorphisms with a gene-environment and gene-gene interaction in steroid hormone metabolizing enzymes. Inherited alterations in the activity of any of these enzymes hold the potential to define differences in cancer risk associated with estrogen carcinogenesis like breast cancer.

Breast cancer is one of the most common and important diseases affecting women. Epidemiological studies have indicated that environmental xenobiotics or their metabolites, some with estrogenic or androgenic agonist and antagonist activities, may also play a significant role in the development of breast cancer [7].

The cytochrome P450 CYP1B1 *(CYP1B1)* is responsible for the hydroxylation of estrogens to the 2-hydroxy estrogen (2-OH HE) and 4-OH HEs, as well as a number of polycyclic aromatic hydrocarbons (PAH) and aryl amines, including several that are potent mammary gland carcinogens in rodents. At least seven different SNPs in *CYP1B1* have been described, of which one in exon 3, encodes the heme-binding domain, at codon 432 (Val→Leu) *(CYP1B1*3)*. The CYP1B1*3 product, a lower Km value for both 2- and 4-hydroxylation has been observed when compared to differ significantly from *CYP1B1*1* [8]. The *CYP1B1*1 and CYP1B1*3* alleles were detected by minor modifications of the methods described by Fritsche et al. [9]. The *CYP1B1*3* allele was associated with a significantly increased susceptibility of breast cancer [the adjusted OR for age, age at menarche, age at first full-term pregnancy, BMI and smoking status; 2.32 (95% CI 1.26-4.25, p=0.007)]. The comparision of genotype frequencies according to the basis of BMI, susceptibility of breast cancer was almost three-fold increased among women with a BMI greater than 24 kg/m$^2$ [With the exception of the BMI under evaluation, ORs and 95% CIs adjusted for other variables; 2.81 (1.38-5.74)]. The results showed that the *CYP1B1*3* variant which is predicted to be associated with higher activity was positively related to the susceptibility of breast cancer and was specific to women with a BMI greater than 24 kg/m$^2$.

Catechol-*O*-methyltransferase (*COMT*; E.C.2.1.1.6.) is one of several phases II enzyme, which is responsible for the detoxification of catecholamine including 2-CE and 4-CE by *O*-methylation. The level of *COMT* activity is controlled by a common genetic polymorphism being homozygous for a low activity allele termed *COMT-L* (V158M) [10]. Reduced *COMT* activity might increase the risk of cancer due to accumulation of CE, which causes oxidative DNA damage [6]. The *COMT-H* and *COMT-L* alleles were detected by minor modifications of the methods described by Lachman et al. [10]. In the case of *COMT*, the allele frequency of high activity *COMT-H* allele and low activity *COMT-L* allele was found to be 0.58 and 0.42 in the cases. There was no significant difference in susceptibility for breast cancer development between patients with *COMT-L* (V158M) and *COMT-H* alleles [the adjusted OR; 0.86 (95% CI 0.46-1.60, p=0.63)], and susceptibility was not affected by menopausal status, BMI, and other susceptibility factors.

Manganese containing superoxide dismutases (*MnSOD;* EC 1.15.1.1), the only known superoxide scavenger in mitochondria, may be particularly important for antioxidant defense and hence production of reactive oxygen radicals (ROS) [6]. A one base pair transition (T→C) leads to a Val→Ala amino acid change at codon 16 in the –9 position of signal sequence of *MnSOD*, produces a conformational change in the helical structure of the protein. This change may decrease the efficiency of transport into mitochondria [11]. Because *MnSOD* is a major enzyme involved in the scavenging of free radicals, ROS generated by estrogens and their metabolites may be involved in breast cancer etiology. The *MnSODVal* and *MnSODAla* alleles were detected by minor modifications of the methods described by Shimoda-Matsubayashi et al. [11]. The frequencies of Val / Val, Val / Ala, Ala / Ala genotypes were found to be 0.33, 0.45, 0.22, respectively in cases. There was no significant difference in the frequency of the *MnSOD Ala* allele between cases and controls [the adjusted OR; 0.86 (95% CI 0.43-1.72, p=0.67].

The analysis of susceptibility of breast cancer associated with the *MnSOD* genotypes stratified by *COMT-L* and *CYP1B1* alleles was performed. When *MnSOD Ala* allele was combined with either *COMT-HL* and *COMT-LL* or *CYP1B1*1/*3* and **3/*3* genotypes and all gene-gene interaction together, the risk for developing breast cancer was not significantly increased OR 1.04 (95% CI= 0.95-1.26), OR 1.38 (95% CI= 0.92-2.85) and OR 0.90 (95% CI= 0.68-1.19), respectively. The postmenopausal breast cancer susceptibility was increased in patients with *MnSOD Ala*, *CYP1B1*1* and *COMT-L* variants OR: 1.26 (95% CI= 0.93-1.72). However, the susceptibility for developing breast cancer approaches significance in patients with a BMI greater than 24 kg /m$^2$ [OR: 1.42 (95% CI= 1.04-1.93)], when *MnSOD Ala* was combined with either *CYP1B1*1* and *COMT-L* genotypes. This finding suggests that *MnSOD Ala*, *CYP1B1*1* and *COMT-L* variants are involved in the susceptibility to breast cancer in certain women.

Many arylamine and hydrazine drugs, as well as for a number of known carcinogens aromatic and the heterocyclic amines present in the diet, cigarette smoke and the environment can be either detoxified by arylamine *N*-acetyltransferase (*NAT2;* EC 2.3.1.5) and eliminated from the body or bioactivated to metabolites that have the potential to cause toxicity and or cancer [12]. Thirteen different SNPs in *NAT2* gene occurring single or in combination define numerous alleles (15-20) associated with decreased expression, low activity, enzyme instability and biochemical phenotypes ranging from slow to fast acetylators [13]. The *NAT2* genotyping was detected by a modification of the methods of Bell et al. [12]. The *NAT2*5A*, **5B*, **5C*, **6* and **7* alleles were detected. The rapid acetylators of *NAT2* allele frequencies in cases and controls were 50%-43.7% and slow acetylators of *NAT2* allele frequencies were 50%-56.3%, respectively. The frequency of rapid genotype in cases was slightly more common than controls, although there was no significant difference in the genotype frequency of the *NAT2* rapid allele between two groups [the adjusted OR; 0.78 (95% CI 0.44-1.38, p=0.39)]. The most common slow allele was the *NAT2*5B* allele in both cases (38.1%) and controls (38.9%). Among the **5B* slow alleles in cases (25%), **5B/*6* was slightly higher than among controls (21.4%), whereas *5B/*5B* slow allele in cases (9.5%) was slightly lower than among controls (14.6%). The second most frequent slow allele, *NAT2*6/*6* was the same as among cases (9.5%) and controls (10.7%). Slightly less controls (43.7%) than cases (50%) had the wild allele *NAT2*4*. Some 8.3% of the cases and 6.8% of all controls were homozygous rapid allele carriers (**4/*4*). Only four and one case subject were found to be *4/*5C, **12A/*12C*, respectively, whereas no control subject was found. Only three controls were found to be *6/*7, whereas no case subject was found.

When *NAT2 slow* allele was combined with either *COMT-HL* and *COMT-LL* or *CYP1B1*1/*3* and **3/*3* genotypes and all gene-gene interaction together, the risk for developing breast cancer was not significantly increased OR 1.30 (95% CI= 0.71-2.37), OR 0.93 (95% CI= 0.52-1.65) and OR 1.28 (95% CI= 0.55-2.96), respectively.

Most of the information has been collected up to now on the effect of genetic polymorphism on the individual ability to activate and deactivate estrogens and xenobiotics, whereas no information is available on inter-individual variability of to *CYP1B1*, *COMT*, *MnSOD* and *NAT2* genotypes and the influence of these genotypes on onset of menarche / menapause in Turkish healthy women. These genes were examined stratified different risk factors in 103Turkish women. In all genotypes, only the case of COMT, the *COMT-L* (*COMT*2*) allele was more frequent among postmenopausal women associated with a significantly increased ($X^2$=3.820, p=0.05). However, in the cases of other genes, the *CYP1B1*3*, *MnSOD Ala* and *NAT2 slow* alleles, there were no significant differences in the frequency of premenopausal and postmenopausal women [$X^2$=0.360, p=0.55; $X^2$=0.026, p=0.87; $X^2$=0.653, p=0.42; respectively]. The frequencies of *CYP1B1*3* allele (0.27), *COMT-L* allele (0.39) *MnSOD* Ala allele (0.56) and *NAT2 slow* allele (0.56) determined in Turkish healthy women were found to be similar with Caucasian population-based studies. The comparision of genotype frequencies according to the basis of BMI, there were not any significant differences in each genes. However, *CYP1B1*3* and *COMT*1* genotypes were related to increased risk among women with a BMI greater than 27 kg/m$^2$ (Fisher's exact test, p=0.044 ). *COMT*, *MnSOD* genotypes stratified for according to *CYP1B1* genotypes (gene-gene interaction), age and menopausal status; *CYP1B1*3*, *COMT-L* and *MnSOD Ala* alleles does approach significance in older than 45 years and postmenopausal women with compared with the combination of the low risk genotypes (Fisher's exact test, p=0.012). The correlation between genotype and early age at menarche was significant in women who carried both *CYP1B1*3* and *COMT-L* alleles than the women who carried wild alleles ($X^2$ =4.57, p=0.032). Although the small sample size of each combination of estrogen metabolizing, the results suggest that the *CYP1B1*3* and *COMT-L* alleles influence age at menarche in healthy Turkish women.

The study of the relationship among human genetic polymorphisms, cancer susceptibility will undoubtedly have increasingly important implications for risk assessment and the prevention, early diagnosis, and intervention of clinical disease and cancer. There is evidence for the existence of polymorphism in each of the genes encoding these enzymes and it is possible that combinations of polymorphic enzymes may be better predictors of cancer risk than polymorphisms in one or two genes alone. Also, the variety of exogenous and endogenous exposures that may influence the development of carcinoma warrants further investigation of genetic polymorphisms at xenobiotic metabolizing loci and analysis of gene-gene and gene-environment interactions in large series of patients.

## References


[1] http://www.snp.cshl.org

[2] Ingelman-Sundberg, M. (2001) Genetic variability in susceptibility and response to toxicants. Toxicol Letts 120, 259-268

[3] Nebert, D.W. et al. (1999) Genetic epidemiology of environmental toxicity and cancer susceptibility: human allelic polymorphisms in drug-metabolizing enzyme genes, their functional importance, and nomenclature issues. Drug Metabol. Reviews 31, 467-487.

[4] Miller, M.C. et al. (2001) Genetic variability in susceptibility and response to toxicants. Toxicol. Letts. 120, 269-280

[5] Gemignani, F. et al. (2002) A catalogue of polymorphisms related to xenobiotic metabolism and cancer susceptibility. Pharmacogenetics 12, 459-463

[6] Thompson, P.A. and Ambrosone, C. (2000) Molecular epidemiology of genetic polymorphisms in estrogen metabolizing enzymes in human breast cancer. J Natl. Cancer Inst. Monographs 27, 125-134.

[7] Yager, J.D. and Liehr, J.G. (1996) Molecular mechanisms of estrogen carcinogenesis. Ann. Rev. Pharmacol. Toxicol. 36, 203-232.

[8] http://www.imm.ki.see/CYPalleles /cyp1b1.htm

[9] Fritsche, E. et al. (1999) Detection of Cytochrome P450 1B1 Bfr I polymorphism: genotype distribution in healthy German individuals and in patients with colorectal carcinoma. Pharmacogenetics 9, 405-408.

[10] Lachman, H.M. et al. (1996) Human Catechol-O-methyltransferase pharmacogenetics: description of a functional polymorphism and its potential application to neuropsychiatric disorders. Pharmacogenetics 6, 243-250

[11] Shimoda-Matsubayashi, S. et al. (1996) Structural Dimorphism in the Mitochondrial Targeting Sequence in the Human Manganese Superoxide Dismutase Gene. Biochem. Biophysical Res. Commun. 226, 561-565.

[12] Bell, D.A. et al. (1993) Genotype/ phenotype discordance for human arylamine N-acetyltransferase (NAT2) reveals a new acetylator allele common in African-Americans. Carcinogenesis 14, 1689-1692.

[13]. _http://www.louisville.edu /medschool/pharmacology/NAT.html

*Essays in Bioinformatics*
*D.S. Moss et al. (Eds.)*
*IOS Press, 2005*


# Bioinformatics approaches in Molecular Systematics: the case of *Silene* section *Siphonomorpha* Otth (*Caryophyllaceae*)

Helena COTRIM[1], M. Salomé PAIS[1], Michael F. FAY[2] and Mark W. CHASE[2]

[1]*Plant Molecular Biology and Biotechnology Laboratory, ICAT, Ed. ICAT, Faculty of Sciences, University of Lisbon, Campo Grande P-1749-016 Lisbon, Portugal, email: hcotrim@icat.fc.ul.pt/hmcotrim@fc.ul.pt.*
[2] *Jodrell Laboratory, Royal Botanic Gardens, Kew, Richmond, Surrey, TW9 3DS, UK.*

**Abstract.** The primary goal of computational molecular biology, like molecular biology itself, is to understand the meaning of the genomic information and how this information is expressed. Molecular systematics makes phylogenetic inferences from molecular data using computacional methods. The systematics of *Silene* section *Siphonomorpha* Otth was approached from three different perspectives, the first analysing global relationships within the section, the second studying two pairs of taxa with problematic species boundaries, and the third using one of the species to study rarity at ecological and genetic level.

## 1. Introduction

*Silene* (Caryophyllaceae) is a large plant genus with more than seven hundred species found in the Northern Hemisphere. The genus includes many rare species, and 80% of those referred to as rare or endangered [1, 2] belong to section *Siphonomorpha* Otth. The section includes circa twenty-five taxa, two of which are a widely distributed in Europe (*Silene nutans* and *S. italica*) and several regional or local endemics occurring mainly in the Iberian Peninsula. Their taxonomy has undergone several modifications since Otth. Morphological separation is difficult, especially in the *S. italica* complex. Breeding barriers are essentially absent in the group, which makes species boundaries a crucial subject. Moreover the extreme morphological similarity of some of the taxa hinders their conservation.

## 2. Material and Methods

Molecular and morphological methods were applied to trace evolutionary relationships within section *Siphonomorpha*. Nuclear ITS and plastid *trn*L-F DNA were sequenced for eighteen taxa [3]. Plastid *trn*L-F microsatellites and nuclear AFLPs were also used. The micromorphology of *Silene longicilia* ssp. *cintrana* and *S. rothmaleri* was studied using scanning electron microscopy. Karyological characterisation of both species was made with silver staining and in situ hybridisation techniques [4]. The relationships between *S. longicilia* ssp. *longicilia* and *S. longicilia* ssp. *cintrana* were analysed with AFLPs. *Silene rothmaleri*, one of the numerous rare taxa of this section, previously considered extinct [5], was used as a case study of rarity. Ecological traits of this taxon were studied and populations characterised. The within and between population genetic variability was assessed by RAPD across the entire range of the taxon [6].

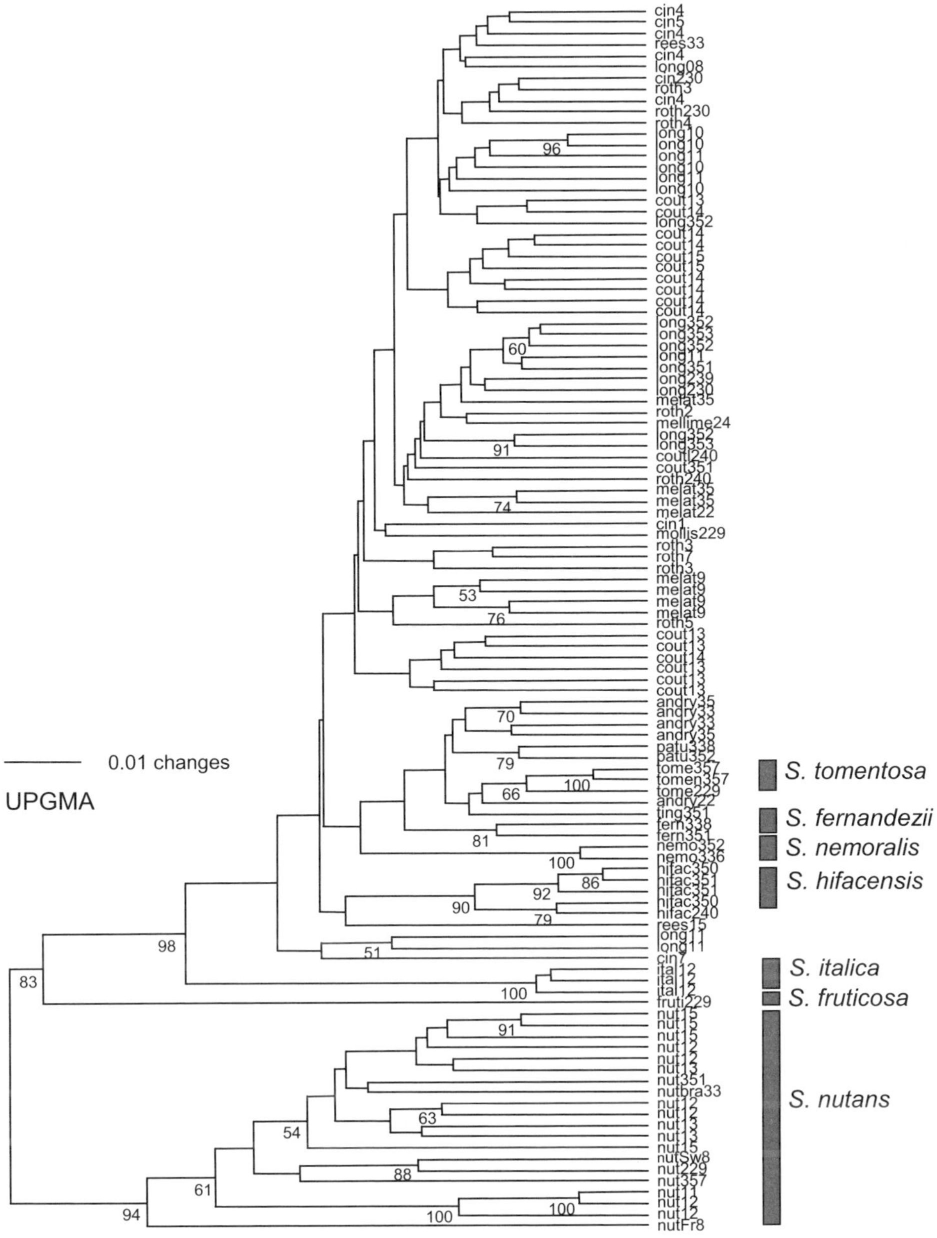


Figure 1- UPGMA analysis of 121 AFLP genotypes of *Silene* section *Siphonomorpha*. Branch lengths are proportional and bootstrap percentages greater than 50% are indicated below branches. See [3] for legend key of the taxa.

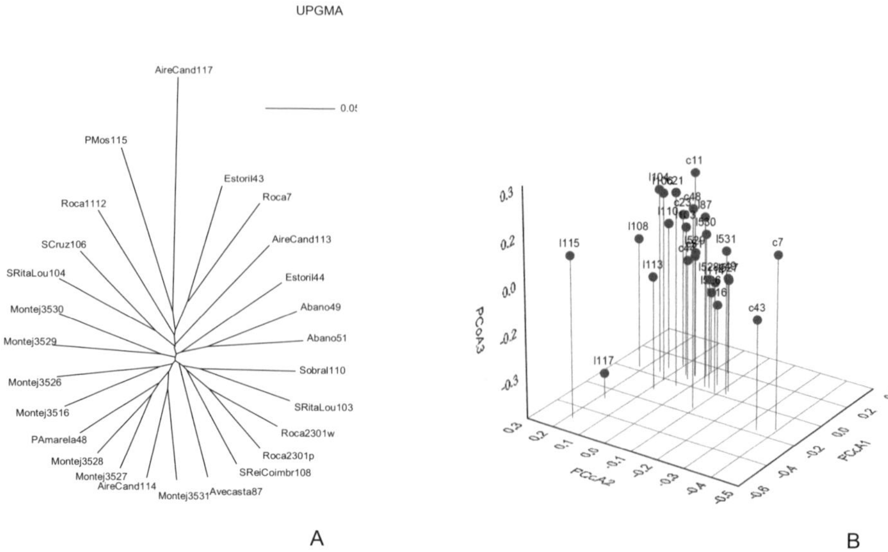


Fig. 2. **A**: Unrooted dendogram of UPGMA for 26 individuals of *Silene longicilia* ssp. *cintrana* and *S. longicilia* ssp. *longicilia* (Nei & Li distance). Legends refer to the geographic origin of the material, cf. [3] for details. **B**: Principal coordinates analysis of twenty six individual genotypes (AFLPs) of *Silene longicilia* ssp. *longicilia* and *S. longicilia* ssp. *cintrana*, for eigenvalues and percentage of variance retained by axis see [3].

For all the topics mentioned a bioinformatics approach was used either by aligning and editing sequences with Sequence Navigator and Autoassembler (PE Applied Biosystems, Inc.) or performing parsimony analysis (PAUP 4.0 β for Macintosh, Swofford, 1998). Genetic fingerprints comparison and analysis was done using Genescan and Genotyper software (PE Applied Biosystems Inc.). Also other bioinformatics methods were applied like Neighbour Joining (PAUP 4.0 β) and multivariate methods (UPGMA and PCoA).

## 3. Results and Discussion

Global relationships within the section analysed with AFLPs are depicted in Fig.1. Molecular and morphological methods demonstrated that *Silene* section *Siphonomorpha* is a group of closely related species, with *S. nutans* being the most distinct taxon [3]. Thus, these species were considered better treated as two instead of one section: section *Italicae* and section *Siphonomorpha*, the former including thirteen of the species analysed and the latter including two species, *S. nutans* and *S. viridiflora*.

The first section corresponds to a group of entities probably not behaving as biological species, but including several endemics contributing to global biodiversity. The low level of divergence within section *Italicae* and frequently common plastid markers are evidence of a recent shared common origin.

Study of *S. longicilia* ssp. *cintrana* and *S. rothmaleri* revealed differences in trichome length and density and hilum cell morphology. In both species the 18S-5.8S-26S rDNA probe labelled four sites on the short arms of two submetacentric chromosomes and one locus was labelled with 5S rDNA probe [4]. The species differed in the physical position of this 5S rDNA locus. The NOR activity analysed by Ag-staining in metaphase cells also showed differences between the species.

AFLPs analysis revealed that there is no genetic differentiation between *Silene longicilia* ssp. *longicilia* and *S. longicilia* ssp. *cintrana*. The data showed a high degree of genetic diversity although lacking population structure (Fig.2A and B), indicating gene flow and a panmictic population. Neither morphological nor genetic discontinuities were detected within the species, thus indicating that both named entities correspond to *S. longicilia*, with no reason for the maintenance of the subspecific rank

The study of *S. rothmaleri*, one of the rare species of the group, revealed a non-specific micro-habitat coloniser [5]. The species presence in its habitat was related to the availability of the micro-habitat (deposits, step, fissure, scree) on the southwestern Portuguese coast either in sea cliff or mountain *facies*. The species analysed with RAPD displayed a high degree of genetic diversity mainly distributed within populations rather than between populations and a considerable degree of differentiation with population-specific markers [6] as depicted in Fig. 3.

Conservation implications and future prospects of this work include the need for revision of the conservation status of *S. longicilia*, the study of the rare *taxa* belonging to *S. rosulata* from North Africa, the evaluation of the relationships between *S. coutinhoi* and *S. mellifera* and a better understanding of genetic variability in *S. nutans* and *S. italica*.

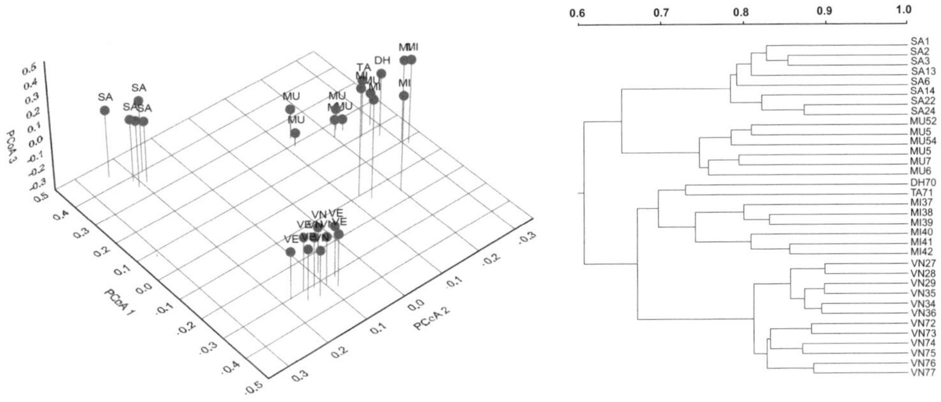


Figure 3-**A:** PCoA of 34 plants of *Silene rothmaleri.* Axis 1 represents 25% of the variance, axis 2 (12%) and axis 3 (8%). **B**: PCoA of 34 plants of *Silene rothmaleri.* Axis 1 represents 25% of the variance, axis 2 (12%) and axis 3 (8%). cf.[6] for details.

## References


[1] EEC 92/43, Council directive of 21 May 1992 on the conservation of natural habitats and of wild fauna and flora. O. J. L206, 22.07.92. 1992

[2] K. S. Walter and H. J. Gillett (editors), *1997 IUCN Red List of Threatened Plants*. Compiled by the World Conservation Monitoring Centre. IUCN – The World Conservation Union, Gland, Switzerland and Cambridge, UK. lxiv + 862. 1998

[3] H. M. C. Cotrim, Molecular Systematics of *Silene* section *Siphonomorpha* Otth – a conservation perspective. PhD thesis, Faculty of Sciences of the University of Lisbon, Portugal. 2001

[4] O. Pontes, H. M. C. Cotrim *et al.*, Physical mapping, expression patterns and interphase organisation of rDNA *loci* in Portuguese endemic *Silene cintrana* and *Silene rothmaleri*. *Chromosome Research* 8(4) (2000): 313-317.

[5 H. M. C. Cotrim, M. J. Pinto, Population distribution and habitat colonisation pattern in Southwest Portuguese endemism *Silene rothmaleri* P. Silva. XIV Jornadas de Fitossociologia, Bilbao 1994.

[6] H. M. C. Cotrim *et al.*, *Silene rothmaleri* P. Silva (Caryophyllaceae) a rare, fragmented but genetically diverse species. *Biodiversity and Conservation* 12(2003):1083-1098.

## Volume contributors:

ÁGOSTON, Vilmos; Bioinformatics Group, Biological Research Center, Hungarian Academy of Sciences, Temesvári krt. 62, 6726 Szeged, Hungary; Telephone: +36-62-599-766; Fax: +36-62-423-576; E-mail: vilagos@nucleus.szbk.u-szeged.hu

ALMEIDA, Maria Gabriela -CQFB, Departamento de Química, Faculdade de Ciências e Tecnologia, Universidade Nova de Lisboa, 2829-516; Monte de Caparica, Portugal; Telephone: +351-21-2948550, ext.10957; Fax: +351-21-2948345; E-mail: mga@dq.fct.unl.pt

AYGÜN KOCABAS, Neslihan; Department of Toxicology, Faculty of Pharmacy, Gazi University, 06330 Etiler-Ankara, Turkey; Telephone: +90 312 2154468/1104; GSM: +905324232865; E-mail: neslihan@gazi.edu.tr; neslihanak@hotmail.com

BALDÉ, Aladje; Plant Molecular Biology and Biotechnology Laboratory, ICAT-FCUL, University of Lisbon, Edifício ICAT, FCUL, Campo Grande, P-1749-016 Lisboa, Portugal; Telephone: +351-217500163; Fax: +351- 217500172; E-mail: abalde@fc.ul.pt

BAIROCH, Amos; Swiss Institute of Bioinformatics, CMU - 1, rue Michel Servet, CH-1211 Geneva 4, Switzerland. Telephone: +41-22-3795050; Fax: +41-22-3795858; E-mail: swiss-prot@expasy.org

BOECKMANN, Brigitte; Swiss Institute of Bioinformatics, CMU - 1, rue Michel Servet, CH-1211 Geneva 4, Switzerland. Telephone: +41-22-379-5859 ; Fax: +41-22-379-5858; E-mail: Brigitte.Boeckman@isb-sib.ch

BRYANT, Stephen H.; National Center for Biotechnology Information (NCBI), National Library of Medicine, National Institutes of Health, 8600 Rockville Pike, Bethesda, MD 20894 USA; Telephone: +1 301 496-2475; Fax: +1 301 480-9241; E-mail: bryant@ncbi.nlm.nih.gov

CARUGO, Oliviero, I.; International Centre for Gernetic Engineering and Biotechnology, AREA Science Park, Padriciano 99, I-34012 Trieste, Italy. Telephone: +39-040-3757340; Fax: +39-040-226555; E-mail: o.carugo@icgeb.org

CARVER, Tim J.; The Wellcome Trust Sanger Institute, Wellcome Trust Genome Campus, Hinxton, Cambridge, CB10 1SA, UK; Telephone: +44 1223 834244; Fax: +44 1223 494919; E-mail: tjc@sanger.ac.uk

CEMAZAR, Masa.; International Centre for Gernetic Engineering and Biotechnology, AREA Science Park, Padriciano 99, I-34012 Trieste, Italy. Telephone: 61-7-33462328; Fax: +61-7-33462029; E-mail: m.cemazar@imb.uq,edu.au

CHASE, Mark W.; Molecular Systematics Section, Jodrell Laboratory, Royal Botanic Gardens, Kew, Richmond, Surrey TW9 3DS, United Kingdom; Telephone: +44-20-8332-5364; Fax: +44-20-8332-5310; E-mail: m.chase@kew.org

COTRIM, Helena M. C. ; Plant Molecular Biology and Biotechnology Laboratory, ICAT-FCUL, University of Lisbon, Edificio ICAT, FCUL, Campo Grande, P-1749-016 Lisboa, Portugal; Telephone: +351-217500163; Fax: +351-217500172; E-mail: hcotrim@icat.fc.ul.pt/hmcotrim@fc.ul.pt

COTTAGE, Amanda.; MRC Rosalind Franklin Centre for Genomic Research, Genome Campus, Hinxton, Cambridge, CB10 1SB, UK; Telephone: +44-1223-494500; Fax: +44-1223-494512; E-mail: acottage@rfcgr.mrc.ac.uk

de HAAN, Jorn R.; Laboratory of Analytical Chemistry, Radboud University Nijmegen, Toernooiveld 1, 6525 ED Nijmegen, the Netherlands; Telephone: +31 24 3653179; Fax: +31 24 3652653; E-mail: J.deHaan@science.ru.nl

EDWARDS Yvonne J.K.; Comparative Genomics & Bioinformatics, School of Biological and Chemical Sciences, Queen Mary, University of London, Mile End Road, London E1 4NS, UK; Telephone: +44 20 7882 3717; Fax: +44 207 882 5556; E-mail: y.j.edwards@qmul.ac.uk

ELGAR, Greg.; MRC Rosalind Franklin Centre for Genomic Research, Genome Campus, Hinxton, Cambridge, CB10 1SB, UK; Telephone: +44 1223 494562; Fax: +44 1223 494512; E-mail: gelgar@rfcgr.mrc.ac.uk

FERRO ROJAS, Serenella; Swiss Institute of Bioinformatics, CMU - 1, rue Michel Servet, CH-1211 Geneva 4, Switzerland Telephone: +41-22-3795050; Fax: +41-22-3795858; E-mail: swiss-prot@expasy.org

FAY, Michael F.; Section of Genetics, Jodrell Laboratory, Royal Botanic Gardens, Kew, Richmond, Surrey, TW9 3DS United Kingdom; Telephone:+44-20-8332-5315; Fax:+44-20-8332-5310; E-mail: m.fay@kew.org

FOGTMAN Anna, Institute of Biochemistry and Molecular Biology, University of Wroclaw, 50-137 Wroclaw, Tamka 2, Poland; Telephone: +48-71-3752-393; Telefax: +48-71-3752-608; E-mail: fogi@grid.icm.edu.pl

FORTES, A. Margarida; Unit of Plant Molecular Biology and Biotechnology, ICAT, FCUL, Campo Grande, 1749-016 Lisboa, Portugal. Telephone: +351-21-7501063; Fax: +351-21-7501072; E-mail: amfortes@fc.ul.pt

GASTEIGER, Elisabeth; Swiss Institute of Bioinformatics, CMU - 1, rue Michel Servet, CH-1211 Geneva 4, Switzerland. Telephone: +41-22-3795050; Fax: +41-22-3795858; E-mail: swiss-prot@expasy.org

GONÇALVES, Luisa L. Faculty of Pharmacy, Room #514, 19 Russel Street, Toronto, Ontario M5S 2S2, Canada. Telephone: +1-416-978-5061;Fax: +1-416-978-8511; E-mail: luisamlima@netc.pt

GOUVEIA, Manuela; Departamento de Botânica, Universidade da Madeira, Largo do Município, P-9050, Funchal, Portugal;Telephone: +351 291705387, Fax: +351 291705399; E-mail: mgouveia@uma.pt

HEGEDÜS, Zoltán; Bioinformatics Group, Biological Research Center, Hungarian Academy of Sciences, Temesvári krt. 62, 6726 Szeged, Hungary; Telephone: +36-62-599-766; Fax: +36-62-423-576; E-mail: Hegedus@nucleus.szbk.u-szeged.hu

HELLEN, Elizabeth.; MRC Rosalind Franklin Centre for Genomic Research, Genome Campus, Hinxton, Cambridge, CB10 1SB, UK; Telephone: +44-1223-494500; Fax: +44-1223-494512; E-mail: ehellen@rfcgr.mrc.ac.uk

HRANUELI, Daslav; Faculty of Food Technology and Biotechnology, Department of Biochemical Engineering, Section for Bioinformatics, University of Zagreb, Pierottijeva 6, 10000 Zagreb, Croatia; Telephone: +385-1-4826252; Fax: +385-1-4836083; E-mail: hranueli@rudjer.irb.hr

JUDGE, David P.; Department of Genetics, University of Cambridge, Tennis Court Road, Cambridge CB2 3EH, UK. Telephone: +44-1223-333614; Fax: +44-1223-333992; E-mail: dpj10@mole.bio.cam.ac.uk

KAJÁN, László; International Centre for Gernetic Engineering and Biotechnology, AREA Science Park, Padriciano 99, I-34012 Trieste, Italy; Telephone: +39-040-3757340; Fax: +39-040-226555; E-mail: kajan@icgeb.org

KOCABAS, Fahri S.; Informatics Institute, Medical Informatics Department, Middle East Technical University, Inönü Bulvari, 06531, Ankara, Turkey; Telephone : +90 312 4022214; Fax: +90 312 4250813; GSM : +905356205589; E-Mail : fkocabas@tsk.mil.tr

LAMPREIA, Jorge; CQFB, Departamento de Química, Faculdade de Ciências e Tecnologia, Universidade Nova de Lisboa, 2829-516 Monte de Caparica, Portugal. Telephone: +351-21-2948352; Fax: +351-21-2948550; E-mail: jorge.lampreia@dq.fct.unl.pt

LEUNISSEN, Jack A.M. Laboratory of Bioinformatics, Wageningen University and Research Centre, Dreijenlaan 3, 6703 HA Wageningen, the Netherlands; Telephone: +31-317-482036; Fax: +31-317-483584; E-mail: jack.leunissen@wur.nl

MACKAY, Alan L.; School of Crystallography, Birkbeck College, Malet St. London WC1E 7HX UK; Telephone: +44 207 631 6800 Fax +44 207 631 6803. E-mail: a.mackay@mail.cryst.bbk.ac.uk

MILES, Andrew. J.; School of Crystallography, Birkbeck College, Malet St. London WC1E 7HX UK; Telephone: +44 207 631 6800; Fax +44 207 631 6803. E-mail: a.miles@mail.cryst.bbk.ac.uk

MIZRACHI, Ilene; National Center for Biotechnology Information, National Library of Medicine, National Institutes of Health, 8600 Rockville Pike, Bethesda, MD 20894 USA; Telephone: +1 301 496-2475; Fax: +1 301 480-9241; E-mail: mizrachi@ncbi.nlm.nih.gov

MOSS, David S.; School of Crystallography, Birkbeck College, Malet St., London WC1E 7HX UK; Telephone: +44-207-631 6800; Fax +44-207-631-6803. E-mail: d.moss@bbk.ac.uk

MOURA, José; CQFB, Departamento de Química, Faculdade de Ciências e Tecnologia, Universidade Nova de Lisboa, 2829-516 Monte de Caparica, Portugal. Telephone: +351-21-2948382 #8345; Fax: +351-21-2948550; E-mail: jose.moura@dq.fct.unl.pt

MULLAN, Lisa J.; European Bioinformatics Institute, Genome Campus, Hinxton, Cambridge, CB10 1SD, UK; Telephone: +44-1223-494448; Fax: +44-1223-494468; E-mail: lisa@ebi.ac.uk

PAIS, Maria S; Unit of Plant Molecular Biology and Biotechnology, ICAT, FCUL, University of Edificio ICAT, FCUL, Campo Grande, P-1749-016 Lisboa, Portugal; Telephone: +351-21-7501063; Fax: +351-21-7501072; E-mail: maria.pais@fc.ul.pt

PILJAC, Jasenka; Department of Molecular Biology, Institute Ruđer Bošković; Bijenička c. 54, PO Box 180, 10002 Zagreb Croatia; Telephone: +385-1-4560-987; Fax: +385-1-4561-117; E-mail: jpiljac@irb.hr

SANSOM, Clare E.; School of Crystallography, Birkbeck College, Malet St. London WC1E 7HX UK; Telephone: +44-207-631-6800; Fax +44-207-631-6803. e-mail: c.sansom@mail.cryst.bbk.ac.uk

SAYERS, Eric W; National Center for Biotechnology Information, National Library of Medicine, National Institutes of Health, 8600 Rockville Pike, Bethesda, MD 20894 USA; Telephone: ++1 301-402-4039; Fax: +1 301 480-9241; E-mail: sayers@ncbi.nlm.nih.gov

PAIS, Maria S.; Plant Molecular Biology and Biotechnology Laboratory, ICAT-FCUL, University of Lisbon, Edificio ICAT, FCUL, Campo Grande, P-1749-016 Lisboa, Portugal; Telephone: +351 217500006; Fax: +351 217500172; E-mail: maria.pais@fc.ul.pt

PATEL, Sunil.; 334, Cambridge Science Park, Milton Rd, Cambridge, Cambridgeshire CB4 UK; Telephone: +44-1223-228500; Fax: +44-1223-228501; E-mail: spatel@accelrys.com

PONGOR, Sándor; International Centre for Genetic Engineering and Biotechnology, AREA Science Park, Padriciano 99, I-34012 Trieste, Italy. Telephone: +39-040-3757300; Fax: +39-040-226555; E-mail: pongor@icgeb.org

PORTELA, Miriam B.D.; MRC Rosalind Franklin Centre for Genomic Research, Genome Campus, Hinxton, Cambridge, CB10 1SB, UK; Telephone: +44-1223-494500; Fax: +44-1223-494512; E-mail: mportela@rfcgr.mrc.ac.uk

VAVOURI, Tanya.; MRC Rosalind Franklin Centre for Genomic Research, Genome Campus, Hinxton, Cambridge, CB10 1SB, UK; Telephone: +44-1223-494556; Fax: +44-1223-494512; E-mail: tvavouri@rfcgr.mrc.ac.uk

VLAHOVICEK, Kristian; International Centre for Genetic Engineering and Biotechnology, AREA Science Park, Padriciano 99, I-34012 Trieste, Italy. Telephone: +39-040-3757340; Fax: +39-040-226555; E-mail: kristian@icgeb.org

WALLACE, Bonnie A.; School of Crystallography, Birkbeck College, Malet St. London WC1E 7HX UK; Telephone: +44-207-631-6800 Fax +44-207-631-6803. e-mail: b.wallace@mail.cryst.bbk.ac.uk

## Course participants

## Organisers:

MOSS, David S.; School of Crystallography, Birkbeck College, Malet St. London WC1E 7HX UK; Telephone: +44 207 631 6800; Fax +44 207 631 6803. E-mail: d.moss@bbk.ac.uk

JELASKA, Sibila; Department of Molecular Biology, Faculty of Science, University of Zagreb, Rooseveltov trg 6, 10 000 Zagreb, Croatia, Tel. +385 1 48 77 34 and +385 1 48 26 261, Fax. 48 26 260, e-mail: sibila@hazu.hr

## Lecturers:

HRANUELI, Daslav; Faculty of Food Technology and Biotechnology, Department of Biochemical Engineering, Section for Bioinformatics, University of Zagreb Pierottijeva 6, 10000 Zagreb, Croatia; Telephone: +385-1-4826252; Fax: +385-1-4836083; E-mail: hranueli@rudjer.irb.hr

JUDGE, David P.; Department of Genetics, University of Cambridge, Tennis Court Road, Cambridge CB2 3EH, UK. Telephone: +44-1223-333614; Fax: +44-1223-333992; E-mail: dpj10@mole.bio.cam.ac.uk

LEUNISSEN, Jack A.M.; Laboratory of Bioinformatics, Wageningen University and Research Centre, Dreijenlaan 3, 6703 HA Wageningen, the Netherlands; Telephone: +31-317-482036; Fax: +31-317-483584; E-mail: jack.leunissen@wur.nl

MACKAY, Alan L.; School of Crystallography, Birkbeck College, Malet St., London WC1E 7HX UK; Telephone: +44-207-631-6800; Fax +44-207-631-6803;. E-mail: a.mackay@mail.cryst.bbk.ac.uk

MOSS, David S.; School of Crystallography, Birkbeck College, Malet St. London WC1E 7HX UK; Telephone: +44 207 631 6800; Fax +44 207 631 6803. E-mail: d.moss@bbk.ac.uk

PONGOR, Sándor; International Centre for Genetic Engineering and Biotechnology (ICGEB), AREA Science Park, Padriciano 99, I-34012 Trieste, ITALY. Telephone: +39-040-3757300; Fax: +39-040-226555; E-mail: pongor@icgeb.org

SANSOM, Clare E.; School of Crystallography, Birkbeck College, Malet St. London WC1E 7HX UK; Telephone: +44-207-631- 6800; Fax +44 207 631 6803. e-mail: c.sansom@mail.cryst.bbk.ac.uk

VLAHOVICEK, Kristian; International Centre for Genetic Engineering and Biotechnology (ICGEB), AREA Science Park, Padriciano 99, I-34012 Trieste, ITALY. Telephone: +39-040-3757340; Fax: +39-040-226555; E-mail: kristian@icgeb.org

**Students:**

AHMED, Mohamed Tawfic; Suez Canal University, Ismailia, Egypt; Fax: +20 2 4186049; E-mail: motawfic@tedata.net.eg

AMBRIOVIC RISTOV, Andreja; Division of Molecular Biology, Ruder Boskovic Institute, Bijenicka 54, 10000 Zagreb, CROATIA. Telephone: +385-1-4571240; Fax: +385-1-4561177; E-mail: andrea@irb.hr

AYGÜN KOCABAS, Neslihan; Department of Toxicology, aculty of Pharmacy, Gazi University, 06330 Etiler-Ankara, Turkey; Telephone: +90 312 2154468/1104; GSM : +905324232865; E-Mail : neslihan@gazi.edu.tr; neslihanak@hotmail.com

BALDÉ, Aladje; Plant Molecular Biology and Biotechnology Laboratory, ICAT-FCUL, University of Lisbon, Edificio ICAT, FCUL, Campo Grande, P-1749-016 Lisboa, Portugal; Telephone: +351 217500163; Fax: +351 217500172; E-mail: abalde@fc.ul.pt

CANATAN, Halit; Department of Medical Biology and Genetics, Faculty of Medicine, Firat (Euphrates) University, Elazig 23119, Turkey; Telephone: +90-424-237000, ext.6712; Fax: +90-424-2379138; E-mail:canatan.2@osu.edu, halitcanatan@yahoo.com

CARDOSO, Eduardo; Chemical & Biochemical Engineering; University of Maryland, Baltimore County 1000 Hilltop Circle, Baltimore, MD 21250 USA; Email: xininhas@hotmail.com

COTRIM, Helena M. C.; Plant Molecular Biology and Biotechnology Laboratory, ICAT-FCUL, University of Lisbon, Edificio ICAT, FCUL, Campo Grande P-1749-016 Lisboa, Portugal; Telephone: +351 217500163; Fax: +351 217500172; E-mail: hcotrim@icat.fc.ul.pt/hmcotrim@fc.ul.pt

FERENAC, Marina; Laboratory for Experimental Cancerology, Department of Molecular Biology, Ruder Boskovic Institute, Bijenicka cesta 54, 10000 Zagreb, CROATIA. Telephone: +385-1-4561093; fax: +385-1-4561177; E-mail: marinaf@irb.hr

FOGTMAN Anna; Institute of Biochemistry and Molecular Biology, University of Wroclaw, 50-137 Wroclaw, Tamka 2, Poland, tel: +48 (71) 3752-393; fax: +48 (71) 3752-608; E-mail: fogi@grid.icm.edu.pl

FORTES, A. Margarida; Unit of Plant Molecular Biology and Biotechnology (ICGEB), ICAT, FCUL, Campo Grande, 1749-016 Lisboa, Portugal. Telephone: +351-21-7501063; Fax: +351-21-7501072; E-mail: amfortes@fc.ul.pt

FRANJEVIC, Damjan; Zoological Department, Faculty of Natural Science, University of Zagreb, Rooseveltov trg 6, 10000 Zagreb, CROATIA. Telephone: +385-1-4877757; Fax: +385-1-4826260; E-mail: damianf@zg.biol.pmf.hr

GONÇALVES, Luisa L.; Faculty of Pharmacy, Room #514, 19 Russel Street, Toronto, Ontario M5S 2S2, Canada; Telephone: +1-416-978-5061;FAX: +1-416-978-8511; E-mail: luisamlima@netc.pt

KOCABAS, Fahri S.; Informatics Institute, Medical Informatics Department, Middle East Technical University, Inönü Bulvari, 06531, Ankara, Turkey; Telephone : +90 312 4022214; Fax: +90 312 4250813; GSM : +905356205589; E-mail : fkocabas@tsk.mil.tr

KLAJN, Rafal; Deparment of Chemistry, University of Warsaw, 1 Pasteur str, 02093 Warszawa, Poland. E-mail: rklajn@hotmail.com. Present address: Department of Chemical and Biological Engineering, Northwestern University, 2145 Sheridan Road, Evanston, 60208 IL, USA. Telephone: +1-847-491-3969; Fax: +1-847-491-3728; E-mail: rafal@northwestern.edu

MRAVINAC, Brankica; Department of Molecular Biology, Ruder Boskovic Institute, Bijenicka cesta 54, 10002 Zagreb, CROATIA. Telephone: +385-1-4561083; Fax: +385-1-4561177; E-mail: brankica@irb.hr

NOGUIERO, Eugenia; Nonaqueous Solvents Biocatalysis Laboratory, Instituto de Tecnologia Qumica e Biolgica (ITQB-Oeiras), Universidade Nova de Lisboa, Lisbon Portugal; Tel: +351962871745; Email: evgenia@aeiou.pt

NOROOZI, Nelson; Outpatient's Clinic for Dentist Surgery, Johannes Gutenberg University Mainz,, Augustusplatz 2, 55131 Mainz, Germany; Telephone: +49-6131/-989737; E-mail: noroozia@web.de

PETROVIC, Vlatka; Division of Molecular Biology, Ruder Boskovic Institute, Bijenicka 54, 10000 Zagreb, CROATIA.Telephone: +385-1-4561083; Fax: +385-1-4561177, E-mail: vpetrov@irb.hr

PILJAC, Jasenka; Molecular Biology Department, Rudjer Boskovic Institute, Bijenicka c. 54, PO Box 180, 10002 Zagreb, CROATIA. Telephone: +385-1-4560987; Fax: +385-1-4561177; E-mail: jpiljac@irb.hr

RADU, Ioan; Department of Molecular Cell Biology, Faculty of Medicine, Transilvania University, Brasov, Romania; Telephone: +40-0216346831; Fax: 0040213323361; E-mail: raduic@k.ro

RAINALDI, Mario; Department of Organic Chemistry, University of Padova, Via Marzolo, 1 35131, Padova, Italy; Telephone: +39-049-827-5266;; Fax. 049 827 5239; E-mail mario.rainaldi@unipd.it

SEMOVSKI, Serge V.; Limnological Institute, Section of Biology, Russian Academy of Science; P.O.Box 4199 Irkutsk, 664033; Russia; Fax: 7-3952-425405; Email: semovsky@lin.irk.ru

VUKELIC, Ana; Mathematics Depatment, Faculty of Food Technology and Biotechnology, University of Zagreb, Pierottijeva 6, 10000 Zagreb, CROATIA. Telephone: +385-1-4605005; Fax: +385-1-4836083; E-mail: avukelic@pbf.hr

# Author Index